The Atmospheric Sciences Entering the Twenty-First Century

Board on Atmospheric Sciences and Climate
Commission on Geosciences, Environment, and Resources
National Research Council

NATIONAL ACADEMY PRESS
Washington, D.C.   1998

NATIONAL ACADEMY PRESS • 2101 Constitution Avenue, N.W. • Washington, DC 20418

NOTICE: The project that is the subject of this report was approved by the Governing Board of the National Research Council, whose members are drawn from the councils of the National Academy of Sciences, the National Academy of Engineering, and the Institute of Medicine. The members of the committee responsible for the report were chosen for their special competences and with regard for appropriate balance.

Support for this project was provided by the Department of Agriculture, the Department of Energy, the Environmental Protection Agency, the Office of Naval Research of the Department of Defense, the Air Force Office of Scientific Research, the National Aeronautics and Space Administration, the National Oceanic and Atmospheric Administration, and the National Science Foundation under Grant No. ATM-9526208. Any opinions, findings, and conclusions or recommendations expressed in this publication are those of the author(s) and do not necessarily reflect the views of the above-mentioned agencies.

Library of Congress Cataloging-in-Publication Data

The atmospheric sciences : entering the twenty-first century /
Board on Atmospheric Sciences and Climate, Commission on
Geosciences, Environment, and Resources, National Research Council.
    p. cm.
Includes bibliographical references and index.
ISBN 0-309-06415-5
    1. Atmospheric physics. 2. Atmospheric chemistry. 3. Atmospheric physics—Research—United States. I. National Research Council (U.S.). Board on Atmospheric Sciences and Climate.
  QC861.2 .A88 1998
    551.5—ddc21   98-40083

The Atmospheric Sciences Entering the Twenty-First Century is available from the National Academy Press, 2101 Constitution Avenue, N.W., Box 285, Washington, DC 20418 (1-800-624-624; http://www.nap.edu).

Copyright 1998 by the National Academy of Sciences. All rights reserved.

Printed in the United States of America

CURRENT BOARD ON ATMOSPHERIC SCIENCES AND CLIMATE

ERIC J. BARRON, (Co-chair), Pennsylvania State University, University Park
JAMES R. MAHONEY, (Co-chair), International Technology Corporation, Washington, D.C.
SUSAN K. AVERY, Cooperative Institute for Research in Environmental Sciences, University of Colorado, Boulder
LANCE F. BOSART, State University of New York, Albany
MARVIN A. GELLER, State University of New York, Stony Brook
DONALD M. HUNTEN, University of Arizona, Tucson
JOHN IMBRIE, Brown University, Providence, Rhode Island
CHARLES E. KOLB, Aerodyne Research, Inc., Billerica, Massachusetts
THOMAS J. LENNON, Weather Services International Corp., Billerica, Massachusetts
MARK R. SCHOEBERL, NASA Goddard Space Flight Center, Greenbelt, Maryland
JOANNE SIMPSON, NASA Goddard Space Flight Center, Greenbelt, Maryland
NIEN DAK SZE, Atmospheric and Environmental Research, Inc., Cambridge, Massachusetts

Staff

ELBERT W. (JOE) FRIDAY, JR., Director
H. FRANK EDEN, Senior Program Officer
LOWELL SMITH, Senior Program Officer
DAVID H. SLADE, Senior Program Officer
LAURIE GELLER, Staff Officer
PETER SCHULTZ, Staff Officer
TENECIA A. BROWN, Senior Program Assistant
DIANE GUSTAFSON, Administrative Assistant

BOARD ON ATMOSPHERIC SCIENCES AND CLIMATE THAT PREPARED THIS REPORT

JOHN A. DUTTON (*Chair*), Pennsylvania State University, University Park
ERIC J. BARRON, Pennsylvania State University, University Park
WILLIAM L. CHAMEIDES, Georgia Institute of Technology, Atlanta
CRAIG E. DORMAN, Office of Naval Research, Arlington, Virginia
FRANCO EINAUDI, Goddard Space Flight Center, Greenbelt, Maryland
MARVIN A. GELLER, State University of New York, Stony Brook
PETER V. HOBBS, University of Washington, Seattle
WITOLD F. KRAJEWSKI, The University of Iowa, Iowa City
MARGARET A. LEMONE, National Center for Atmospheric Research, Boulder, Colorado
DOUGLAS K. LILLY, University of Oklahoma, Norman
RICHARD S. LINDZEN,* Massachusetts Institute of Technology, Cambridge
GERALD R. NORTH, Texas A&M University, College Station
EUGENE M. RASMUSSON, University of Maryland, College Park
ROBERT J. SERAFIN, National Center for Atmospheric Research, Boulder, Colorado

Staff

DAVID H. SLADE, Senior Program Officer and Study Director
DORIS BOUADJEMI,† Administrative Assistant

GREGORY H. SYMMES, Acting Director
WILLIAM A. SPRIGG,† Director
H. FRANK EDEN, Senior Program Officer
KENT L. GRONINGER,† Senior Program Officer
PETER SCHULTZ, Staff Officer
LAURIE S. GELLER, Staff Officer
ELLEN F. RICE, Reports Officer
TENECIA A. BROWN, Senior Program Assistant
KELLY NORSINGLE,† Senior Project Assistant
ANDREW E. EVANS,† Program Summer Intern

---

*Did not participate in the preparation of this report.
†Denotes past staff members who were active during the preparation of this report.

COMMISSION ON GEOSCIENCES, ENVIRONMENT, AND RESOURCES

GEORGE M. HORNBERGER (*Chair*), University of Virginia, Charlottesville
PATRICK R. ATKINS, Aluminum Company of America, Pittsburgh, Pennsylvania
JERRY F. FRANKLIN, University of Washington, Seattle
B. JOHN GARRICK, PLG, Inc., Newport Beach, California
THOMAS E. GRAEDEL, Yale University, New Haven, Connecticut
DEBRA S. KNOPMAN, Progressive Foundation, Washington, D.C.
KAI N. LEE, Williams College, Williamstown, Massachusetts
JUDITH E. MCDOWELL, Woods Hole Oceanographic Institution, Massachusetts
RICHARD A. MESERVE, Covington & Burling, Washington, D.C.
HUGH C. MORRIS, Canadian Global Change Program, Delta, British Columbia
RAYMOND A. PRICE, Queen's University at Kingston, Ontario
H. RONALD PULLIAM, University of Georgia, Athens
THOMAS C. SCHELLING, University of Maryland, College Park
VICTORIA J. TSCHINKEL, Landers and Parsons, Tallahassee, Florida
E-AN ZEN, University of Maryland, College Park
MARY LOU ZOBACK, United States Geological Survey, Menlo Park, California

Staff

ROBERT M. HAMILTON, Executive Director
GREGORY H. SYMMES, Assistant Executive Director
JEANETTE A. SPOON, Administrative Officer
SANDI S. FITZPATRICK, Administrative Associate
MARQUITA S. SMITH, Administrative Assistant/Technology Analyst

The National Academy of Sciences is a private, nonprofit, self-perpetuating society of distinguished scholars engaged in scientific and engineering research, dedicated to the furtherance of science and technology and to their use for the general welfare. Upon the authority of the charter granted to it by the Congress in 1863, the Academy has a mandate that requires it to advise the federal government on scientific and technical matters. Dr. Bruce M. Alberts is president of the National Academy of Sciences.

The National Academy of Engineering was established in 1964, under the charter of the National Academy of Sciences, as a parallel organization of outstanding engineers. It is autonomous in its administration and in the selection of its members, sharing with the National Academy of Sciences the responsibility for advising the federal government. The National Academy of Engineering also sponsors engineering programs aimed at meeting national needs, encourages education and research, and recognizes the superior achievements of engineers. Dr. William A. Wulf is president of the National Academy of Engineering.

The Institute of Medicine was established in 1970 by the National Academy of Sciences to secure the services of eminent members of appropriate professions in the examination of policy matters pertaining to the health of the public. The Institute acts under the responsibility given to the National Academy of Sciences by its congressional charter to be an adviser to the federal government and, upon its own initiative, to identify issues of medical care, research, and education. Dr. Kenneth I. Shine is president of the Institute of Medicine.

The National Research Council was organized by the National Academy of Sciences in 1916 to associate the broad community of science and technology with the Academy's purposes of furthering knowledge and advising the federal government. Functioning in accordance with general policies determined by the Academy, the Council has become the principal operating agency of both the National Academy of Sciences and the National Academy of Engineering in providing services to the government, the public, and the scientific and engineering communities. The Council is administered jointly by both Academies and the Institute of Medicine. Dr. Bruce M. Alberts and Dr. William A. Wulf are chairman and vice chairman, respectively, of the National Research Council.

Preface

The atmospheric sciences have progressed in the twentieth century from a fledgling discipline to a global enterprise providing considerable benefits to individuals, businesses, and governments. Through research and applications, the atmospheric sciences provide information that contributes to protection of life and property, agriculture, economic and industrial vitality, management of air quality, battlefield decisions, and national policies concerning energy and environment.

This report sets forth recommendations intended to strengthen atmospheric science and services and to enhance benefits to the nation. It is thus intended for those who share the responsibility for maintaining the pace of improvement in the atmospheric sciences, including leaders and policy makers in the public sector, such as legislators and executives of the relevant federal agencies; decision makers in the private sector of the atmospheric sciences; executives of other economic endeavors whose activities are dependent on atmospheric information, and of course, university departments that include atmospheric science.

Today the activities of the atmospheric sciences extend from the search for fundamental understanding to a wide range of specific applications in weather, climate, air quality, and other environmental issues. Moreover, the Board on Atmospheric Sciences and Climate (BASC) believes that new alliances between government, the private sector, and academe are developing rapidly and will advance the atmospheric sciences and services. Nevertheless, the federal government has a key and continuing role in supporting research to ensure that weather forecasts and warnings will improve, that uncertainties about a changing climate or air quality will be reduced, and that future atmospheric impacts and benefits will be identified early enough to ensure the safety and vitality of the nation.

This study was supported by the Department of Agriculture, the Department of Energy, the Department of Defense, the Environmental Protection Agency, the National Aeronautics and Space Administration, the National Oceanic and Atmospheric Administration, and the National Science Foundation. It began with a poll of leaders in science and engineering to obtain their views of issues and priorities for the atmospheric sciences.

The Board then requested its continuing committees for chemistry, solar-terrestrial, and climate research and two ad hoc teams of experts in atmospheric physics and dynamics/weather forecasting to assess for BASC the scientific challenges facing their disciplines, each discipline's contribution to the national well-being, and the research needed to face the challenges. These five technical reports, called "Disciplinary Assessments," were prepared for consideration by BASC. They are published here because they contain valuable ideas and suggestions that could interest research workers and federal agencies.

BASC then used the Disciplinary Assessments, together with input received from a variety of scientific sources as the basis for its appraisal of the major changes facing the atmospheric sciences as a whole. Some remarkably consistent themes emerged across the five Disciplinary Assessments, themes that permitted BASC to develop its vision of the future for the atmospheric sciences.

Thus, the Board's conclusions and recommendations for atmospheric sciences and services are a summary and a synthesis of the Disciplinary Assessments and recommendations and are presented as Part I of this report. In Part I the Board also points out some opportunities and challenges that derive from its own broad survey of the state and future of the atmospheric sciences. Although the major part of this report focuses on science issues, Part I points to other key elements of a national agenda for atmospheric sciences and services.

Part II, "Disciplinary Assessments," contains the five assessments, each devoted to a major research area within the atmospheric sciences. These areas and the chairs of the study groups follow: Atmospheric Physics, William A. Cooper; Chemistry, William Chameides; Dynamics and Weather Forecasting, Kerry Emanuel; Upper Atmosphere and Near-Earth Space, Marvin Geller; and Climate and Climate Change, Eric J. Barron. The Board is indebted to all who contributed to this study. Their names appear in the appropriate chapters of Part II. Portions of the Disciplinary Assessments included in Part II of this report have been abstracted and used as input to a forthcoming NRC report *Global Environmental Change: Research Pathways for the Next Decade*.

The Twenty-First Century report has been reviewed by individuals chosen for their diverse perspectives and technical expertise, in accordance with procedures approved by the NRC's Report Review Committee. The purpose of this independent review is to provide candid and critical comments that will assist the authors and the NRC in making the published report as sound as possible and to ensure that the report meets institutional standards for objectivity, evidence, and responsiveness to the study charge. The content of the review comments and draft manuscript remain confidential to protect the integrity of the deliberative

PREFACE

process. We wish to thank the following individuals for their participation in the review of this report:

> Bruce Albrecht, University of Miami
> Richard A. Anthes, University Corporation for Atmospheric Research
> Eugene W. Bierly, American Geophysical Union
> John S. Chipman, Department of Economics, University of Minnesota
> Ralph J. Cicerone, University of California, Irvine
> Paul J. Crutzen, Max-Planck-Institut für Chemie
> Richard M. Goody, Jet Propulsion Laboratory, California Institute of Technology
> Thomas E. Graedel, Yale University
> John Hallet, Desert Research Institute
> Dennis L. Hartmann, University of Washington
> D.A. Henderson, School of Hygiene and Public Health, Johns Hopkins University
> James R. Holton, University of Washington
> Donald Hornig, Harvard School of Public Health (emeritus)
> Donald R. Johnson, University of Wisconsin
> Richard S. Lindzen, Massachusetts Institute of Technology
> Syukuro Manabe, Institute for Global Change Research Program, Tokyo, Japan
> Marcia M. Neugebauer, Jet Propulsion Laboratory, California Institute of Technology
> Edward S. Sarachik, University of Washington
> Joanne Simpson, Goddard Space Flight Center, National Aeronautics and Space Administration
> George Siscoe, retired
> Robert M. White, President, Washington Advisory Group

Although the individuals listed above have provided many constructive comments and suggestions, responsibility for the final content of this report rests solely with the authoring committee and the NRC.

The Board and I are grateful to David H. Slade, Senior Program Officer and Study Director, whose acumen and energy contributed much to the organization and writing of this report, and to Doris Bouadjemi, Administrative Assistant, who directed its publication with skill and dedication. The Board and I are also grateful to William A. Sprigg, former director of BASC, whose energy, dedication and innovation contributed much to the work of the Board and to its achievements in recent years.

John A. Dutton
Chair

Contents

SUMMARY 1

PART I
1 INTRODUCTION 13
    Four Centuries of Progress, 14
    The Atmospheric Sciences and Other Disciplines, 14
    Looking Forward to the Twenty-First Century, 15

2 CONTRIBUTIONS OF THE ATMOSPHERIC SCIENCES TO THE
   NATIONAL WELL-BEING 17
    Protection of Life and Property, 17
      Need for Forecasts and Warnings, 18
      Progress in Weather Services, 21
    Maintaining Environmental Quality, 22
      Chlorofluorocarbons and Ozone, 22
      Greenhouse Gases and Global Change, 22
      Aerosols, 23
      Role of Atmospheric Sciences in Environmental Issues, 23
    Enhancing National Economic Vitality, 24
      Benefits of Weather and Climate Information, 24
    Strengthening Fundamental Understanding, 26

3 SCIENTIFIC IMPERATIVES AND RECOMMENDATIONS
   FOR THE DECADES AHEAD 28
   Atmospheric Science Imperative 1: Optimize and
     Integrate Observation Capabilities, 29
       New Observing Opportunities, 30
       Requirements for Optimizing and Integrating
         Observing Systems, 31
       Observing System Simulation Experiments, 33
     Atmospheric Science Imperative 2: Develop
       New Observation Capabilities, 33
       Water in the Atmosphere, 34
       Wind Obsrvations, 35
       Observations in the Stratosphere, 36
       Observations in Hear-Earth Space, 37
     Atmospheric Research Recommendation 1: Resolve
       Interactions at Atmospheric Boundaries and Among
       Different Scales of Flow, 37
       Surface Properties, 38
       Long-Term Interactios with the Oceans, 38
       Clouds and Their Consequences, 39
       Aerolsols and Atmospheric Chemistry, 40
       The Fundamental Problem of Nonlinearity, 40
     Atmospheric Research Recommendation 2: Extend a Disciplined
       Forecast Process to New Areas, 41
     Atmospheric Research Recommendation 3: Initiate Studies
       of Emerging Issues, 43
         Climate, Weather and Health, 44
         Water Resources, 44
         Rapidly Increasing Emissions to the Atmoshere, 45

4 LEADERSHIP AND MANAGEMENT CHALLENGES IN THE
   DECADES AHEAD 46
     Leadership and Management Recommendation 1: Develop a
       Strategy for Providing Atmospheric Information, 46
         A Changing System for Providing Weather Services, 47
         Prospects for Atmospheric Information, 48
         Implications of Distributed Atmospheric Information Services, 49
     Leadership and Management Recommendation 2: Ensure
       Access to Atmospheric Information, 50
     Leadership and Management Recommendation 3: Assess
       Benefits and Costs, 51
         Federal Funding of Atmospheric Research and Services, 52
     Leadership and Management Planning, 58

CONTENTS   *xiii*

PART II DISCIPLINARY ASSESSMENTS

1   ATMOSPHERIC PHYSICS RESEARCH ENTERING THE
    TWENTY-FIRST CENTURY   63
        Summary, 63
            Major Scientific Goals and Challenges, 63
            Key Components of the Scientific Strategy, 64
            Initiatives to Support the Strategies, 65
            Expected Benefits and Contributions to the National
                Well-Being, 65
            Recommended Atmospheric Physics Research, 66
        Introduction, 68
            Mission, 68
            Major Research Themes and Past Accomplishments, 69
            Perspective for the Future, 70
        Scientific Challenges and Questions, 71
            Atmospheric Radiation, 71
            Cloud Physics, 74
            Atmospheric Electricity, 77
            Boundary Layer Meteorology, 79
            Small-Scale Atmospheric Dynamics, 81
        Disciplinary Research Challenges, 83
        Contributions to National Goals, 106

2   ATMOSPHERIC CHEMISTRY RESEARCH ENTERING THE
    TWENTY-FIRST CENTURY   107
        Summary, 107
            Major Scientific Questions and Challenge, 108
            Overarching Research Challenges, 109
            Disciplinary Research Challenges, 110
            Infrastructural Initiatives, 110
            Expected Benefits and Contribution to the National Well-Being, 111
        Introduction and Overview, 111
            The Mission, 112
            Insights of the Twentieth Century, 114
            Disciplinary Research Challenges, 121
            Overarching Research Challenges, 132
            Infrastructural Initiatives, 135
            Conclusion, 140
        The Environmentally Important Atmospheric Species:
            Scientific Questions and Research Strategies, 140

Stratospheric Ozone, 140
Atmospheric Greenhouse Gases, 147
Photochemical Oxidants, 157
Atmospheric Aerosols, 162
Toxics and Nutrient, 166

3 ATMOSPHERIC DYNAMICS AND WEATHER FORECASTING RESEARCH ENTERING THE TWENTY-FIRST CENTURY   169
    Summary, 169
       Emerging Research Opportunities, 170
       Key Recommendations, 173
    Introduction, 175
       Basic Research Foci, 175
       Technique Developments, 187
       Technological Developments, 193
    Conclusion, 197

4 UPPER ATMOSPHERE AND NEAR-EARTH SPACE RESEARCH ENTERING THE TWENTY-FIRST CENTURY   199
    Summary, 199
       Major Scientific Goals and Challenges, 200
       Key Components of the Scientific Strategy, 200
       Scientific Requirements for the Coming Decade(s), 200
       Expected Benefits and Contributions to the National Well-Being, 202
       Upper-Atmosphere and Near-Earth Space Research Tasks, 202
    Introduction, 204
       The Sun, 204
       Interplanetary Space, 204
       The Magnetosphere, 205
       The Ionosphere-Upper Atmosphere, 206
       The Middle Atmosphere, 206
       Cosmic Rays, 208
       Research Priorities 208,
    Stratospheric Processes Important for Climate and the Biosphere, 211
       Stratospheric Ozone, 213
       Volcanic Effects, 218
       Solar Effects, 219
       Quasi-Biennial Oscillation Effects, 220
       Atmospheric Effects of Aircraft, 220
       The Role of the Stratosphere in Climate and Weather Prediction, 223

Key Initatives, 223
Measures of Success, 225
Space Weather, 225
Scientific Background, 228
Critical Science Questions, 237
History and Current Research Activities, 241
Key Initiatives 242
Middle-Upper Atmosphere Global Change, 245
Scientific Background, 245
Critical Science Questions, 249
Key Initiatives, 250
Contributions to the Solution of Societal Problems, 254
Measures of Success, 255
Solar Influences, 256
Solar Energy Output over a Solar Cycle, 257
Separating Solar and Anthropogenic Effects, 259
Solar Influences on the Earth's Upper and Middle Atmosphere, 263
Physical Basis of the Solar Activity Cycle, 265
Long-Term Changes in Solar Behavior: Solar-Type Stars, 268
Key Initiatives, 270
Contributions to the Solution of Societal Problems, 271

5   CLIMATE AND CLIMATE CHANGE RESEARCH
    ENTERING THE TWENTY-FIRST CENTURY   272
    Summary, 272
    Introduction, 276
    Mission Statement, 278
    Perspectives for the Twenty-First Century, 279
      Insights of the Twentieth Century, 279
      The Scientific Questions, 296
      Key Drivers for Research in the Twenty-First Century, 297
    Objectives and Requirements for Climate Research, 302
      Objective 1, 302
      Objective 2, 307
      Objective 3, 309
      Objective 4, 310
      Objective 5, 311
      Objective 6, 314
      Objective 7, 316
    Priorities for Climate Research, 318
      Build a Permanent Climate Observing System, 319
      Extend the Instrumented Climate Record Through
        Development of Integrated Historical and Proxy Data Sets, 320

Continue and Expand Diagnostic Efforts and
  Process Study Research to Elucidate Key
  Climate Variability and Change Processes, 320
Construct and Evaluate Climate Models That Are
  Increasingly Comprehensive, Incorporating
  All Major Components of the Climate System, 321
Cross-Cutting Requirements, 322
  Education, 322
  Institutional Arrangements, 323
Contributions to National Goals and Needs, 324

REFERENCES   325

APPENDIX A   Acronyms and Abbreviations   341

APPENDIX B   Listing of Reports by the Committee on
Atmospheric Sciences and the Board on Atmospheric
Sciences and Climate Since 1958   346

INDEX   349

The Atmospheric Sciences
Entering the Twenty-First Century

Summary

The atmospheric sciences have developed an impressive capability over the past century to help society anticipate atmospheric phenomena and events. Progress continues today as improved observational and remote sensing capabilities provide more accurate resolution of atmospheric processes. In addition, enhanced physical understanding, new modeling strategies, and powerful computers combine to provide improved atmospheric simulations and predictions. This progress leads the Board on Atmospheric Sciences and Climate (BASC) to the following vision for the atmospheric sciences entering the twenty-first century:

> *Improvements in atmospheric observations, further understanding of atmospheric processes, and advances in technology will continue to enhance the accuracy and resolution of atmospheric analysis and prediction. As a consequence, society will enjoy greater confidence in atmospheric information and forecasts and will be able to act more decisively and effectively.*

THE ATMOSPHERIC SCIENCES AND OTHER DISCIPLINES

This report, by design, focuses on the atmosphere, but recognizes that the atmosphere interacts intimately with other parts of the Earth and its environment. The fluxes of materials and energy between the atmosphere and the oceans, the Earth's surface, the ecosystems of the planet, and the near-space environment shape the structure and evolution of atmospheric processes and events. The report emphasizes atmospheric research, but it also focuses on the societal implications of atmospheric science, including the benefits of improved knowledge of atmospheric processes for the health and welfare of the environment and society.

OPPORTUNITIES AND IMPERATIVES FOR THE ATMOSPHERIC SCIENCES

As BASC studied the progress and future of the atmospheric sciences, it became evident that a critical contemporary theme involves the increasing importance of integrated observation systems and new observations of critical variables; thus, these constitute the subject of the Board's two highest-priority recommendations, designated as "Imperatives."

Atmospheric Science Imperative 1:
Improve Observation Capabilities

The atmospheric science community and relevant federal agencies should develop a specific plan for optimizing global observations of the atmosphere, oceans, and land. This plan should take into account requirements for monitoring weather, climate, and air quality and for providing the information needed to improve predictive numerical models used for weather, climate, atmospheric chemistry, air quality, and near-Earth space physics activities. The process should involve a continuous interaction between the research and operational communities and should delineate critical scientific and engineering issues. Proposed configurations of the national and international observing systems should be examined with the aid of observing system simulation experiments.

In addition, new opportunities for advances in research and services lead to the second imperative:

Atmospheric Science Imperative 2:
Develop New Observation Capabilities

The federal agencies involved in atmospheric science should commit to a strategy, priorities, and a program for developing new capabilities for observing critical variables, including water in all its phases, wind, aerosols, and chemical constituents and variables related to phenomena in near-Earth space, all on spatial and temporal scales relevant to forecasts and applications. The possibilities for obtaining such observations should be considered in studying the optimum observing systems of Imperative 1.

Contemporary numerical computer models of the atmosphere are sufficiently varied and powerful that they can predict or simulate a range of phenomena such as climate change and air pollution episodes, as well as forecasting the weather. However, observations of critical variables on time and space scales relevant to forecasts are essential to improving such numerical simulations and predictions.

ATMOSPHERIC RESEARCH RECOMMENDATIONS

Common themes in the five Disciplinary Assessments presented in Part II of this report lead to two quite different sets of recommendations: one concerned directly with atmospheric research and related issues and a second with leadership and management.

Atmospheric Research Recommendation 1:
Resolve Interactions at Atmospheric Boundaries
and Among Different Scales of Flow

The major weather, climate, and global observation programs supported by the federal government and international agencies should put high priority on improved understanding of interactions of the atmosphere with other components of the Earth system and of interactions between atmospheric phenomena of different scales. These programs, including the U.S. Weather Research Program, the U.S. Global Change Research Program, and other mechanisms for supporting atmospheric research, require observational, theoretical, and modeling studies of such interactions.

Atmospheric studies are shaped today by the recognition that contemporary approaches must seek to understand, model, and predict the components of the Earth's environment as coupled systems. Critical scientific questions focus on the exchanges of energy, momentum, and chemical constituents between the troposphere and the surface below, as well as with the layers above.

Atmospheric Research Recommendation 2:
Extend a Disciplined Forecast Process to New Areas

A strategy and implementation plan for initiating experimental forecasts and taking advantage of a disciplined forecasting process should be developed by appropriate agencies and the scientific community for climate variations, key chemical constituents, air quality, and space weather events.

The impact of a disciplined process of forecasting (observe, predict, assess accuracy, improve methods) has intensified with the advent of numerical simulation of atmospheric processes because precise quantitative comparisons between forecasts and actual observations can be made easily and because proposed modifications in computer models can be applied to difficult cases retrospectively. Several of the atmospheric sciences are developing capabilities for making quantitative forecasts and improving capabilities by applying the discipline of forecasting. The opportunities include climate forecasting, atmospheric chemistry, and space weather.

Atmospheric Research Recommendation 3: Initiate Studies of Emerging Issues

The research community and appropriate federal agencies should institute interdisciplinary studies of emerging issues related to (1) climate, weather, and health; (2) management of water resources in a changing climate; and (3) rapidly increasing emissions to the atmosphere.

The emerging issues identified in the recommendation require the attention of atmospheric scientists in a wide range of disciplines and collaborators concerned with human dimensions.

RECOMMENDATIONS FOR LEADERSHIP AND MANAGEMENT IN THE ATMOSPHERIC SCIENCES

BASC recommends that leadership and management issues related to planning and priorities should be addressed forthrightly by the entire atmospheric science community, including federal agencies, professional societies, and the academic and private sectors.

Leadership and Management Recommendation 1: Develop a Strategy for Providing Atmospheric Information

The Federal Coordinator for Meteorological Services and Supporting Research should lead a thorough examination of the issues that arise as the national system for providing atmospheric information becomes more distributed. Key federal organizations, the private sector, academe, and professional organizations should all be represented in such a study and should help develop a strategic plan.

Rapid changes in the national weather information system are occurring because of the following:

1. Quantitative information on global weather and climate data, visualizations, and predictions are readily available on global information networks.
2. Computer-to-computer communication enables weather-dependent enterprises to incorporate atmospheric information more readily in their decision making.

Leadership and Management Recommendation 2: Ensure Access to Atmospheric Information

The federal government should move forthrightly and aggressively to protect the advance of atmospheric research and services by maintaining the free and open exchange of atmospheric observations among all countries and by preserving the free and open exchange of data among scientists.

The increasing dependence on distributed capabilities has significant implications for access to atmospheric data and information, especially since some nations and some industries advocate schemes for limiting access to electronic data. Two principles have long governed the traditional U.S. view of international atmospheric data:

1. Data acquired for public purposes with public funds should be publicly available.
2. The free and open exchange of atmospheric observations will enhance atmospheric research, understanding, and services for all nations.

Leadership and Management Recommendation 3: Assess Benefits and Costs

The atmospheric science community, through the collaboration of appropriate federal agencies and advisory and professional organizations, should initiate interdisciplinary studies of the benefits and costs of weather, climate, and environmental information services.

Reasons for examining the benefits and costs of maintaining the atmospheric enterprise are to ensure that funds invested in atmospheric research and services are highly leveraged in serving national interests and to help determine which new directions in atmospheric research and services will provide the most benefit to the public and private sectors.

LEADERSHIP AND MANAGEMENT PLANNING

BASC believes that a national research environment requires a strong disciplinary planning mechanism. This view is reinforced by an obvious contemporary reality: opportunities for progress and service in the atmospheric sciences are far more plentiful than resources. Thus, the efforts of the discipline must be guided by an overall vision and by reasoned priorities. All partners in the atmospheric science enterprise—those in government, universities, and a wide range of private endeavors—must join together as an effective team focused on the future. For this to come to pass, there must be clear responsibilities for priorities and progress, for resources and results.

DISCIPLINARY ASSESSMENTS

To assess the state of the science and look to the future, BASC asked three of its continuing committees and two ad hoc groups of experts to prepare separate assessments that analyze critical scientific issues, identify major opportunities and initiatives for each discipline, and recommend a scientific and programmatic agenda for the decade or two ahead.

These assessments focus both on improvements in fundamental science and on service to society. For the near future, they emphasize forecasts of atmospheric phenomena with significant societal impacts and propose efforts to predict important aspects of seasonal climate variability, chemical processes, and space weather phenomena. For the longer term, they emphasize the resolution of climate variability on the scale of decades to centuries and the possibilities of projecting climate variations.

Atmospheric Physics Research

Atmospheric physics seeks to understand physical processes in the atmosphere, including atmospheric radiation, the physics of clouds, atmospheric electricity, boundary layer processes, and small-scale atmospheric dynamics. The complexity and interactive nature of the physical processes in the atmosphere will usually defy prediction unless supplementary and organizing principles can be found.

Three scientific strategies are recommended:

1. Develop and verify a capability to predict the influence of small-scale atmospheric physical processes on large-scale atmospheric phenomena.
2. Develop a quantitative description of the processes and interactions that determine the observed distribution of water substance in the atmosphere.
3. Improve capabilities for making critical measurements in support of studies of atmospheric physics.

Pursuing these strategies requires specific research efforts related to the interactions among a variety of processes, including those of atmospheric radiation, clouds and other components of the hydrological cycle, aerosols, atmospheric electricity, boundary layer meteorology, and small-scale dynamics. Such studies will require new measurement and analysis techniques based on contemporary concepts and technology.

Atmospheric Chemistry Research

The "Environmentally Important Atmospheric (chemical) Species," by virtue of their radiative and chemical properties, affect climate, key ecosystems, and all living organisms. The challenge for atmospheric chemistry research in the coming decades is to develop the tools and scientific infrastructure necessary to document and predict the concentrations and effects of these species on local, regional, and global spatial scales and on daily to decadal time scales. These Environmentally Important Atmospheric species are stratospheric ozone, greenhouse gases, photochemical oxidants, atmospheric aerosols, toxics, and nutrients.

The recommended strategies for atmospheric chemistry research include the following:

1. Document the chemical climatology and meteorology of the atmosphere through the development of monitoring networks.
2. Develop and evaluate predictive models for atmospheric chemistry and air quality.
3. Provide assessments of the efficacy of environmental management activities by gathering and assessing air quality data.
4. Develop holistic and integrated understanding of the environmentally important atmospheric species and the chemical, physical, and biological interactions that couple them.

The infrastructure necessary to advance atmospheric chemistry research and applications includes:

- global observing systems,
- surface exchange measurement and ecological exposure systems,
- environmental management systems,
- instrument development and technology transfer programs, and
- facilities for studying condensed-phase and heterogeneous chemistry.

Atmospheric Dynamics and Weather Forecasting Research

Progress in the study of atmospheric interactions that shape weather phenomena has created opportunities to make major advances that will lead directly to improved weather warnings and predictions. The recommended research includes the following efforts:

1. Optimize observing systems, by better collection and utilization of data over the oceans and determination of optimal combinations of available and new observations.
2. Develop adjoint techniques, that target specific regions of the atmosphere for special observations that will lead to greatly reduced forecast error.
3. Maintain in situ observations, especially by halting deterioration of the global rawinsonde network and other in situ measurements.
4. Emphasize land-atmosphere interactions, for which better observations and understanding may be the key to improved forecasts of convection, precipitation, and seasonal climate.
5. Improve water vapor observations, including more accurate and higher-resolution measurements to improve a wide variety of forecasts.
6. Emphasize seasonal forecasts, which require deeper understanding of the relative importance of internal atmospheric variability and interactions with longer-scale phenomena in the oceans and land.
7. Emphasize tropical cyclone motion and intensity, especially research on the physics of tropical cyclone motion and changes in intensity, research on interactions with the upper ocean layers, and research to delineate optimal combinations of measurement systems for hurricane forecasting.

Upper-Atmosphere and Near-Earth Space Research

Human-induced and natural changes in the upper atmosphere and near-Earth space now portend increasing impacts on the global environment and societal activities. Four areas of scientific research are essential for understanding and mitigating these impacts; thus, upper-atmosphere and near-Earth space research should emphasize the following:

1. Stratospheric processes that affect climate and the biosphere, including the effects of stratospheric aircraft, ozone-depleting chemicals, volcanic emissions, and solar variability.
2. Space "weather," the short-term variability of the near-Earth space environment that has important effects on satellite performance, human health in space, and communication systems and power grid operation.
3. Global changes in the middle and upper atmosphere in response to natural and anthropogenic influences that have significant effects on the lower atmosphere.
4. The effects of solar variability on the global climate system, which may be significant but must be differentiated from other natural influences and from climate effects associated with human activity.

Research to enable operational prediction of space weather should be given high priority and should emphasize the causes of solar variability and models of the solar-terrestrial system.

Climate and Climate Change Research

Climate research aims to understand the physical and chemical basis of climate and climate change in order to predict climate variability on seasonal to decadal and longer time scales, to assess the role of human activities in affecting climate, and to determine the role of climate change in affecting human activities and the environment. The three dominant scientific goals in climate change are:

1. Understand the mechanisms of natural climate variability on time scales of seasons to centuries, and assess their relative significance.
2. Develop climate change prediction, application, and evaluation capabilities,
3. Project changes in the climate system and relate them to human activities.

The highest priority strategies for pursuing these goals are the following:

- Create a permanent climate observing system.
- Extend the observational climate record through the development of integrated historical and proxy data sets.
- Continue and expand diagnostic efforts and process study research to elucidate key climate variability and change processes.
- Construct and evaluate models that are increasingly comprehensive, incorporating all the major components of the climate system.

Part I

Overview and Recommendations for the Atmospheric Sciences Entering the Twenty-First Century

PART I

1

Introduction

We live in an atmosphere that shapes our activities and states of mind, an atmosphere whose storms threaten our lives and property, and whose climate and composition influence the nature and vitality of our societies. Today we see accelerating progress as the atmospheric sciences develop improved understanding of the atmosphere and thus an increasing capability to help society anticipate atmospheric events. Observational technology and strategies are advancing to provide improved understanding of interactions between the atmosphere, ocean, and land. Enhanced physical and mathematical understanding allows use of increasingly powerful computers to organize observations and convert them into predictions of weather, climate, and air quality. These advances will enable the atmospheric sciences to develop notable new and improved capabilities for serving society these and lead to a vision for the science entering the twenty-first century:

> *Improvements in atmospheric observations, further understanding of atmospheric processes, and advances in technology will continue to enhance the accuracy and resolution of atmospheric analysis and prediction. As a consequence, society will enjoy greater confidence in atmospheric information and forecasts and will be able to act more decisively and effectively.*

Realizing this vision requires a focus on the observational and modeling infrastructure, on research tasks with the greatest promise, and on mechanisms to ensure that federal investments in atmospheric research and operations are effective and produce results significant to the nation in the early decades of the twenty-first century.

FOUR CENTURIES OF PROGRESS

The theoretical bases of the contemporary atmospheric sciences have been established over the past four centuries, beginning in the seventeenth century with the study of gases and the formulation of Newton's laws of motion and continuing in the late nineteenth century with the formalization of thermodynamics and radiation theory. In North America, the study of meteorology and climate began in the colonial period with Benjamin Franklin's investigation of the passages of large-scale fronts and storms on the East Coast and the relationship of lightning to electricity.

Early in the twentieth century, telegraphic transmission of observations and the demands of aviation, agriculture, and other activities stimulated meteorological research and the development of conceptual models of storm and frontal systems that made crude short-range predictions possible. Steady progress ensued, with important advances often arising from technological developments. Contemporary technologies being applied to atmospheric analysis and prediction include remote sensing satellites, radars that measure precipitation and wind, laser systems, and powerful digital computers. The classical cycle continues: observations stimulate theory, and, in turn, theory and new understanding stimulate new observations and new capabilities for understanding and prediction.

In recent decades, the emphases in the atmospheric sciences have moved simultaneously in two directions: toward the smaller space and time scales involved with physical processes and toward the larger scales involved in the evolution and prediction of climate and environmental change. Moreover, there is an increasing emphasis on atmospheric chemistry and on the prediction of upper atmosphere processes. In all cases, observations, theory, and computer models combine to provide new understanding and prediction.

THE ATMOSPHERIC SCIENCES AND OTHER DISCIPLINES

To appreciate the key role of the atmospheric sciences in understanding our world, imagine a planet without an atmosphere. The surface is rough and torn, battered by meteor impacts. Ultraviolet radiation and streams from solar flares impinge freely on the surface, making life there impossible. The temperature contrasts between day and night, between equator and pole, are astonishingly large. The differences between this stark planet and our own blue and green Earth arise from the presence of a substantial atmosphere.

The atmosphere shelters life on Earth from the hazards of space and provides the global transport system that maintains the resources necessary for life. In addition to addressing their own scientific challenges, the atmospheric sciences illuminate important issues in a range of disciplines, while in turn drawing on these disciplines to create more realistic portraits of the forces and constraints that shape atmospheric behavior.

Oceanography is perhaps the closest scientific partner of the atmospheric sciences. The ocean surface is a critical boundary for the atmosphere, which in turn provides a critical boundary for the oceans. Across their common interface, the atmosphere and ocean exchange energy, momentum, and some important chemical constituents, most notably water in various forms. Because the ocean and atmosphere interact so intimately in shaping and controlling the planetary environment, many major initiatives today concerned with climate and climate change are collaborative efforts between the two sciences.

Improved understanding of atmospheric and oceanic circulations has opened new areas of research for the geological sciences by providing independent, physically based climate estimates that stimulate new ideas and validate conclusions. Research in atmospheric chemistry provides new understanding of chemical reactions, sources, and sinks that enhance understanding of the chemical evolution of the planet. Climate conditions shape the distribution of life and must be taken into account by scientists defining optimum conditions and limits of flora and fauna, as species or as ecosystems.

Today, human health responses to weather and climate are becoming apparent. Direct connections related to heat stress and respiratory problems are augmented by climate variations that modify disease vectors and affect carriers of infectious diseases. Moreover, both the occurrence of skin cancer and the vigor of immune systems appear to be related to the intensity of ultraviolet radiation reaching the Earth's surface and hence are subject to variations in atmospheric chemical constituents.

In 1963, the noted physicist Alvin M. Weinberg argued that the relative scientific merit of disciplines could be assessed by their impact on other disciplines, saying "that field has the most scientific merit which contributes most heavily to and illuminates most brightly its neighboring scientific disciplines." On this basis, the atmospheric sciences stand high among the disciplines that study the Earth and its biological systems.

LOOKING FORWARD TO THE TWENTY-FIRST CENTURY

As the century turns, new areas of emphasis are enriching the atmospheric sciences and sharpening their contributions to society. At the same time, fundamental and worldwide changes in economic activity and public policy are being driven at unprecedented rates by the information revolution. Some aspects of contemporary society are increasingly sensitive to atmospheric events. Thus, timely and accurate information to support critical decisions is increasingly valuable. Moreover, as the Earth's population grows and its economic engines accelerate, human activities may force the atmosphere-ocean-land system in ways and directions that we do not yet completely comprehend.

As atmospheric issues become more complex and interdisciplinary, the implications more critical, and federal funds more scarce, it becomes especially pressing to set priorities for the atmospheric sciences so that they simultaneously develop fundamental understanding and serve national needs effectively. The improved benefits of atmospheric science achieved through the synergy of increasing capability and greater user confidence will be the measure of success as the twenty-first century unfolds.

PART I
2

Contributions of the Atmospheric Sciences to the National Well-Being

The responsibilities of the atmospheric sciences include the advance of fundamental understanding, the prediction of weather and climate change, and the identification of environmental threats. This section examines four ways in which the atmospheric sciences contribute to national well-being and the achievement of national goals through protection of life and property, maintaining environmental quality, enhancing economic vitality, and strengthening fundamental understanding. Today, they contribute to a broad range of decisions concerning individual actions, business and economic strategies, and public policy.

Atmospheric information is valuable when it helps to clarify the advantages and risks of alternative courses of action for private and public decision makers, but assessing the influence of atmospheric information on public safety and economic activities is not straightforward. Atmospheric information is usually only one of many factors that influence a decision, and the value of a forecast often depends on such factors as lead time, forecast resolution, and expected accuracy and on user constraints such as the ability to respond or to assess realistically the costs versus benefits of various courses of action.

PROTECTION OF LIFE AND PROPERTY

The United States experiences a great diversity of weather, some of it the most severe in the world. As a consequence, this nation has been a leader in developing the understanding and technological capability to provide forecasts of the continually changing weather and warnings of severe weather events such as floods, tornadoes, and hurricanes. Atmospheric information and forecasts are provided in the United States through a four-way partnership:

1. The government acquires and analyzes observations and issues forecasts and warnings.

2. The government, newspapers, radio, and television all participate in dissemination of weather forecasts and severe weather warnings.

3. Private-sector meteorological firms use government data and products to provide weather information for the media and special weather services for a variety of industries and activities.

4. Government, university, and private-sector scientists develop improved understanding of atmospheric behavior and help to turn this advancing understanding into new capabilities and technology for observing and predicting atmospheric events. This four-way partnership has served the nation well; it could be nurtured to generate even greater benefit.

Although we can do little to change the atmosphere or the weather, we can do much to anticipate atmospheric events such as severe weather and thus provide opportunities for protecting lives and property. These capabilities are expanding rapidly, for both traditional weather impacts and new ones, including applications to environmental quality, to solar events that affect satellites in Earth orbit, communications, to power transmission and changes in weather patterns associated with El Niño events. Today, the nation is reaping significant benefits from its investments in atmospheric observations and prediction capabilities.

Need for Forecasts and Warnings

The benefits of weather forecasts and warnings as measured by lives saved, injuries avoided, or property that has not been damaged cannot be estimated easily. Determining the economic benefits of long-term forecasts, such as those associated with El Niño, is even more difficult. Nevertheless, casualties produced by unusual weather events are substantial, both in an absolute sense and relative to other natural disasters.

Long-term fatality statistics for tornadoes and hurricanes are shown in Tables I.2.1 and I.2.2, using data that go back to the 1930s and 1900s. The data show remarkable progress. More detailed information on weather-related fatalities and damage in the United States for 1991-1995 is given in Table I.2.3. The number of fatalities attributable to weather is typically 300-400 per year. However, one large event, such as the extreme temperatures of 1995, can add as many as 1,000 fatalities to that total. Similarly, Hurricane Andrew (1992) and the floods in 1993 each caused more damage than is attributable to all weather events in some other years. Changnon et al. (1997) present an analysis of the effects of recent weather events on the U.S. insurance industry.

Table I.2.4 provides worldwide statistics on deaths owing to civil strife, natural disasters, and other environmental causes. It shows that more than 25 percent of deaths are due to drought, famine, and severe weather and that these

TABLE I.2.1 Reported Fatalities from Tornadoes in the United States

Period	Fatalities
1930s	1,945
1940s	1,786
1950s	1,419
1960s	945
1970s	998
1980s	522
1986-1995	485

SOURCE: Aviation Weather and Storm Prediction Center, National Centers for Environmental Prediction, National Weather Service.

TABLE I.2.2 Reported Fatalities from Representative Hurricanes in the United States (1900-1995)

Year	Rank[a]	Location	Fatalities
First Half of Twentieth Century			
1900	1	Galveston	>8,000
1909	6	Louisiana and Mississippi	406
1915	4	Texas and Louisiana	550
1919	9	Florida, Texas, Louisiana	287
1928	2	Florida	1,836
1935	5	Florida	414
1938	3	Southern New England	600
Second Half of Twentieth Century			
1961	25	Texas	46
1965	16	Southern Louisiana	75
1969	11	Southern States to West Virginia	256
1972	14	Florida to New York	122
1979		Caribbean Islands to New York	22
1980		Texas Coast	2
1989	20	Carolinas	56
1992		Florida and Louisiana (Andrew)	23

[a]Among top 30 deadliest mainland U.S. hurricanes 1900-1995.

SOURCE: Hebert et al., 1996.

TABLE I.2.3A Weather-Related Fatalities in the United States, 1991-1995

Weather Event	Fatalities[a]					
	1991	1992	1993	1994	1995	Total
Extreme temperatures	49	22	38	81	1,043	1,233
Convective storms[b]	144	93	99	155	153	644
Floods	61	62	103	91	80	397
Other[c]	75	40	62	20	51	248
Snow, ice, avalanches	45	64	67	31	17	224
Hurricanes	13	27	2	9	17	68
Marine storms	4	0	0	1	1	6
Total	391	308	371	388	1,362	2,820
Annual Average						564

TABLE I.2.3B Weather-Related Damage in the United States, 1991-1995

Weather Event	Damage ($ millions)					
	1991	1992	1993	1994	1995	Total
Hurricanes	1,164	33,611	15	426	5,932	41,148
Floods	874	690	21,288	921	1,250	25,023
Other[c]	1,878	1,932	5,019	893	359	10,081
Convective storms[b]	1,527	1,580	1,086	1,001	2,638	7,832
Snow, ice, avalanches	516	28	602	1,143	111	2,400
Extreme temperatures	224	479	416	52	1,120	2,291
Marine storms	45	31	1	3	2	82
Total	6,228	38,351	28,427	4,439	11,412	88,857
Annual average						17,771

SOURCE: Office of Meteorology, National Weather Service.

[a]The fatalities represent only those deemed to be directly attributable to weather and floods. Number of fatalities in which weather was a contributing factor would be much larger.
[b]Term includes tornadoes, thunderstorms, lightning, and hail.
[c]Includes drought, dust storms, rain, fog, strong winds, fire weather, and mud slides.

events account for more than 92 percent of all people affected by all the disasters combined.

Weather fatalities and damage could be mitigated by improved design and construction standards for buildings and critical systems; relocation of residents from hazard prone locations; and earlier, more accurate, and more focused warnings of severe weather. Although it is difficult to estimate accurately the relative effectiveness of implementing these three strategies, it seems that improving severe weather warnings, despite requiring investments in observational technology and forecast capabilities, might be the least expensive and most feasible of

TABLE I.2.4 Worldwide Disaster Statistics: Annual Averages for 1960-1989

Classification of Disaster	Number of Events	Percent	Deaths	Percent	Persons Affected	Percent
Civil strife	4.5	5.8	97,087	62.2	3,916,454	5.6
Drought and famine	10.3	13.1	21,220	13.6	36,185,464	51.8
Weather (storms and floods)	37.0	47.1	17,894	11.5	28,182,075	40.4
Earthquakes, volcanoes	10.5	13.4	16,583	10.6	1,400,787	2.0
Fires and epidemics	16.2	20.6	3,228	2.1	160,371	0.2
Totals	78.5	100.0	156,012	100.0	69,845,151	100.0

SOURCE: Office of U.S. Foreign Disaster Assistance (1991); adapted from Bruce (1994).

the three options and might generate the increased public confidence that would lead to even greater response to warnings. It is noteworthy that early warnings appear to have been effective in limiting to 23 the number of deaths attributed to Hurricane Andrew in 1992, even though the direct damage reached a record total of more than $25 billion (Hebert et al., 1996).

Progress in Weather Services

In the United States, daily weather maps and forecasts were first provided by the Army Signal Corps, starting in the 1870s. Radiosonde networks established over much of the world in the late 1940s and 1950s, practical numerical weather prediction initiated in the late 1950s, radar data networks also established in the 1950s, and weather satellites deployed in the 1960s all led to measurable improvements in forecasts.

The National Oceanic and Atmospheric Administration (NOAA) and the National Weather Service (NWS) are now completing a historically unique technological development cycle. New facilities include a national Doppler weather radar network, major weather satellite improvements, the enhanced Automated Surface Observation System (ASOS) surface network, and a computerized network of weather display and prediction systems (Automated Weather Interactive Processing System—AWIPS) to be distributed at forecast centers around the country. In addition to improving weather forecasts, these new capabilities will be used for the study of climate and of atmospheric chemistry and physics, and for a variety of applications and special purposes. However, the full benefits of these new technological systems will be realized only if there is a sustained and substantial commitment by the government to support research efforts such as the U.S. Global Change Research Program (USGCRP) and the U.S. Weather Research Program.

MAINTAINING ENVIRONMENTAL QUALITY

The human habitability of the Earth depends on certain basic prerequisites, including the availability of clean air and water, food, shelter, and security from natural hazards. Natural processes sometimes create adverse environments, and local anthropogenic activities sometimes create life-threatening conditions. Examples include the lethal episodes of smog in London and in Pennsylvania valleys apparently caused by coal burning during the nineteenth and early twentieth centuries. Local anthropogenic influences on air quality and other aspects of the environment continue to be major problems for individual cities, states, and countries. When the pollution caused by one jurisdiction interferes with the economic interests of another, governments become involved in complex issues.

Chlorofluorocarbons and Ozone

Recent concerns have focused on threats posed by certain trace gases that are long-lived and mix throughout the entire global atmosphere. Important examples are the manufactured chlorofluorocarbon (CFC) gases that were widely used in spray cans and refrigeration systems in the middle to late twentieth century. These gases have lifetimes of years because they have no natural removal mechanisms in the lower atmosphere. CFCs released at the Earth's surface diffuse in a few years to the upper atmosphere where they are exposed to energetic solar radiation. After a series of chemical transformations, CFCs lead to a decrease in the ozone concentration in the stratosphere, which in turn permits an increase in dangerous ultraviolet radiation at the Earth's surface. As described later, scientific research produced an explanation of the processes involved. When the cause-and-effect relationship became known and accepted, the governments of many nations developed and signed agreements (the "Montreal Protocol") to limit products deemed harmful to the ozone layer.

Greenhouse Gases and Global Change

A more complex problem is posed by the increasing concentrations of so-called greenhouse gases, primarily carbon dioxide released by the combustion of coal, natural gas, and petroleum. Some of the carbon dioxide is absorbed in the terrestrial system, but a globally dispersed residual has been accumulating in the atmosphere at a rate of about 0.5 percent per year.

Theory suggests that increasing concentrations of greenhouse gases will result in a warming of the planetary surface and that the direct effects may be amplified by feedback mechanisms, including absorption and re-emission of infrared radiation by atmospheric water vapor, itself a greenhouse gas. The observed concentrations of carbon dioxide (and other greenhouse gases such as methane) have increased since the beginning of the industrial revolution in the eighteenth century in parallel with the consumption of fossil fuels. Moreover, there appears to have been an increase in the Earth's surface temperature over the last century. Although all of the complex interactions are not fully understood, there is enough evidence that the climate research community has accelerated efforts to understand the links and interactions among components of the climate system. The federal government has responded to these concerns with the U.S. Global Change Research Program that brings together the efforts of many agencies in a coordinated and comprehensive study (see USGCRP, 1997). Other governments are also concerned and have turned to the international scientific community for advice. In its most recent report, the Intergovernmental Panel on Climate Change (IPCC, 1996) has stated, "The balance of evidence suggests that there is a discernible human influence on global climate."

Understanding and taking action on the effects of greenhouse gases are each challenging because of the difficulty of modeling the entire climate system and the complexity of assessing economic, social, and political implications. Atmospheric scientists will have to collaborate with disciplines that can delineate other dimensions of these issues to develop an understanding that may be useful to governments and society. Moreover, the level of public confidence in climate change predictions will directly influence the urgency with which the nation undertakes measures to mitigate, or adapt to, the predicted change.

Aerosols

Anthropogenic aerosols created by the burning of some fossil fuels are another example of environmental consequences of human activities. These tiny particles can limit visibility, cause lung problems, foul delicate machinery, reduce the intensity of the sunlight reaching the Earth's surface, and contribute to acid rain. Aerosols influence the formation of clouds and hazes. Screening of sunlight and changes in cloud properties have potential climate effects since they could disturb the energy balance of the Earth-atmosphere system. An important question is the extent to which anthropogenically produced aerosols are offsetting, delaying, or altering the nature of global warming attributed to increasing concentrations of greenhouse gases. Years of research will be required in order to develop the quantitative understanding necessary to advise policy makers about the consequences of increasing concentrations of aerosols (see NRC, 1996a).

Role of the Atmospheric Sciences in Environmental Issues

Atmospheric scientists often respond to an emerging environmental problem with dynamical and chemical studies and with analyses of the associated scientific issues. Sometimes such studies reveal that scientific advances are required to explore and then understand the new problem and its relation to atmospheric processes or events.

Studies focused solely on the scientific aspects of an environmental issue are rarely either sufficient or satisfactory. Environmental issues and their potential consequences must be addressed in the context of their human implications, including economic, social, and political aspects. Thus, the atmospheric sciences become most effective as partners in interdisciplinary collaborations aimed at resolving the intertwined scientific and human aspects of an environmental issue.

To maintain the confidence of society, atmospheric researchers must maintain high standards of scientific integrity and, in either purely scientific or applied efforts, must concentrate on scientific questions and rigorous analysis of causes and effects. They must remain neutral with respect to the economic and political considerations that swirl around all issues related to maintaining environmental quality. Through this strategy, atmospheric scientists will help clarify for society the consequences of a range of alternative actions and policy options.

ENHANCING NATIONAL ECONOMIC VITALITY

A wide range of activities that contribute to national economic vitality are sensitive to weather and climate variations. The atmospheric sciences have a long and successful history of assisting these activities in two ways:

1. The public and private weather information sectors provide predictions of weather and its consequences—some focused on specific activities—that allow government and industry to reduce the economic loss and disruption of activities owing to adverse weather or to take advantage of favorable conditions for action.

2. The atmospheric sciences develop and disseminate a wealth of information that contributes to reducing long-term vulnerability or sensitivity to atmospheric conditions and climate variations and, in some cases, specifies envelopes of opportunity for certain activities.

The atmospheric sciences thus contribute to decision making in both public and private domains and help lubricate the national economic engine. A more effective contribution to enhancing national economic vitality will require improved collaboration between providers and users of atmospheric and environmental information to ensure that needs, decision processes, capabilities, and constraints on alternative actions are fully explored.

Benefits of Weather and Climate Information

Contributions to the gross domestic product (GDP) of activities that are weather sensitive to some extent are shown in Table I.2.5. Other specific examples are available, including the estimate in Table I.2.3 that damage by adverse weather averaged $17.7 billion per year in 1991-1995. As another example, the Federal Aviation Administration assesses the economic cost of weather-related

TABLE I.2.5 Categories of U.S. Activities That Display Sensitivity to Weather and Climate

Industry	Contribution to Gross Domestic Product, 1996 ($ billion)	Percent of Gross Domestic Product
Industries Sensitive to Weather and Climate		
Agriculture, Forestry, and Fisheries	115.5	1.9
Construction	222.1	3.7
Transportation and Public Utilities	529.3	8.8
Retail Trade	557.5	9.3
Finance, Insurance, and Real Estate	1106.1	18.4
Subtotal	2530.5	42.1
Industries Generally Not Sensitive to Weather and Climate		
Mining	85.2	1.4
Manufacturing	1063.0	17.7
Wholesale Trade	394.4	6.5
Services	1182.7	19.7
Government	755.7	12.6
Subtotal	3481.0	57.9
Gross Domestic Product	6011.5	100.0

SOURCE: Bureau of Economic Analysis, Department of Commerce.

delays of U.S. airline traffic at $1 billion per year; improved forecasts of winds aloft would lead to substantial savings in airline costs for fuel (see NRC, 1994a). Governments spend considerable sums on removing snow from highways and repairing highway damage from weathering. Agricultural activities are sensitive to weather events and air quality. The benefits of weather information to the construction, retail, or tourism industries are large but would be difficult to specify precisely.

In attempting to estimate the benefits of weather, climate, and air quality information, it is essential to distinguish those effects that can be mitigated with timely, accurate information from those that cannot. Although the accurate warnings of hurricanes can reduce the loss of life, they provide little protection against the destruction of buildings by hurricane winds or tidal surges. Even though forecasts of precipitation rates can assist in managing flood control facilities, the damage from severe and widespread floods is largely independent of forecast accuracy. Agricultural damage from an extended drought may be severe, even though both long-term outlooks and short-term forecasts were accurate. However, forecasts of severe winter storms allow individuals and industries to make suitable advance preparations. Accurate measurements and forecasts of air pollution could be used to mitigate pollution through temporary decreases in emission rates. Improved understanding, modeling, and prediction of the interactions between solar phenomena and near-Earth space will help reduce damage to satellites in orbit and perhaps reduce disruptions of communications and electrical power networks.

Although weather and climate forecasts may not provide for mitigation of all harmful effects, the use of climate data and extreme-value statistics, along with impact assessments, design studies, and possibly appropriate codes or standards, can significantly reduce the adverse consequences of severe weather events. Moreover, long-term climate records provide information critical to urban planning, land-use planning, agricultural strategies, and air quality standards. As skill develops in seasonal climate forecasting on a regional scale, both governments and industry can begin to realize further benefits in shaping strategies to expected weather conditions. Forecasts of El Niño are proving useful in agricultural planning in several Latin American countries (NRC, 1996b).

STRENGTHENING FUNDAMENTAL UNDERSTANDING

The following examples illustrate how enhanced fundamental understanding of the atmosphere leads to practical benefits and can stimulate progress.

In the 1930s, Professor Carl-Gustav Rossby was trying to understand the large-scale patterns of the middle and upper troposphere being observed from 3 to 10 km above the surface by newly developed balloon-borne instrument packages. He developed a highly simplified version of the equations of atmospheric motion and predicted a periodic wave structure (now known as Rossby waves) that corresponds to some of the observations. A fundamental aspect of large-scale, midlatitude flow was described by just a few symbols. This work foreshadowed and contributed to more detailed understanding of large-scale atmospheric flow and was also a significant stimulus for the development of numerical weather prediction and thus for the increased success of contemporary weather forecasts and climate models.

Evidence that state-of-the-art numerical models can reproduce complex atmospheric processes was provided by Joseph Klemp and Robert Wilhelmson in the mid-1970s. Their numerical simulation successfully modeled the three-dimensional structure and dynamics of the powerful thunderstorms common to the Great Plains and other U.S. locations. Their work provided a foundation for the present understanding of severe storm dynamics and perhaps for a method of numerically predicting such events in the future.

The discovery in the last decade of the processes by which the release of the manufactured chlorofluorocarbon gases used in spray cans and refrigeration can damage the protective ozone layer of the stratosphere involved laboratory experiments, theoretical process analysis, ground-based and satellite observations, and for verification, instrumented aircraft measurements over the South Pole. The success of the endeavor, and its importance to human life and safety, was recognized by award of the 1995 Nobel Prize in Chemistry to Sherwood Rowland, Mario Molina, and Paul Crutzen.

A widely recognized contribution to fundamental understanding by a meteorologist is chaos theory, pioneered by Edward N. Lorenz beginning in the 1960s. Professor Lorenz explored the properties of a simplified system of equations describing convection. He discovered, through numerical experimentation, that the evolving solutions of these equations were aperiodic and ultimately unpredictable, even though they were clearly deterministic in the sense that they were governed by the equations of the system. Such chaotic behavior of nonlinear systems is now known to be common rather than rare, and this discovery has resulted in a new paradigm for phenomena occurring in almost every field of science. It has also resulted in a widely accepted theory of atmospheric predictability and has led to a deeper understanding of the mathematical structure of atmospheric motion and the nature of strategies required to predict the statistics that describe climate. Of even greater significance, perhaps, is the fact that the understanding of chaos and nonlinear dynamics that stemmed from basic research in meteorology has now illuminated phenomena studied in many scientific disciplines.

PART I

3

Scientific Imperatives and Recommendations for the Decades Ahead

To accelerate progress in the first decades of the twenty-first century, the atmospheric sciences must advance understanding of processes through which the atmosphere evolves. To do so requires organizing and optimizing interactions between conceptual and technological advances through a comprehensive integration of observational, modeling, and other research efforts.

Part II of this report presents Assessments of five disciplines within the atmospheric sciences:

1. Atmospheric physics
2. Atmospheric chemistry
3. Atmospheric dynamics and weather forecasting
4. Upper atmosphere and near-Earth space
5. Climate and climate change

These Disciplinary Assessments, prepared by three standing committees and two ad hoc groups of the Board on Atmospheric Sciences and Climate (BASC), emphasize understanding and predicting atmospheric phenomena and processes. For the near future, they emphasize forecasts of atmospheric phenomena with significant societal impacts, including seasonal climate variability, chemical processes and air quality, and space weather events. For the longer term, they emphasize the resolution of climate variability on the scale of decades to centuries and the possibility of projecting climate variations. Each assessment analyzes critical scientific issues, identifies the major opportunities and initiatives

> **Box I.3.1**
> **Atmospheric Science Recommendations**
>
> ***Imperatives for Atmospheric Research***
>
> 1. Optimize and integrate atmospheric and other Earth observation, analysis, and modeling systems.
> 2. Develop new observation capabilities for resolving critical variables on time and space scales relevant to forecasts of significant atmospheric phenomena.
>
> ***Recommendations for Atmospheric Research***
>
> - Resolve and model interactions at the boundaries between the atmosphere and other Earth system components, and interactions within the atmosphere among phenomena of different scales;
> - Apply the discipline of forecasting in atmospheric chemistry, climate, and space weather research in order to advance knowledge, capabilities for prediction, and service to society.
> - Develop collaborative studies of three emerging issues: (1) climate, weather, and health; (2) climate change implications for water resource management; and (3) implications of rapidly increasing atmospheric emissions.

for the discipline, and recommends a scientific and programmatic agenda for the decade or two ahead based on an evaluation of priorities within the discipline.

Each of these Assessments cites the critical role of comprehensive and integrated observations in research and applications, leading to BASC's identification of two imperatives concerning observations as the highest-priority endeavors for the atmospheric sciences and services. In addition, three recommendations for research also emerge from these studies. They are stated in brief form in Box I.3.1.

This section discusses the highest-priority imperatives and recommendations for atmospheric research in greater detail, first stating the imperative or recommendation and then providing its justification. Section 4 of Part I addresses some of the leadership and management issues that should be considered in shaping the financial and infrastructural agendas for atmospheric sciences.

ATMOSPHERIC SCIENCE IMPERATIVE 1: OPTIMIZE AND INTEGRATE OBSERVATION CAPABILITIES

The atmospheric science community and relevant federal agencies should develop a specific plan for optimizing global observations of the atmosphere, oceans, and land. This plan should take into account requirements for monitoring weather, climate, and air quality and for providing the information needed to improve predictive numerical mod-

els used for weather, climate, atmospheric chemistry, air quality, and near-Earth space physics activities. The process should involve a continuous interaction between the research and operational communities and should delineate critical scientific and engineering issues. Proposed configurations of the national and international observing system should be examined with the aid of observing system simulation experiments.

Atmospheric observation, modeling, and prediction systems are increasingly interdependent; thus, the components of the atmospheric information system should be optimized as part of an end-to-end system. Moreover, there is an increasing need in both research and applications to integrate observations of the atmosphere, ocean, and land surface.

As a consequence of the increasing synergy of observation and modeling systems, four-dimensional[1] data bases portraying the evolution of actual and predicted conditions are replacing the traditional synoptic snapshots of atmospheric conditions. Made possible by new capabilities in observational, computational, and communications technology, these four-dimensional data bases containing predictions of traditional variables will be the input to distributed computer procedures that prepare application-specific forecasts formulated in terms of key impact variables,[2] decision aids, and recommendations for action.

Somewhat different data formats and strategies may also be required. Thus, the effort to optimize and improve national and global observing systems should examine commonalities and differences among the needs of weather prediction, climate monitoring and projection, atmospheric chemistry and air quality prediction, near-Earth space physics, and other environmental disciplines.

New Observing Opportunities

New opportunities for acquiring atmospheric observations suggest that the observing system of the future may be dramatically different from that of today. Three examples illustrate this point:

1. ***Commercial aircraft observations:*** Sensors carried by commercial air transport aircraft are now producing thousands of observations of winds and

---

[1] A four-dimensional data base is one in which data are stored according to four independent variables, or coordinates: longitude, latitude, height above mean sea level, and time.

[2] Impact variables are descriptions of atmospheric conditions that directly affect a given operation, in contrast to the meteorological variables used in theory and modeling, such as pressure and temperature. For aviation, as an example, such impact variables include cloud height, visibility, and intensity of icing or turbulence.

temperatures daily over the contiguous United States; the use of humidity sensors is currently being evaluated. The ascent and descent of these aircraft at major air terminals can provide a dense set of atmospheric soundings in addition to enroute observations (NRC, 1994a; Fleming, 1996).

2. *Global positioning system:* Signals from the Global Positioning System (GPS) satellites intended for use in navigation can also, through occultation techniques, produce profiles of the atmospheric index of refraction (Ware et al., 1996) which can be used to derive temperature and humidity profiles or can be assimilated directly into models. With these GPS measurements, global fields of important variables could be measured relatively inexpensively and continuously for weather and climate applications.

3. *Adaptive strategies:* High-resolution measurements in specific locations may sometimes enhance prediction of significant phenomena, suggesting that some observations can be made adaptively in response to empirical evidence or modeling techniques that identify sources of errors. For example, operational models predicting severe weather or tropical storms might be configured to specify locations where higher-resolution initial conditions would enhance accuracy (Burpee et al., 1996). Then, observations could be obtained in these regions from satellites or remotely piloted scientific aircraft.

Requirements for Optimizing and Integrating Observing Systems

The integrated portrayal of atmospheric evolution and interactions in four-dimensional data bases containing both existing and some critical new observations is essential for the advances envisioned in this report. The following requirements must be met to achieve this objective:

- *Integration with modeling efforts:* Efforts to optimize observing systems must take into account the analysis and prediction models that will assimilate the data, whether they are weather, climate, upper-atmosphere, or chemical models. Thus, rigorous end-to-end analyses—from observations through models to accuracy of predictions—are required to assess the benefits and costs of overall strategies and specific observation schemes. For example, issues such as whether radiosonde observations can be replaced with satellite data must take account of many considerations, including forecast accuracy, operational costs, and especially the integrity of the climate record.
- *Increases in computing power:* Advances in computing power will be necessary to conduct fully coupled simulations of the entire Earth system—including the atmosphere, oceans, land surface, ice, and chemical and biological subsystems—in order to study climate and its variability on various scales. High-resolution weather forecasts, along with ensemble approaches to assessing confidence in the prediction will similarly require greater computer power. Both activities may impose changing requirements on optimization criteria.

- *Assimilation of new forms of data:* Innovative assimilation schemes for incorporating new forms of data into analysis and prediction models may enable significant advances. For example, recent research has shown that using model variables to calculate the model radiance fields and then iterating the model to achieve agreement with radiance fields as observed by satellites is superior to attempting to convert satellite observations into traditional variables such as temperature and humidity. Significant benefits from other new forms of data, including those from commercial aircraft and air quality networks, may derive from similar innovation and analysis. Earth Observing System (EOS) satellites will provide global observations focused on understanding climate variability and evolution. In addition, EOS and other research satellite efforts will provide new streams of data and new instruments that may be adapted for operational use.
- *Multiple uses of data bases:* As data from surface and upper-air observations, operational radar, geostationary and polar-orbiting satellites, and other sources such as air quality networks are incorporated into the data base, they can be shared by the weather, climate, and applications communities. Because resources are limited, it will be critical to strive for the maximum benefit for all from the incremental investments required to optimize and integrate the observational and analysis systems. Indeed, the goal should be to integrate atmospheric data bases with those developed by other disciplines to represent ocean and land surface conditions.
- *Information organizing systems:* A requirement in forecasting is to identify the limited subset of data that is critical for the accuracy of a prediction. Machine-learning concepts, statistical models relating key variables or decision parameters to computer forecasts, expert systems, and the concepts of fuzzy logic all have potential for organizing observed data and computer simulations in ways that strengthen atmospheric information services.
- *International collaboration:* Four-dimensional data bases will transcend national boundaries. It is essential that the long tradition of sharing weather and climate data globally be maintained. The proposed integration of satellite observations between the United States and European space and meteorological agencies is an important harbinger of progress; in contrast, attempts to limit the availability of local data are serious impediments to the cooperation required.

Many of the issues involved in improved observations and improved weather services are being addressed in the federal interagency effort to create a future North American Atmospheric Observing System (NOAA, 1996) as a composite observing system for the twenty-first century. Until it has been established rigorously that new approaches are fully adequate alternatives, it is essential to retain the full capabilities of the worldwide radiosonde network because it is the foundation of the present atmospheric observation and prediction system.

Observing System Simulation Experiments

Getting the highest return from investments in atmospheric information and forecasts is the clear motivation for integrating observing systems. Assessing alternatives is critical as automated systems replace human observers in an attempt to reduce ongoing operational costs by making capital investments in systems that have lower overall life-cycle costs.

Observing system simulation experiments enabled by contemporary computers and numerical models can help to determine the optimum configurations of observing system components and numerical models relative to forecast accuracy and overall costs, including capital investments, operational costs, and the costs of disseminating and archiving data. Such experiments allow rigorous and quantitative examination of a wide range of strategies and can indicate whether new resources will produce demonstrable benefits. Simulations of observing and prediction systems can thus be a critical mechanism for managing the advance of the discipline and its service to society.

ATMOSPHERIC SCIENCE IMPERATIVE 2: DEVELOP NEW OBSERVATION CAPABILITIES

The federal agencies involved in atmospheric science should commit to a strategy, priorities, and a program for developing new capabilities for observing critical variables, including water in all its phases, wind, aerosols and chemical constituents, and variables related to phenomena in near-Earth space, all on spatial and temporal scales relevant to forecasts and applications. The possibilities for obtaining such observations should be considered in studying the optimum observing systems of Imperative 1.

Contemporary numerical atmospheric computer models are sufficiently varied and powerful that they can predict or simulate a range of phenomena such as climate change or air pollution episodes, as well as the course of the weather. However, observations of critical variables on time and space scales relevant to forecasts are essential to improving such numerical simulations and predictions. Some of the required observations present significant technological challenges.

In some cases, higher-resolution observations of traditional meteorological variables such as water vapor are required to portray important processes accurately. In others, both research and applications require some new observations, for example, of aerosols and trace chemicals and their vertical transports in both clear and mixed-phase environments.

The required improvements will arise in part from new instruments and new technology, as in new remote sensing techniques, and in part from new observational strategies. For example, automation, adaptive observations, and international cooperation will all contribute to improved observations, as demonstrated by the Tropical Ocean Global Atmosphere-Tropical Atmospheric Ocean array, a network of buoys stretched across the equatorial Pacific.

Water in the Atmosphere

Water in all its phases and changes in these phases have a pervasive role in weather, climate, and atmospheric chemistry; thus, observing the phases and amounts of water in the atmosphere is important to understanding atmospheric events on all scales. The energy released by changes in the phase of water substance is of major importance in driving global wind systems.

Water Vapor

On the storm scale, prediction of convective precipitation is limited by uncertainty in the distribution of water vapor in the atmosphere and the amount of water in the soil. On the longest time scale, water vapor is the most important greenhouse gas, and significant uncertainty in climate models can be traced to inadequate understanding of the water budget and water-induced radiation feedback. Water vapor in the atmosphere varies markedly over small scales vertically and horizontally; thus, its variability is significant at scales not resolvable by the radiosonde network.

Today, water vapor profiling technology and its integration with other measurements and models to improve resolution are evolving rapidly. Notable advances are being made with remote sensing techniques. In situ water vapor measurements are essential to resolve fine structure and to constrain remotely sensed data. Radiosonde observations satisfy these needs only partially because of large spatial and temporal gaps. Greater coverage along commercial air routes and more soundings near airports will be made possible through the Commercial Aviation Sensing Humidity (CASH) program [sponsored by the Federal Aviation Administration (FAA) and the National Oceanic and Atmospheric Administration (NOAA) Office of Global Programs], which is installing humidity sensors on air transport aircraft; these sensors will have sufficient resolution to obtain profiles during aircraft ascent and descent. However, improvements are needed because current devices cannot measure humidity accurately on an aircraft flying through clouds.

Clouds

Clouds are the visible evidence of a host of complex processes involved with the change of water vapor to both liquid and solid forms. Many critical aspects of the cloud processes involved in the formation of precipitation and in interactions between water drops and ice particles are still not well understood and require detailed observational and modeling studies.

These processes also have important impacts on chemical reactions in aqueous and mixed-phase environments and must be better understood in order to improve models of local and global chemical cycles.

As described later, clouds are an important control on the planetary energy budget, inducing significant modifications of the incoming streams of solar radiation and the outgoing streams of infrared radiation. Moreover, latent heat released during the formation of liquid water is a significant source of energy for both severe weather and large-scale synoptic systems.

Precipitation

The global distribution of precipitation is tied to the dynamics and energetics of the atmosphere, the coupling of atmosphere to ocean, and ocean currents themselves. Precipitation observations thus reflect both regional and local conditions and are critically needed for climate studies, verification of global models, and forecasts of severe weather and flood conditions. Truly global observations can be made effectively only by satellites; progress with precipitation observations is being made with the Special Sensor Microwave/Imager (SSM/I). Another satellite system, the Tropical Rainfall Measurement Mission (TRMM), launched on November 27, 1997, promises to improve precipitation estimates and estimates of energy released over the tropics, using a combination of high-resolution radar, a passive microwave radiometer, and measurements in the visible to infrared. Variability in the location of tropical rainfall has been shown to be a critical link in the changes leading to El Niño events.

Precipitation estimates over the United States have been greatly enhanced by deployment of NOAA's WSR-88D radars, and these improvements are being incorporated into river basin flood models. Further development is necessary to (1) improve the accuracy of rainfall estimates by upgrading WSR-88D radars to include multiparameter capability, (2) to improve rain gauge reliability, and (3) to strengthen the ability to estimate rainfall from satellites in space. As in the case of water vapor, improved precipitation-estimation algorithms must be based on combinations of relevant data. An expert system might be developed to produce precipitation estimates from input information including radar, satellite, and rain gauge data, as well as season of the year and type of cloud system.

Wind Observations

Owing to modern technology, the density of wind observations has significantly increased over the United States and some other continental areas in recent years. In the central United States, radar wind profilers, sensitive to the motions of refractive index fluctuations in clear air, are now demonstrating their usefulness as forecast model input and research tools. WSR-88D Doppler radars, although used mainly to document precipitating weather systems, can measure winds in clear air to altitudes of several kilometers over land during the day. Some civilian transport aircraft provide wind observations along air routes and during ascent and descent.

Wind information over the open ocean is sparse. Radiosondes are launched from islands and from some ships, but many of the remote sites are being abandoned or becoming unreliable owing to fiscal pressures in other nations. Winds inferred from satellite observations of cloud motion can be useful but are sometimes inaccurate.

More accurate and comprehensive wind observations are needed for nearly all the endeavors of atmospheric science. Adaptive and episodic observations, obtained on demand in data-sparse regions where information is critical, could help. Presently, aircraft are deployed to gather data in and around tropical cyclones through penetrations and dropsondes; future strategies could also incorporate remotely piloted scientific aircraft (Holland et al., 1992) and drifting balloons that rise or fall in response to remote control. Satellite-borne lidar systems for measuring wind could become a reality in the next decade, with several promising techniques now under development. Satellite scatterometer measurements, such as those from the European Research Satellite ERS-1, and microwave brightness from SSM/I, when constrained by models or independent observations, should prove useful for estimates of surface winds over the oceans. New airborne radars developed and operated by the National Center for Atmospheric Research and the NOAA Aircraft Operations Center are providing unprecedented detail on the fine-scale structure of weather systems, including severe storms and hurricanes.

Observations in the Stratosphere

The stratosphere affects the world's weather through interactions and exchanges with the troposphere and is an integral part of the global climate and chemical system. Sampling meteorological variables, trace gases, and aerosols in the stratosphere will require a combination of ground-based balloon and remote sensing, satellite measurements, and piloted and remotely piloted aircraft—all blended with numerical models. Satellite systems should overlap in time, to provide continuity of record as well as to provide confidence in comparison of studies separated by several years. In situ aircraft measurements will also be used to evaluate new remote sensing devices. Remotely piloted aircraft may become the platform of choice because of their lower cost and high-altitude capability, provided that safety issues are resolved and the technology proves to be operationally feasible. Both space-borne and airborne measurements will benefit from miniaturization of remote sensing and in situ instruments.

Observations in Near-Earth Space

Phenomena created in the near-Earth environment by charged particles and varying solar magnetic fields are referred to collectively as "space weather" and have a variety of significant effects on human activities, as discussed later. The prediction of space weather phenomena requires observations that can be obtained in part through present and planned satellite programs. For example, the GPS radio occultation technique (Ware et al., 1996) gives vertical profiles of electron density and total electron content in the ionosphere.

Because space weather phenomena are forced by solar variability, three sets of solar observations should be emphasized: (1) coronal mass ejections, (2) magnetic fields including those in the corona, and (3) near-Sun solar wind properties. Moreover, measurements of the interaction of solar particles and the solar magnetic field with the Earth's magnetosphere and ionosphere are important for basic understanding and development of numerical models.

ATMOSPHERIC RESEARCH RECOMMENDATION 1: RESOLVE INTERACTIONS AT ATMOSPHERIC BOUNDARIES AND AMONG DIFFERENT SCALES OF FLOW

The major weather, climate, and global observation programs supported by the federal government and international agencies should put high priority on improved understanding of interactions of the atmosphere with other components of the Earth system and of interactions between atmospheric phenomena of different scales. These programs, including the U.S. Weather Research Program, the U.S. Global Change Research Program, and other mechanisms for supporting atmospheric research, require observational, theoretical, and modeling studies of such interactions.

Atmospheric studies are shaped today by the recognition that interactions with neighboring systems are critical to improved understanding and prediction. It is no longer sufficient to study the individual components of the Earth system in isolation—attention must now turn to understanding, modeling, and predicting them as coupled systems. In many cases the relevant coupled systems are the actual physical subsystems; in others, the interacting subsystems are more abstract—for example, when small-scale eddies interact with the largest scales of atmospheric flow. Critical scientific questions focus on the exchanges of energy, momentum, and chemical constituents among the atmosphere and the surface below as well as the layers above leading to near-Earth space.

In the initial study of a dynamical system, interactions with the exterior environment can sometimes be ignored. This approach served the numerical weather forecasting community well in its initial efforts. However, as the length of the forecast period increased, the need to account for the flows of dynamically significant quantities across system interfaces became evident. In climate dynamics, knowledge of this interaction between the components of the Earth system is essential to progress. For example, it is now recognized that the El Niño-Southern Oscillation (ENSO) is driven by interactions within the combined tropical ocean-atmosphere system.

Surface Processes

Processes occurring at the boundary between the atmosphere and the land and sea are vital elements in weather, climate, atmospheric chemistry, and global change. Atmospheric interfaces with the ocean and the land are similar because both represent the intersection of an atmospheric boundary layer with a boundary layer on the surface below. The ocean interface is better understood except under high winds; the land interface merits greater emphasis.

Surface fluxes over the ocean are important in early cyclogenesis (Kuo et al., 1991), and the churning up of cold ocean water by hurricanes (Shay et al., 1992) helps determine their evolution and motion (Bender et al., 1993). The weather and climate are particularly sensitive to air-sea interaction over the tropical South Pacific, where the warmest open-ocean surface temperatures coincide with Earth's greatest annual precipitation (Webster and Lukas, 1992).

Surface temperature, moisture, and fluxes are as strongly linked to land surface properties as they are to atmospheric variables. Satellites provide estimates of surface temperature, vegetation, and moisture characteristics that can be converted into estimates of fluxes but require further calibration against in situ measurements of local characteristics (Kogan, 1995).

Methods of estimating fluxes over the ocean using satellites and buoys are perhaps at a more advanced state than over land, with the major uncertainties associated with light and strong winds, wave spectra, and precipitating convection. The TOGA-TAO array is providing a long time series of atmosphere and ocean data useful for understanding sea-air interaction in the equatorial Pacific. Continued developments in remote sensing of surface wind, temperature, and radiative fluxes are needed, along with adaptive sampling techniques for in situ and remotely sensed data in cyclones.

Long-Term Interactions with the Oceans

On decadal to centennial time scales, the interaction between the upper and the deeper parts of the world ocean is believed to be a primary control on the natural variability of surface temperature on the planet. Moreover, the flow of heat between ocean layers sets the time scale for significant response to forcing such as the greenhouse effect. Better quantification and understanding of these interactions will help in assessing the difference between natural fluctuations of the large-scale surface temperature and the response to external forcing.

Vertical transport of heat in the oceans is mediated by currents that are often of very small horizontal scale compared to basin-wide gyres. Flows connecting the ocean surface to deep water are thought to occur in only a few locations on the planet. For example, North Atlantic deep water is formed in very small regions in the Norwegian Sea. Local phenomena such as the intermittent formation of sea ice can have significant effects in modulating flow from the cold briny surface waters to the very bottom of the world ocean.

Surface interactions are also important. Accurate simulation of the exchange of heat and fresh water at the ocean surface is a demanding contemporary challenge for coupled ocean-atmosphere climate models. Sea ice variation also is only poorly understood, although it is clearly important in long-term climate change because of its high reflectivity and its inhibiting effects on thermal exchange between the atmosphere and ocean.

The coupling of the atmosphere and ocean in the tropics gives rise to the ENSO cycle of alternating extremes of warm and cold sea surface temperatures with periods of three to seven years that affect the global climate system. The information between system components is communicated through wind stress, sea surface temperature, radiation, and precipitation. Although current models exhibit solutions with ENSO-like oscillations that suggest useful predictive capability, they are not always realistic. Improvement in computation of the fluxes across the tropical atmosphere-ocean boundary should lead to notable improvements in understanding and practical application.

Clouds and Their Consequences

Clouds and cloud systems are relatively small but numerous and, in sum, play a key role in shaping weather and climate. For example, the reflection of sunlight by the planet is determined largely by the cloud cover, and a significant fraction of the infrared radiation flowing from the surface is absorbed by clouds and reradiated to the surface as part of the greenhouse effect. Hence, clouds are strong controls on the amount of solar energy absorbed, the planetary energy budget, and the surface temperature. Success in modeling a changing climate will require correct specification of the effects of clouds and the way in which these effects will themselves change as climate evolves.

The effects of clouds on radiation streams are determined by their horizontal and vertical distribution and the type, size, and concentration of their constituent particles. The properties of cloud particles are a consequence of the initial concentrations of aerosols and thermodynamic properties, subsequent vertical motion and mixing, and encounters with particles having different properties.

Deep, precipitating convection interacts with motions of larger scales in other important ways as well. For decades, theoretical studies have explored the effects of convective heating (the latent heating associated with convective rain) on the growth of synoptic- to global-scale waves in the tropics. More recently, it has been demonstrated that the effects of ENSO on weather in other parts of the world are determined largely by the atmospheric response to heating by deep convection over the equatorial Pacific, which is in turn modulated by the sea surface temperature distribution.

Today, numerical models that can resolve properties of individual clouds and cloud ensembles are being used to study the response of clouds to larger-scale forcing and to develop convective parameterization schemes for climate models whose resolution is too coarse to represent clouds explicitly. Toward this end, the results of cloud-resolving models are presently being compared to available observations, and cloud-resolving models are being coupled with ocean models to increase the realism of the simulation.

Aerosols and Atmospheric Chemistry

Suspensions in the atmosphere of tiny liquid or solid particles are known as aerosols and can cause macroscopic changes in air quality, atmospheric heating, and air chemistry. Aerosols have both short-term and long-term effects in the atmosphere: cloud particles form on them; they absorb and scatter radiation passing through the atmosphere, thus altering the local heat balance; and they are sites of chemical reactions with atmospheric trace gases. Aerosols should be the focus of much research in the coming decades, including gathering more data on aerosol chemistry and its evolution in time and space, and improved understanding of the light-scattering proportions of different types of aerosols (NRC, 1996a).

Trace gas concentrations in the atmosphere depend on sources and sinks are usually at the surface of the Earth and on processes within the atmosphere. Atmospheric transport and modification of chemical constituents in mesoscale flow include reaction, advection and diffusion, and interactions with water in its various phases. Improved understanding is clearly important for air quality modeling and prediction.

The Fundamental Problem of Nonlinearity

Many of the key challenges in atmospheric sciences embody the fundamental problem of all geophysical fluid flow—the nonlinear interaction between phenomena with various length and time scales in the flow and phenomena of different length and time scales in the boundary conditions, external forcing, and within the flow itself.

Most geophysical flow problems involve interactions with a boundary and thus develop turbulent boundary layers. Energy, momentum, and other properties are passed across these boundary layers from the fixed surfaces to the larger-scale flows. Geophysical problems rarely allow simple closures; some interactions must usually be retained, even if in somewhat simplified form. In fact, it is typical in atmospheric boundary layers for scales varying as much as eight orders of magnitude in both space and time to be involved in fluxes from the surface. Crude treatment of these interactions limits the accuracy of long-range forecasts.

Yet even without significant boundary effects or internal interactions with water processes, nonlinear interactions between flow components of different scales has been a continuing challenge for more than a century. Significant progress has been made, as reflected by contemporary understanding of turbulence, predictability, and the phenomena of chaos. Nevertheless, we have yet to develop ways to describe chaos mathematically by finding analytic methods for specifying the nature of attractors and the statistical characteristics of resulting flows (Dutton, 1992). Of particular interest is the extent to which long-term climate simulations can determine global or local statistics accurately, even though the numerical solutions are surely not deterministic.

Until the difficulties created by nonlinearity are dramatically reduced, they will remain pre-eminent challenges in the attempt to understand geophysical flows and predict them quantitatively.

ATMOSPHERIC RESEARCH RECOMMENDATION 2: EXTEND A DISCIPLINED FORECAST PROCESS TO NEW AREAS

A strategy and implementation plan for initiating experimental forecasts and taking advantage of a disciplined forecasting process should be developed by appropriate agencies and the scientific community for climate variations, key chemical constituents and air quality, and space weather events.

Much of the effort in the atmospheric sciences is aimed, either explicitly or implicitly, at extending the range and improving the accuracy of forecasts of atmospheric phenomena—weather and air quality on the short term, climate variation on the longer term. As practiced for more than a century, weather prediction involves the traditional steps of the scientific method:

- Collect and analyze observations of present conditions.
- Develop and use subjective or quantitative methods and models to infer future conditions from these observations.
- Assess the accuracy of the forecast with observations of actual conditions.
- Analyze forecast results to determine how methods and models can be improved.

Through this process, weather prediction imposes a demanding discipline on both individual forecasters and all of the atmospheric sciences. The accuracy of the forecast every day, in a wide variety of locations, is a continuing measure of the success and progress in this field—and of the integration of theory, technology, and practical methods.

The impact of this discipline has intensified with the advent of operational numerical weather prediction because precise quantitative comparisons between forecasts and actual observations can be made easily. The inclusion of computers in the forecast cycle also facilitates improvement because proposed modifications in the computer model can be applied to difficult cases retrospectively.

As shown in Part II, several of the atmospheric sciences are now developing capabilities for making quantitative forecasts and will benefit substantially by establishing experimental operational forecast procedures and thus engaging the discipline of forecasting.

Climate forecasting presents one such opportunity, on both the seasonal and the decadal scales. The TOGA program has stimulated considerable progress in monitoring and predicting seasonal to interannual climate fluctuations such as El Niño. In the early 1980s, observations in the tropical Pacific Ocean were so limited that the 1982-1983 El Niño was in progress for several months before its magnitude and implications were known. The Tropical Atmosphere Ocean observing system is now operational and includes 70 moored ocean buoys to observe surface meteorological properties and the upper-ocean thermal structure, surface drifting buoys that measure sea surface temperature, expendable bathythermographs from ships of opportunity that probe the ocean temperature field to 700 m, and island tide gauges that monitor sea level variability. Research and operational satellites complement the in situ measurements. This system allowed the evolution of the extreme El Niño event of 1997-1998 to be monitored on a day-to-day basis. Moreover, these observations have supported the development of coupled ocean-atmosphere models of the tropical Pacific Ocean that can predict tropical Pacific sea surface temperatures with confidence months to a year in advance. These developments in the observing system, together with related advances in coupled forecast models, have led to the implementation at NOAA's National Centers for Environmental Prediction of a routine, model-based, short-term climate forecast system for the United States. Similarly, the newly formed International Research Institute for Climate Prediction, created by NOAA, has taken on the complementary responsibility for the rest of the world.

Successful forecasts of climate variations over a decade or more would be of great value to a variety of industrial and government activities. These forecasts would be statistical in nature and thus focused perhaps on seasonal departures from means; as statistics, they might include an estimate of expected accuracy. Verifying such climate forecasts poses special problems because of the length of time necessary to accumulate a meaningful ensemble of cases.

Another opportunity is created by the notable advances in atmospheric chemistry over the past 20 years. The rapidity with which the scientific community established the cause of the Antarctic ozone hole only a few years after its discovery and impressive advances in both global and mesoscale chemical modeling demonstrate that the necessary links between dynamics and chemical processes are increasingly well understood and that operational models are feasible. The discipline of forecasting applied to air quality modeling will lead to improved understanding of the factors that produce fluctuations in chemical constituent concentrations. As the forecasting process matures, it will provide continuing benefits in managing the environment of urban areas and assessing the consequences of chemical perturbations induced by pollutants, fires, volcanoes, and other environmentally significant events.

A third opportunity arises with space weather—the collection of phenomena associated with emissions of energy and mass from the Sun and instabilities in the Earth's magnetosphere. Space weather events can produce considerable disturbances in the ionosphere, which in turn induce communications disruptions and produce transient induced currents that lead to failures of electrical power networks. The streams of high-energy radiation can also result in satellite malfunction and may be lethal for astronauts in space without sufficient protection. Today, the increasing societal dependence on more than 250 satellites in geosynchronous and low-Earth orbits for communication, navigation, and Earth observations creates a new urgency for understanding and predicting space weather. This urgency will increase as hundreds of new satellites are placed in orbit for global cellular phone and other information transfer services.

Space weather phenomena have been predicted for many years using statistical methods, primarily by the military. Today, advances in observations of solar phenomena and the solar wind can be coupled with increasingly quantitative research models of the solar-terrestrial system to develop an operational space weather forecasting system.

In these three examples, as with prediction of the weather a few days in advance, the relentless discipline of forecasting can be expected to stimulate the interplay between improvements in observation, theory, and practice required to develop predictive capabilities of broad value to society.

ATMOSPHERIC RESEARCH RECOMMENDATION 3: INITIATE STUDIES OF EMERGING ISSUES

The research community and appropriate federal agencies should institute interdisciplinary studies of emerging issues related to (1) climate, weather, and health; (2) management of water resources in a changing climate; and (3) rapidly increasing emissions to the atmosphere.

A number of interdisciplinary scientific issues are emerging as important candidates for attention by atmospheric scientists.

Climate, Weather, and Health

Both threats to human health or other biological systems and infectious diseases have been linked to climate variability (Shope, 1991; IPCC, 1996; Patz et al., 1996). Temperature, rain, sunshine, wind, humidity, and soil moisture may affect the emergence and spread of infectious diseases (Landsberg, 1969; Colwell and Huq, 1994; Epstein, 1995; Morse, 1995; Patz et al., 1996). Climatic factors provide limiting conditions for the distribution of vector-borne diseases; weather events can determine the timing, outbreak, and spread of disease. For example, air temperature controls the latitude and altitude distribution of mosquitoes that are vectors for encephalitis, meningitis, dengue, and yellow fever. In the tropics, rainfall controls the emergence of the anopheles mosquito, the vector for malaria.

The increasing frequency of water-borne diseases, such as typhoid, hepatitis, and bacillary dysentery, is associated with flooding. A direct threat to human well-being arises because the Earth's protective stratospheric ozone shield has been decreasing (WMO, 1995), leading to detectable increases of ultraviolet (UV-B) radiation at the Earth's surface (Herman et al., 1996). Thus, health officials expect a consequent increase in skin cancer, weakened immune systems, and other health-related concerns (Taylor et al., 1988; Cooper et al., 1992; International Agency for Research on Cancer, 1992; Johnson and Tinning, 1995).

Understanding the links between weather, climate, and various diseases and their vectors will require collaboration among a number of disciplines, particularly epidemiology and entomology. As this understanding increases, multidisciplinary observation technologies, sophisticated numerical modeling techniques, timely weather and climate analyses and predictions, and international cooperation in environmental research will combine to offer new tools in combating disease.

Water Resources

The design and operation of water resource management systems have long taken account of expected extremes in both weather and climate conditions. Now, however, population expansion coupled with the possibility of climate change mandates that systems for water supply or flood control and mitigation be designed for evolving, rather than stationary, weather and climate conditions. The effects of possible climate change on the spatial and temporal distribution of floods and droughts, as well as on patterns of precipitation, temperature, and wind, must be better understood to improve water management designs and strategies. Thus, the stability of climate and possible changes in the statistical structure of extreme weather events are important new issues for water resource management.

Moreover, the operation of water management systems is evidently linked to both weather and climate forecasts; thus, issues of predictability become important. Forecast lead times and operational flexibility in managing water systems interact to determine the range of possible actions and the possible benefits.

Atmospheric involvement in water resources extends to certain aspects of the nation's ground water supply as well. Emerging research problems involve aspects of multiphase ground water flow and coupled mass-biochemical contaminant transport that come into play as rainwater is absorbed by the ground and transported laterally through underground rock strata. Robust description and understanding of these complex phenomena involve laboratory and field experiments and the development of mathematical and numerical models capable of simulating unsteady processes with time constants varying over many orders of magnitude.

With the atmosphere providing the rapid transport portion of the hydrological cycle, it is evident that atmospheric science, in collaboration with hydrology and soil science, must focus more sharply on the complex issues associated with understanding, predicting, and managing the fluxes of water on which society depends.

Part of the World Climate Research Program, the Global Energy and Water Cycle Experiment (GEWEX), was initiated in 1988 to observe and model the hydrological cycle and energy fluxes in the atmosphere, at the land surfaces and in the upper oceans. GEWEX will significantly increases our understanding of the water-energy cycle and thus provide the basis for a more sophisticated water management system.

Rapidly Increasing Emissions to the Atmosphere

Beginning with the Industrial Revolution, environmental quality has been increasingly threatened. With the expansion of worldwide industry, new threats to environmental quality pose new research questions. One example is the economic development along the western rim of the Pacific Ocean and the concomitant increase in the emission of gases produced by combustion and industrial processes. The rapid expansion of emissions in this region is now, or will soon be, mirrored elsewhere around the world during the twenty-first century, a trend that seems likely to result in chemical and climate impacts on both regional and global scales.

These increasing emissions pose new questions concerning the regional, hemispheric, and global consequences of rapidly intensifying sources of pollution. What will the fate of these materials be as they move across the globe? Will they reach the Arctic? Will they wash out and fertilize specific regions of the ocean? How and at what rate will gaseous sulfur convert to particulates? Will dispersion and conversion rates differ in El Niño years?

The potential impacts of these emissions, both globally and locally, merit considerable study. To be effective, this study should be carried out in a highly collaborative mode, with social scientists, economists, and others.

PART I

4

Leadership and Management Challenges in the Decades Ahead

Significant leadership and management challenges in the atmospheric sciences accompany the new opportunities and directions for research and service discussed here (see Box I.4.1). Assessing them appropriately is as important as meeting the scientific imperatives if atmospheric science is to achieve its potential for service to society over the coming decades.

The need for coordination and collaboration of the atmospheric sciences is increasingly urgent. Atmospheric services as well as observing systems are becoming more distributed, and there are threats to the integrity of the fundamental, worldwide atmospheric observing system. Research is more interdisciplinary and motivated now, in some cases, by issues with potentially serious national and global implications. For these reasons, maintaining the effectiveness of research and services in the governmental, private, and academic components of the atmospheric sciences will require a thoughtful and innovative strategic plan.

LEADERSHIP AND MANAGEMENT RECOMMENDATION 1: DEVELOP A STRATEGY FOR PROVIDING ATMOSPHERIC INFORMATION

The Federal Coordinator for Meteorological Services and Supporting Research should lead a thorough examination of the issues that arise as the national system for providing atmospheric information becomes more distributed. Key federal organizations, the private sector, academe, and professional organizations should all be represented in such a study and should help develop a strategic plan.

> **Box I.4.1**
> **Recommendations for Leadership and Management in the Atmospheric Sciences**
>
> - Develop a strategic viewpoint to shape an increasingly distributed national structure for providing atmospheric information from a variety of governmental and private-sector organizations.
> - Maintain the free and open exchange of atmospheric observations among all countries, and preserve the free and open exchange of data among scientists.
> - Develop a clear understanding of the benefits and costs of weather and climate services.

Two primary consequences of the contemporary information revolution for atmospheric sciences and services are the following:

1. Quantitative information on nearly any topic is readily available on global information networks. Individuals with a modem and a computer have unprecedented resources for examining global weather and climate data, visualizations, and predictions. What was once the province of government supercomputers is now common currency.

2. Computer-to-computer communication enables weather-dependent enterprises to incorporate atmospheric information more readily into their decision making. Four-dimensional data bases containing the classical meteorological variables can be transformed into four-dimensional data bases containing variables of interest to users and critical to their decisions.

The full implications for public and private weather services are not yet clear, but it is obvious that rapid change is in progress.

A Changing System for Providing Weather Services

From the beginning of organized attempts to forecast weather events a century or so ago, nearly all observation networks and both national and global analysis and prediction services have been instituted, funded, and managed by national governments. In the United States, public forecasts and warnings of severe weather are the responsibility of the National Weather Service (NWS). The centralized model has served this and other nations well in many respects, leading to greatly improved observations and the impressive weather prediction capabilities enjoyed today in all developed countries.

As communications capabilities improved, weather information became a potential source of competitive advantage or profit. Private-sector weather forecast firms developed products of special interest to their clients, television stations sought weather presentations that would attract and retain viewers, and *The Weather Channel* created a 24-hour nationwide weather information distribution service using NWS data and supported by advertisers. Universities in the United States created a capability and infrastructure to distribute weather data for academic purposes (Fulker et al., 1997), and the private sector similarly created distribution capabilities to meet the data and information needs of a wide variety of clients. Electronic digital communication made it possible for the government to contract with the private sector to provide aviation weather and flight planning capabilities to pilots who have access to a computer and a modem. Data from each of the more than 100 Doppler Next Generation Radar (NEXRAD) radar units installed as part of the modernization of the National Weather Service are collected on site by four private firms and made available in various forms, including national and regional mosaics, for both private and public purposes. Today, the World Wide Web offers an amazing quantity and diversity of weather information,[1] provided by government agencies, the private sector, academic institutions, and individuals, to those willing to search for it and use it.

Specialized short-range numerical prediction models have been developed at several universities, and in addition to education and research, some are being employed to produce weather predictions available to the public. A survey (Auciello and Lavoie, 1993) showed that 11 NWS Weather Forecast Offices are involved in direct collaboration with universities in such research and service activities; another 8 are associated with federal research organizations; and the Cooperative Program for Operational Meteorology, Education, and Training has supported fifteen collaborative research projects involving NWS forecasters and cooperating researchers.

Despite the richness of the meteorological feast, it is important to keep in mind that all of this information is based on government-financed observations, computer analyses and predictions, and on the research that makes improved approaches possible.

Prospects for Atmospheric Information

Contemporary approaches to atmospheric information focus on user activities, provide more specific local information, are integrated quantitatively into formal decision systems operated by the user, and in some cases take advantage of expert systems and machine learning approaches.

The national weather information partnership is changing at a rapid rate, in part because new approaches and technology favor the development of strong

---

[1] A search of the World Wide Web science and technology category in June 1996 using the keyword "weather" produced a list of 7,211 entries; one of the first listed provided links to a wide variety of sources of current weather information.

relationships between private-sector meteorologists and their clients. Indeed, the private sector is an increasingly significant employer of atmospheric scientists. Furthermore, forecast services focused on particular industries or economic sectors are increasingly likely to be privatized.

The combination of increasing communication bandwidth and increasing computational power in workstations will enable new approaches to regional or local weather or air quality prediction. The key idea is to combine NWS predictions based on global data with the power of workstations to produce local forecasts. Thus, NWS predictions represented as a four-dimensional data base on regional grids for a forecast period in a range of days will be the input data for workstation models tailored to specific activities and locations (NRC, 1994a). Quantitative predictions of variables critical to user activities will then be incorporated into numerical or other decision models that they use to manage their enterprises. The work of many atmospheric scientists will focus on helping users create models of their own activities, controlling the flow of information to them, and assisting in making key operational decisions.

Similar innovations can be expected as current experimental approaches to predicting climate variations on interannual and longer time scales demonstrate success. Successful models and methods will be employed to develop forecasts for specific applications and will operate in nested hierarchies to produce regional and local forecasts of climate variations.

As an extension of these ideas, forecasts tailored to the activities of specific requesters may become available interactively through the Internet, the Web, or other communication systems. In this case, forecast systems might produce scenarios of weather, air quality, climate, or near-Earth space events in response to a user's electronic request and deliver them as a visualization, perhaps in a time-space format.[2] It is conceivable that such capabilities might be provided by advertisers or as a service to customers by firms that have close ties with particular industries.

Implications of Distributed Atmospheric Information Services

The issue before all of the partners in the atmospheric sciences is whether the evolution to a more distributed national atmospheric information system is to occur with or without strategic guidance and some attempt at design of an optimal system.

At one end of the spectrum of possible action, it could be argued that the information revolution enables the emergence of an efficient buyer's market in

---

[2]A prototype of such presentations may be found in the flight weather cross sections that were long ago prepared manually for pilots of cross-country flights. Drawn in height versus distance (or time) along the flight path, these cross sections depicted weather phenomena that the forecaster expected the flight to encounter.

atmospheric data and predictions and that eventually the entire process can be relegated, perhaps with some government support, to the private sector. At the other end of the spectrum, it can be argued that the federal responsibility to provide warnings of severe weather to protect lives and property and to provide atmospheric information and forecasts critical to enhancing safety, health, and economic vitality cannot be delegated.

The model now emerging lies somewhere between these two extremes. The government retains responsibility for warnings and predictions to protect life and property and, as a consequence, retains the responsibility for acquiring and processing the observations necessary to perform this function. Moreover, in support of this mission, the government retains the responsibility for generating state-of-the-science numerical atmospheric predictions that are the basis for predictions of variables relevant to user needs and decision-making processes.

There are important issues here; whose resolution is important to all of the partners. What criteria should govern the design of an optimal atmospheric information system? Should the government seek to recover costs of observations from the public by mechanisms other than taxes? Who is to be responsible for forecasts for critical activities such as agriculture and aviation? Should federal agencies be responsible for supporting research to improve forecasts for such critical activities? What is the appropriate role for academic research, both basic and applied, in such an evolving weather information system, and how should such research be supported so that it remains vigorous and contributes to national goals? The answers to such questions depend in part on financial and political considerations and will require discipline-wide planning and leadership.

LEADERSHIP AND MANAGEMENT RECOMMENDATION 2: ENSURE ACCESS TO ATMOSPHERIC INFORMATION

The federal government should move forthrightly and aggressively to protect the advance of atmospheric research and services by maintaining the free and open exchange of atmospheric observations among all countries and by preserving the free and open exchange of data among scientists.

The increasing dependence on distributed capabilities has significant implications for access to atmospheric data and information. As the capabilities for exchanging information increase, so do the political pressures for seeking local advantage and restricting exchange. Moreover, as electronic data become more valuable to some industries, they will advocate schemes to limit access to such data that would have adverse consequences for atmospheric and other sciences (NRC, 1995a).

Some countries have eschewed a direct responsibility for weather information or warnings and have privatized the national capability (e.g., Japan and New

Zealand) or created independent subsidiaries (e.g., Great Britain). Such approaches are being promoted by some individuals in this country.

Some countries are marketing and selling their weather data and information in order to recover some of the costs of acquiring it, and therefore are restricting the availability to other countries and other weather services that might provide channels for the data to be used in competition with their national service. Obviously, such restrictions run counter to the historical trends that made all weather data available on a global basis in order to support global forecasts that serve all nations. Climate and global air quality research has also become international, requiring the same vigilance in protecting data access, quality, continuity, and comparability.

Two principles have long governed the traditional U.S. view and should be maintained vigorously:

1. Data acquired for public purposes with public funds should be publicly available at no more than the marginal cost of reproduction or transmission.
2. The free and open exchange of atmospheric observations by all countries will enhance atmospheric research and understanding and improve atmospheric services for all nations and their citizens.

The critical point for atmospheric data implicit in the principles cited above is that competitive or economic advantage should be gained with value added to the basic data through analysis, visualization, or prediction methodologies, not by restricting the flow of data themselves. The increasing capabilities of computers and communications have created global markets and global financial venues that are transforming private industry at an astounding speed. They will similarly transform atmospheric data, information, and services throughout the world. Attempts to restrict the flow of meteorological information are not wise in a world that requires a global view for success, health, and prosperity.

LEADERSHIP AND MANAGEMENT RECOMMENDATION 3: ASSESS BENEFITS AND COSTS

The atmospheric science community, through the collaboration of appropriate agencies and advisory and professional organizations, should initiate interdisciplinary studies of the benefits and costs of weather, climate, and environmental information services.

There are a number of reasons for embarking on a thorough examination of benefits and costs across the full range of atmospheric services. First, better understanding of the relationships between benefits and costs of atmospheric information in a wide range of private- and public-sector activities is essential to formulate more effective scientific and service strategies for the atmospheric sciences (Johnson and Holt, 1997). Second, this understanding is required by federal agencies to motivate and justify investments in both research and operations, and to ensure that funds invested in atmospheric research and services are highly leveraged in providing benefits related to national goals.

Another important reason is to identify which new directions in research or services will provide benefits to a wide range of public and private interests. For example, the optimization of observing systems should, in the contemporary environment, proceed past the generic needs of weather and climate analysis and prediction to examine a wide range of specific needs and opportunities in applications such as transportation, health, environmental engineering, and mitigation of flood damage. Furthermore, forecast accuracy, the costs of preparation to mitigate damage, and the costs of damage when no preparation occurs all interact to produce guidelines for optimum strategies that will vary with activity and acceptance of risk. Similar arguments are presented by Pielke and Kimple (1997).

Katz and Murphy (1997) have noted that the difficulty in assessing the costs and benefits of atmospheric information is due partly to its multidisciplinary nature. Besides meteorology, such an undertaking must include the disciplines of economics, psychology, and statistics, as well as the allied fields of management science and operations research. Furthermore, most of the studies of benefits and costs available in the literature either examine specific applications or are formulated as case studies; an exception is a benefit-cost analysis related to the modernization of the NWS (Chapman, 1992).

A comprehensive and rigorous assessment of benefits and costs, involving collaboration among members of the atmospheric information partnership and a number of other disciplines, therefore is required.

Federal Funding of Atmospheric Research and Services

The U.S. government has supported atmospheric observations and data analysis for more than 100 years and atmospheric research for more than 50 years. Various mechanisms for coordinating atmospheric research and services over these years have left a record that allows us to compare progress, funding levels, and coordination schemes. Today, accurate budget information is essential to wise leadership and management of a complex endeavor, in order to assess its effectiveness, balance, and commitment to initiatives and to plan for the future.

Formal federal coordination of atmospheric research began in 1959 when the Federal Council for Science and Technology created the Interdepartmental Committee for Atmospheric Sciences (ICAS), which existed to the end of the Bush administration, when it became known as the Subcommittee on Atmospheric Research (SAR) of the Committee on Earth and Environmental Sciences (CEES), one of several groups covered by the umbrella of the Federal Coordinating Council for Science, Engineering, and Technology (FCCSET) chaired by the President's science adviser. As explained below, this system has been modified substantially in the Clinton administration.

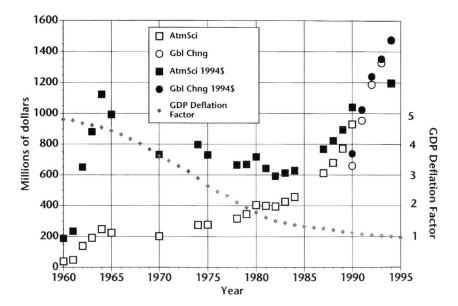

FIGURE I.4.1 Federal funding for atmospheric science and global change research in both current and constant FY 1994 dollars. Estimates for atmospheric science research for FY 1960–1990 are from summaries prepared by the Subcommittee on Atmospheric Research, and FY 1994 is a BASC estimate obtained as described in text from data gathered by the Committee on Environment and Natural Resources. Estimates of global change research are derived from documents prepared by the U.S. Global Change Research Program as part of the President's budget. It should be noted that the global change budget includes research areas other than atmospheric science. The GDP deflation factor used to scale to the 1994 data was obtained from the Gross Domestic Product (GDP) price deflator (e.g., National Science Board, 1996, Appendix 4.1) by shifting to 1994 and then taking the reciprocal to obtain a multiplicative factor.

Funding for Atmospheric Research

The first comprehensive summary of federal expenditures for atmospheric research was published by ICAS in 1960. Similar summaries were assembled, somewhat sporadically, until the most recent one prepared by SAR in 1990.

The evolution of funding for atmospheric research as portrayed by ICAS-SAR summaries is shown in Figure I.4.1, along with an estimate of the 1994 research budget[3] assembled by the Board on Atmospheric Sciences and Climate (BASC) by surveying the agencies. In these data, *some* of the global change

---

[3]These data are now somewhat out of date, but in the absence of a formal coordinating mechanism for atmospheric research, information on federal expenditures can be assembled only by surveying individual agencies. The last comprehensive compilation was the 1994 Committee on the Environment and Natural Resources research inventory used to prepare some of the analyses presented here.

TABLE I.4.1 Definition of Functions for BASC's Summary of the CENR

Function	Definition
Data acquisition management	Acquisition, processing, and management of data from and observing systems or numerical models
Forecasts	Research related to improving forecasts or applications of meteorological, climatological, and environmental information in public and private sectors
Observing systems	Development or operation of individual, project-related observing and data systems to acquired atmospheric observations for research purposes
Observing and data system investments	Development and manufacture of multipurpose observing and data systems for atmospheric research and operations
Process studies	Theoretical, observational, and laboratory studies of atmospheric or related processes at all scales
Theory and modeling	Theoretical studies of atmospheric phenomena and development of numerical models and their research applications

research efforts are *included* in broader atmospheric research categories; thus, the estimates are *not additive*. Moreover, whereas atmospheric research funds are for direct research expenditures and concomitant infrastructural support, the total for global change research includes a variety of efforts in other disciplines, such as ecology, ocean sciences, and social science.

An effort to assemble a national research inventory began in November 1993 when President Clinton established the National Science and Technology Council (NSTC) as a replacement for FCCSET and ordered it "to undertake . . . an across-the-board review of federal spending on research and development." In response, the Committee on the Environment and Natural Resources (CENR) asked each agency to provide narrative and budget material describing environmental R&D programs and activities in FY 1994. CENR agencies produced 509 project descriptions, of which some one hundred described atmospheric science research activities. Some 1,000 descriptions, augmented by National Aeronautics and Space Administration (NASA) data on missions for solar and near-Earth space research, were used to prepare the analyses in this section. Although the difficulties of constructing a meaningful budget summary from the disparate sources are recognized, FY 1994 was the last year for which a considerable body of data is available.

To aid understanding of the distribution of funds within atmospheric science, BASC used the CENR project summaries to allot funds to five functions and to each of the five disciplinary areas represented in Part II of this report; funds allocated to related areas (e.g., societal impact, assessment of indoor air quality) were excluded. Definitions of the functions are given in Table I.4.1, and a

TABLE I.4.2 Atmospheric Research and Infrastructural Investments, 1994 (million dollars)

Category	Storm Dynamics	Climate	Atmospheric Physics	Atmospheric Chemistry	Outer Atmosphere	Total
Research Expenditures						
Data acquisition and management	5	180	27	23	63	298
Forecasts and applications	29	45	8	12	0	94
Observing systems	66	71	54	6	24	221
Process studies	3	45	51	38	16	153
Theory and modeling	16	86	18	31	3	154
Subtotal	119	427	158	110	106	920
Observing and Data Systems Investments						
NWS AWIPS	43					43
NWS ASOS and NEXRAD	263	88				351
NESDIS environmental satellite systems	249	125				374
Defense military satellite program	26					26
EOS data and information system		194				194
EOS flights		255				255
Mission development solar and near-Earth space missions						
Subtotal	581	662			64	1,307
Total Research and Related Activities	700	1,089	158	110	170	2,227

SOURCE: Compiled by BASC from CENR research projects inventory for 1994 and NASA data on solar and near-Earth space missions in progress or development in 1994. Allocation of expenditures reported in the CENR project descriptions to categories in this table was done by BASC, in some cases subjectively.

summary of expenditures is presented in Table I.4.2. Many of the expenditures listed for climate are part of the U.S. Global Change Research Program.

The distribution of funding for atmospheric research by agency is shown in Table I.4.3, which was constructed from CENR inventory data and data on total FY 1994 research funding supplied to BASC by the agencies. A summary of agency estimates from the 1990 SAR analysis is shown for comparison. In some cases, base funding included in agency figures was not included in the CENR inventory data; in other cases, infrastructural expenses were not included in the agency estimate.[4]

---

[4]Note, however, CENR data can be used to provide an independent estimate of total funding for atmospheric research by assuming that only part of the expenditures for data and observing systems should be assigned to research. For example, the CENR total for research projects, added to one-quarter of the total for data and observing systems, gives an estimate of $1,246 million compared to the agency estimate of $1,196 million

TABLE I.4.3 Agency Expenditures for Research and Related Activities, FY 1990 and 1994 (million dollars)

Department or Agency	Agency Reports		Compiled from CENR Data		
	FY 1990 (SAR 1990)	FY 1994 (reported to BASC)	CENR Research Projects	Data and Observing Systems	Agency Total
Commerce	73	254	175	768	943
National Aeronautics and Space Administration	509	506	390	513	903
Energy	45	93	107	0	107
Defense	122	67	71	26	97
Environmental Protection Agency	21	84	84	0	84
National Science Foundation	106	135	77	0	77
Interior	25	15	14	0	14
Agriculture	15	16	1	0	1
Transportation	13	26	1	0	1
Total	929	1,196	920	1,307	2,227

SOURCE: Compiled by BASC from CENR research projects inventory for 1994 and NASA data on solar and near-Earth space missions in progress or development in 1994.

Because of these and other shortcomings in atmospheric sciences budget data, the analyses and summaries presented here involve subjective judgment but still give a rough sense of the magnitude and distribution of federal funding in the atmospheric sciences. Nevertheless, key questions regarding balance, focus, and year-to-year changes in federal funding of atmospheric research cannot be answered because of the lack of substantive budget data and analysis.

Funding for Atmospheric Information Services

The national investment in atmospheric sciences includes federal expenditures for the acquisition and management of atmospheric observations, preparation of forecasts and warnings, and distribution of atmospheric information to a wide variety of users in the private and public sectors. It would be of value to estimate private expenditures to provide and procure atmospheric information, a topic about which little is known.

In sharp contrast to the difficulties in assembling research budget summaries, the federal expenditures for meteorological operations are summarized in detail each year by the Office of the Federal Coordinator for Meteorology (OFCM). The funding history since 1969 is shown in Figure I.4.2, and the distribution of FY 1994 expenditures by agency is given in Table I.4.4.

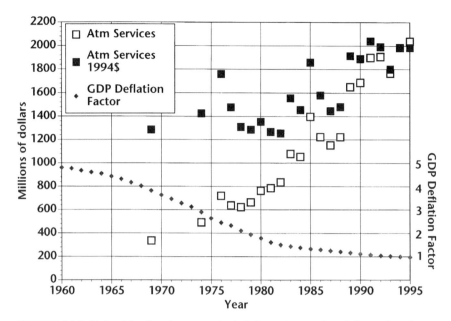

FIGURE I.4.2 Federal funding for atmospheric information services (often referred to as operational meteorology) for FY 1969–1995 as summarized by the Office of the Federal Coordinator for Meteorological Services and Supporting Research, in both current and constant FY 1994 dollars (see legend for Figure I.4.1 for further details).

TABLE I.4.4 Federal Expenditures for Meteorological Operations, FY 1994 (million dollars)

Department or Agency	Budget	
Agriculture		12
Commerce (NOAA)		
National Weather Service	723	
National Environmental Satellite, Data, and		
Information Service	401	1,124
Defense		506
Interior		
Bureau of Land Management		1
Transportation		
FAA	360	
Coast Guard	7	367
National Aeronautics and Space Administration		8
Nuclear Regulatory Commission		<1
Total		2,018

SOURCE: Office of the Federal Coordinator for Meteorology and Supporting Services, with adjustments from data furnished directly by agencies.

It should be understood that the funding for research and related activities (Table I.4.3) and those for meteorological operations (Table I.4.4) are not strictly additive since there are likely to be some overlaps in the reported data.

Summary of Federal Funding

The distribution of federal funding for weather information services is well documented, but funding within key categories of atmospheric research is known only approximately. The question of whether the United States has a balanced and appropriately focused research effort in the atmospheric sciences cannot be answered at present. Obtaining more detailed budgeting information is critical for determining whether important tasks have sufficient support and whether important initiatives are being given appropriate priority.

LEADERSHIP AND MANAGEMENT PLANNING

Many government agencies have interest and involvement in atmospheric research and operations because of the intimate relations between atmospheric phenomena and events and many of the nation's activities. Before it was disbanded, SAR coordinated the research efforts of some 10 agencies. OFCM coordinates operational meteorology through a number of committees and activities. An effective coordinating mechanism for advancing and managing the U.S. Global Change Research Program was developed by CEES under FCCSET and has continued under CENR. The U.S. Weather Research Program is similarly organized and managed to focus on improving understanding and prediction of storm-scale phenomena. A significant component of atmospheric chemistry is coordinated through the North American Strategy for Tropospheric Ozone program and the CENR Subcommittee on Air Quality Research. These interagency interests indicate clearly the breadth of the atmospheric sciences and their importance to the nation.

As evident from this report, the advance of atmospheric science requires that appropriate priorities be determined and implemented so that the research enterprise remains vigorous and focused on activities important in the context of broad national goals.

No One Sets the Priorities; No One Fashions the Agenda

Today, there is reason for considerable concern about planning for atmospheric research. No one sets the priorities; no one fashions the agenda. In part, this is a consequence of the attempt to direct federal research efforts toward a number of strategic initiatives managed by the National Science and Technology Council. In this structure, atmospheric science is viewed as a potential contributor to a number of cross-cutting issues such as global change or natural disasters.

However, for the efforts of atmospheric sciences to serve national needs effectively, they must be integrated into research approaches that serve a number of initiatives simultaneously. Moreover, this integration must recognize that the scientific advances needed to facilitate progress in addressing strategic issues, whatever their interdisciplinary motivation, will occur within the disciplines themselves.

Thus, BASC believes that a national research environment requires a strong disciplinary planning mechanism. This view is reinforced by the very basic contemporary reality for atmospheric sciences and the nation: opportunities for scientific progress and societal service in the atmospheric sciences are far more plentiful than resources. For this reason, the efforts of the discipline must be guided by an overall vision and reasoned priorities.

Therefore, all partners in the atmospheric enterprise—in government, in universities, and in a variety of commercial undertakings—must join together as an effective team focused on the future. For this to come to pass, there must be clear responsibilities for priorities and progress, for resources and results.

Part II

Disciplinary Assessments

PART II

1

Atmospheric Physics Research Entering the Twenty-First Century[1]

SUMMARY

Atmospheric physics seeks to explain atmospheric phenomena that occur on a variety of temporal and spatial scales in terms of physical principles. Areas included in atmospheric physics are atmospheric radiation, aerosol physics, the physics of clouds, atmospheric electricity, the physics of the atmospheric boundary layer, and small-scale atmospheric dynamics.

Major Scientific Goals and Challenges

In each of these areas, we generally have a useful understanding of the physical principles involved at the most fundamental level. However, understanding these physical principles alone does not ensure an adequate understanding of observed atmospheric phenomena because the various realizations of these phenomena are inherently complex and result from complicated interactions among physical processes. Further, these interactions occur across a great range of time and space

---

[1] Report of the Ad Hoc Group on Atmospheric Physics: W.A. Cooper (Chair), National Center for Atmospheric Research; T. Ackerman, Pennsylvania State University; C. Bretherton, University of Washington; S. Cox, Colorado State University; J. Dye, National Center for Atmospheric Research; E. Gossard, Environmental Technology Laboratory, National Oceanic and Atmospheric Administration; D. Lenschow, National Center for Atmospheric Research; V. Ramaswamy, Geophysical Fluid Dynamics Laboratory, National Oceanic and Atmospheric Administration; D. Raymond, New Mexico Institute of Mining and Technology; E. Williams, Massachusetts Institute of Technology.

scales, and many small-scale processes have significant influence on those occurring on larger scales. An understanding must be developed of the collective influence of individual physical processes on larger spatial-scale and longer time-scale phenomena, which are referred to here as organizing principles.

Key Components of the Scientific Strategy

The most critical components of a program to address the scientific issues and challenges are the following:

- To develop and verify an ability to predict the influences of small-scale, atmospheric physical processes, such as turbulence, on large-scale atmospheric phenomena such as thunderstorms.

In many cases, large-scale atmospheric phenomena arise from the collective effects of an ensemble of interactions that occur on much smaller spatial and much shorter temporal scales. Two central impediments to more accurate simulation and prediction of weather and climate, which are comprised of these large-scale events, are the physical understanding of the smaller-scale events and the inability to include them explicitly in models due to computational constraints. The solution lies in developing organizing principles to relate small-scale events to larger-scale phenomena. Progress is being made in this area due to improvements in modeling and computational power, expanded observations from the ground and in situ, and increased communication and collaboration among research scientists.

Considerable attention is being directed toward the use of field observations to verify model results. In the atmospheric sciences, verification is essentially the process of establishing the accuracy of a theory to within some error estimate where the errors include those of both the computational version of the theory (the model) and the observing system. Error estimation is itself often an imprecise quantification in the atmospheric sciences because of a lack of understanding of the propagation of error in nonlinear systems and the inability to create repeatable atmospheric experiments.

- To develop a quantitative description of the processes and interactions that determine the observed distributions of water substance in the atmosphere.

The importance of water, whether vapor, liquid, or solid, in climate and weather processes is self-evident, but there are weaknesses in the current ability to specify the atmospheric water cycle. Among these are poor characterization of upper-troposphere water vapor, uncertainties in surface fluxes and precipitation efficiency, poor representation of ensemble effects of cumulus convection on the transport of water and the characterization of precipitation over oceans, and the absence of a comprehensive understanding of the links between the atmospheric water cycle and other components of the hydrological cycle.

Fortunately, recent improvements appear able to support a comprehensive new approach to these problems. These improvements include, or will include, in situ and remote sensing methods, characterization of precipitation over the oceans, new modeling capability, and comprehensive international programs to study the hydrological cycle at regional scales.

- To improve the capability of making critical measurements in support of studies in atmospheric physics.

Many areas of atmospheric physics instrumentation lag behind what is technically feasible. Platforms suited to atmospheric physics studies are among the most needed facilities. These include new observing satellites and high-altitude, long flight time aircraft. Measurements of water substance, radiation, and trace gases are of maximum importance.

Initiatives to Support the Strategies

Implementation of these research strategies requires the following disciplinary initiatives:

- *Atmospheric Radiation:* to understand the interactions between radiation and components of the hydrologic cycle.
- *Cloud Physics:* to understand water substance interactions and processes, for example, initial droplet formation, cloud chemistry, and the interaction between radiation and clouds, such as is needed in weather and climate research.
- *Atmospheric Electricity:* to enable reduction of fatalities and economic losses due to lightning discharges, and to determine the usefulness of electrical activity as an observational tool for monitoring severe weather and more typical weather as well as climate.
- *Boundary Layer Meteorology:* to understand and make use of the knowledge of boundary layer effects on weather, climate, and human activity.

Expected Benefits and Contributions to the National Well-Being

Many components of the research program recommended here address sources of uncertainty in climate prediction. In addition to expediting the reduction of this uncertainty, other benefits will accrue, such as the improved ability to predict regional and local weather, and will aid in the development of polices to mitigate anthropogenic impacts on the environment and the management of the nation's natural resources.

Recommended Atmospheric Physics Research

Recommended Atmospheric Radiation Research

• Develop and/or test the ability of theory and models of radiation transfer to (1) understand water vapor continuum absorption and in-cloud solar absorption; (2) develop the theory of scattering by nonspherical, including irregular, particles; and (3) understand radiative transfer in cloudy atmospheres.

• Develop observational studies and analyses to (1) better utilize satellite and other remote sensor data, (2) represent the four-dimensional distribution of water vapor, and (3) quantify the direct radiative forcing of climate by trace gases and aerosols.

Recommended Cloud Physics Research

• Develop the ability to predict the extent, lifetimes, and microphysical and radiative properties of stratocumulus and cirrus clouds to (1) resolve stratocumulus-related issues as in atmosphere-ocean coupling, (2) resolve the aerosol-stratocumulus albedo feedback effect, and (3) resolve the role of cirrus clouds in global warming or cooling.

• Improve models of atmospheric radiative transfer; test these models using observations to verify radiative transfer models under different atmospheric conditions and improve parameterizations of these effects in general circulation models (GCMs).

• Increase the attention paid to clouds and their interaction with radiation to (1) permit estimation of the coverage and radiative properties of clouds and (2) improve the theoretical understanding of liquid and solid precipitation formation.

• Conduct critical tests of precipitation mechanism theory with increased attention to dynamic consequences for (1) testing models of warm-rain and ice-phase precipitation processes, (2) evaluating the effects of precipitation production and evaporation on the dynamical evolution of storms, and (3) evaluating the importance of precipitation processes on advertent and inadvertent weather modification.

• Develop the ability to predict size distribution of hydrometeors and aerosol populations to (1) determine their joint influence on the Earth's radiation balance, (2) understand their role in sustaining heterogeneous atmospheric chemical reactions and precipitation formation, and (3) evaluate the influences of microphysical processes on cloud models and the influence of clouds on climate models.

• Investigate the interactions among aerosols, trace chemical species, and clouds; develop and improve the characterization of atmospheric aerosols to (1) characterize cloud condensation nuclei (CCN) activity in chemical global models and (2) develop representations of the radiative effects of aerosols.

Recommended Atmospheric Electricity Research

• Determine mechanisms responsible for charge generation and separation in clouds to understand cloud formation mechanisms and elucidate the fundamental physics of lightning.

• Determine the nature and sources of middle-atmospheric discharges to (1) increase knowledge of these recently discovered phenomena and their possible association with severe weather and (2) explore their effects on radio propagation and atmospheric chemistry.

• Quantify the production of oxides of nitrogen (NO_x) by lightning to better understand upper-troposphere ozone production or loss.

• Investigate the possibility that the global electrical circuit and global and regional lightning frequency might be an indicator of climate change.

Recommended Boundary Layer Meteorology Research

• Understand the structure of cloudy boundary layers to enable characterization of the effects of boundary layer clouds on climate.

• Improve understanding of turbulence and entrainment in the boundary layer to aid in the parameterizations of boundary layers in numerical models and to improve pollution modeling.

• Improve measurements of exchange of water, heat, and trace constituents at the Earth's surface. This is fundamental information for use in most aspects of tropospheric functioning.

• Understand model interactions of the planetary boundary layer, surface characteristics, and clouds for use in analytical and predictive models of daily temperature cycle, hydrologic studies, and pollution prediction.

• Exploit new boundary layer, remote sensors for obtaining a more complete description of three-dimensional boundary layer flow to use in direct comparisons with boundary layer simulations.

Recommended Small-Scale Dynamics Research

• Develop better representations or parameterizations of physical processes occurring on scales smaller than the grid scale in climate models to improve GCM parameterizations.

• Represent the effects of moist convection in large-scale models to improve models of potentially destructive mesoscale convective thunderstorm supercells.

• Improve the dynamical representation of small-scale features in large-scale models to permit better understanding of local severe weather, large-amplitude gravity waves, clear air turbulence, and stratospheric-tropospheric exchange.

Improvement of Capabilities

• Studies in atmospheric physics require new measurement systems such as Doppler infrared radar, polarized lidar, millimeter radar, microwave radiometers, Doppler wind profilers, and polarimetric Doppler radar.

• Develop new analysis techniques such as (1) multispectral algorithms to infer optical depth, cloud liquid water content, and trace gas concentrations from infrared spectrometers, and (2) techniques from other areas that include pattern recognition, intelligent systems, artificial intelligence, chaos theory, and computer visualization.

INTRODUCTION

Mission

Atmospheric physics seeks to explain, in terms of basic physical principles, atmospheric phenomena that occur on a variety of temporal and spatial scales. Thus, atmospheric physics could be interpreted broadly as including all atmospheric phenomena. The field of atmospheric science, however, has traditionally separated large-scale dynamics (meso-, synoptic, and planetary scales) and atmospheric chemistry from atmospheric physics, and this tradition is followed here. Therefore, unavoidable overlap with material in other parts of this document is to be expected.

Areas emphasized here include atmospheric radiation, aerosol physics, the physics of clouds, atmospheric electricity, processes in the planetary boundary layer, and small-scale atmospheric dynamics. In each of these areas, there is generally a useful understanding of the physical principles involved at the most fundamental level, including the laws of motion or electromagnetics. However, understanding these physical principles alone does not ensure an adequate understanding of observed atmospheric phenomena because the various realizations of these phenomena are inherently complex and result from complicated interactions among physical processes. Further, these interactions occur across a great range of time and space scales and many small-scale processes have significant collective influence on large-scale processes. For example, consider the influence that the collective interactions between photons and cloud drops have on cloud life cycle, the influence that cloud life cycle has on synoptic development, and the influence that synoptic development has on climate. Since no large-scale model can hope to include all these processes ab initio, there is a need to develop an understanding of the collective influence of individual physical processes on larger spatial-scale and longer time-scale phenomena, an intellectual process that is called here "the development of organizing principles." Current research in the atmospheric sciences is primarily concerned with finding ways to represent, understand, organize, and predict the results of these complex interactive phenomena.

Major Research Themes and Past Accomplishments

This section reviews some central research themes and recent results in atmospheric physics and identifies as challenges some key problems that impede progress. This review indicates that there have been some notable shifts in the field in the recent past. Most of them are related to the increased attention now being directed to climate studies.

For example, studies of atmospheric radiation have received increased attention as their importance to atmospheric science has become more evident. Growing attention to climate modeling has increased the need to understand radiative transfer in the atmosphere, and recent results demonstrate the influences of radiation on mesoscale weather systems and weather forecasting. The scope of cloud physics has also broadened considerably, with increased attention directed to the roles of clouds in climate, mesoscale meteorology, and atmospheric chemistry. Because of the need to represent small-scale processes in weather and climate models, there has been increased attention to the problem of parameterization, and this is now an active area of research in cloud physics, atmospheric radiation, boundary layer research, and cloud dynamics. Many of the problems being addressed in small-scale dynamics relate to the interactions between this scale and larger scales, and may also involve significant parameterization efforts. Note that "parameterization" is defined here as including two conceptually different processes. It includes using both organizing principles and empirical relationships as a way of incorporating subgrid-scale processes in models. The former implies that the underlying physics of some process is reasonably well understood but cannot be included due to computational constraints so the physics is included via statistical methods or aggregate models. The latter implies that we have a less than adequate physical understanding of the process in question so the process is included by using some appropriate set of observations. Thus, in this context, it is understood that empirical processes include some amount of curve fitting and extrapolation. In practice, most parameterizations include a blend of both of these ideas, which is the view adopted in this document.

Additionally, the simultaneous expansion of computational power and instrument performance in the past two decades has led to an increased focus on the relationship between model output and observational data sets, a process that is referred to as model validation or verification. Validation carries the connotation of establishing the legitimacy of some theory, usually by empirical or logical means. Verification has the connotation of testing the accuracy of some theory, usually by empirical means. Clearly, these concepts are closely allied. Unfortunately, atmospheric phenomena are generally so complex and span so many spatial and temporal scales that logically rigorous validation is impossible; we simply cannot compute or measure all of the relevant quantities.

Verification, as the more limited concept of testing the accuracy of a theory—particularly if it includes testing only some aspects of the theory—is conceptually more tractable. Despite this ambiguity in a precise definition of these terms, however, the concept of model verification by observation is deeply ingrained in current atmospheric physics and has been an important intellectual driver of research in recent years. It is likely to continue to be a strong driver. In this document, the term verification is used to denote the process of establishing the accuracy of a theory within some error estimate. It should be noted further that error estimation is itself often an imprecise quantification in atmospheric science because of a lack of understanding of the propagation of error in nonlinear systems and the inability to create repeatable experiments in the atmosphere.

Studies in atmospheric physics are improving our understanding of radiative transfer, precipitation formation, transport, and other fundamental processes in the atmosphere. Beyond this, the principal benefit to society of studies in atmospheric physics is an improved ability to predict the effects of weather and climate. The present extent of human activity has demonstrable effects on our weather and climate, so it has become imperative to understand these effects and determine their consequences. This adds urgency to the current scientific thrust to understand these phenomena. Because the consequences of weather and climate are so important to us individually and collectively, more reliable predictions of weather and climate would be of great societal and economic value.

Perspective for the Future

Possible components of a program to address the challenges listed below, and to take advantage of the opportunities presented by recent advances, are discussed in the section that follows, which concludes by focusing on three particularly important aspects of the research program recommended—aspects that should be pursued with highest priority:

1. Develop and verify an ability to predict the influences of small-scale physical processes on large-scale atmospheric phenomena.
2. Develop a quantitative description of the processes and interactions that determine the observed distributions of water substance in the atmosphere.
3. Improve capabilities to make critical measurements in support of studies in atmospheric physics.

These imperatives span the needs of all areas in atmospheric physics. Accomplishing the first two would be a significant scientific achievement that would also lead to immediate improvement in our ability to predict weather and climate.

SCIENTIFIC CHALLENGES AND QUESTIONS

Atmospheric Radiation

The discipline of atmospheric radiative transfer and remote sensing has matured greatly in the past decade, and its importance to atmospheric sciences has become increasingly evident. A fundamental interaction that must be understood if we are to develop realistic climate projections is the interactions between radiation and the components of the hydrologic cycle, particularly clouds. A number of studies have shown that radiative processes are important in the development and maintenance of convective systems and therefore have important consequences for weather forecasting and mesoscale meteorology. Satellite remote sensing is providing the meteorological community with a wealth of data on the state of the global atmosphere, while ground-based remote sensing is beginning to supply detailed, temporally continuous data that can be used to understand physical processes in the atmosphere.

Despite the maturation of the discipline, radiative transfer and remote sensing continue to present significant research challenges. Foremost is the continued dichotomy between the modeling of cloud-radiation interactions at the scale of climate models and the understanding of these same processes at the observational and cloud-scale level. Merging of these efforts will be needed before valid parameterizations of radiative transfer in a cloudy atmosphere can be developed. Secondly, the resources devoted to the development of high-quality radiation instrumentation, particularly for in situ and ground-based use, have been minuscule until very recently. At the same time, operational satellite instrumentation has been allowed to expire without adequate replacement. As a result, the quality and quantity of radiation data needed to address current problems are simply not available in many cases. Thirdly, support for basic theoretical and computational research on more realistic cloud-radiation problems has been very limited. This has hampered our ability to understand, for example, the nature of three-dimensional radiative transfer, the scattering of radiation by nonspherical hydrometeors, or the recent and past indications of anomalous absorption by clouds.

At present and in the near future, the following actions are likely to be important aspects of the study of atmospheric radiation. The challenge will be to accomplish them, or to make significant progress in these areas.

Radiation Transfer Models and Observations

The accuracy of fluxes and heating rates given by current models of radiative transfer must be verified by rigorous comparisons of computed quantities with observations at the top and bottom of the atmosphere, and with radiative divergence observations within the atmosphere, especially under cloudy conditions. Recent comparisons of terrestrial infrared model calculations with observations have shown encouraging results for selected cases under clear sky conditions. The verification must be demonstrated for both clear sky and cloudy sky conditions and for both solar and infrared wavelengths. These comparisons are essential in order to address outstanding questions of water vapor continuum absorption and in-cloud solar absorption. A significant part of this challenge is obtaining simultaneous combined data sets characterizing the microphysical and radiative properties of clouds, especially ice clouds, and the distributions of temperature and water vapor.

Radiation Transfer Through a Medium Containing Nonspherical Particles

Although there is a complete and apparently adequate theory for scattering by spherical particles, there is no similar basis for theoretical prediction of the scattering by particles having the complex shapes assumed by ice crystals. Such a theory, spanning the relevant ranges in particle size and wavelength, is needed to assess the radiative effects of cirrus clouds and other clouds containing ice crystals. Given the difficulty of constructing a theory to cover all possible ice crystal shapes, theoretical developments will have to be combined with coupled observations of microphysics and observations to test and extend the theory.

Three-Dimensional Models of Radiative Transfer in Cloudy Atmospheres

Recent research has shown that the macrophysical (three-dimensional) variations in cloud fields may be as important in determining the radiative properties of the cloud field as are the microphysical characteristics of clouds. It is important that these macrophysical effects on radiative transfer be quantified and included in parameterizations of radiation used in weather and climate models.

Innovative Approaches to Analyses of Data from Satellites and Other Remote Sensors

The volume of data produced by satellite and ground-based sensors is difficult to handle by current techniques, and the problem will escalate in the next decade. Automated systems and new techniques will be needed for routine analyses of remotely sensed data from both satellite and ground-based systems. These analyses should produce quantitative results with specified error limits that go well beyond our current capabilities in terms of retrieving geophysical parameters. There is also the opportunity to convert a number of remote sensing instruments previously used in research applications to more routine usage. These instruments include sophisticated lidars, millimeter radars, microwave radiometers, and interferometers. Examples of critical parameters they could monitor are cloud location (base and top), surface radiation budget on scales from small mesoscale to global, profiles of atmospheric radiative heating, hydrometeor size distributions, ice and liquid water contents, and water vapor.

Four-Dimensional Distribution of Water in the Atmosphere

Because radiative transfer in the Earth system is inextricably linked to the components of the hydrologic cycle, understanding the hydrologic cycle and the resulting distribution of water vapor is crucial to a complete understanding of radiative interactions. If climate models are to represent radiative transfer properly, they must incorporate improved representations of the hydrologic cycle on scales ranging from cloud-scale and mesoscale processes to the large-scale circulation. The horizontal and vertical distributions of water vapor play critical roles in determining the radiation fluxes and heating rates, often producing preferred regions for cloud development by their dominating influence on the radiation budget. These distributions may be monitored locally from the surface by Raman scattering or differential absorption lidar (DIAL), infrared interferometers, and rawinsondes; regional and global distributions may be monitored using radiometric or interferometric measurements from satellites or using sensors carried by commercial aircraft.

Improved Understanding of the Roles of Clouds in Climate

Both model simulations and observations have revealed that cloud-radiative interactions play a significant role in climate and climate change. The fundamental physical issues in this regard represent a huge challenge that can be broken down into a number of steps such as (1) developing more physically based cloud-radiation parameterizations; (2) including the explicit treatment of cloud microphysical and macroscopic properties in climate models; and (3) incorporating the dependence and influence of these properties on the large-scale dynamics and thermodynamics of the models. Upon successful completion of these three steps, an accurate representation of the radiative heating of the atmosphere should emerge.

Explicit efforts should be directed toward using the results of process studies, cloud-scale and mesoscale model simulations, and long-term cloud-scale data, as well as large-scale global data. This will improve our understanding of the roles of clouds in climate and the parameterization of these effects in climate models.

Direct Radiative Forcing of Climate by Trace Gases and Aerosols

The potential importance of changes in concentration and distribution of radiatively active species other than the traditional, well-mixed greenhouse gases and clouds, has become more evident in recent years. The effects of species having short lifetimes and pronounced spatial and temporal dependencies (e.g., tropospheric and stratospheric aerosols, ozone) should be evaluated accurately in order to understand their effects on climate.

Interactions Between Radiation and Other Physical Processes Such as Chemistry, Transport, and Transformation Processes

Some trace gases and aerosols have complex life cycles from the time they enter the atmosphere until they are removed. These cycles may be influenced by photochemical, chemical, and microphysical processes. They may also be influenced by local environmental conditions. It is critical to understand these life cycles if the effects of these constituents are to be reliably simulated in climate models.

Cloud Physics

Recent research has led to a substantial broadening of the scope of studies in cloud physics. Studies related to initial droplet formation, growth of rain, ice formation and growth, scavenging of particles, and other topics in traditional cloud microphysics are still being investigated, but in the past few years, increased attention has been directed to the roles of clouds in climate, to cloud chemistry, and to studies of the interactions between radiation and clouds.

At present and in the near future, the actions described below are likely to be important aspects of the study of cloud physics.

Coverage and Radiative Properties of Clouds

Cirrus and stratocumulus clouds have received particular attention over the past decade because of their important roles in the radiation balance of the Earth. Both have been the subject of intensive field campaigns and many numerical simulations. Although these studies have led to improved understanding of the nature of such clouds, they still have left important problems to be resolved. Among these are the causes of stratocumulus breakup, the quantitative factors determining entrainment into stratocumulus clouds, the factors determining ice concentrations and size distributions in cirrus clouds, and the detailed interactions with radiation in both cloud types. Additionally, for clouds that undergo substantial entrainment, cloud dynamics could strongly modulate cloud drop number concentration (CDNC), thereby helping to describe the relationship with height between cloud condensation nuclei and cloud drop number concentration. Solution to these problems appears to be feasible in the coming decade. In addition, these studies require extension to clouds of the middle troposphere, which often have a more complicated, mixed-phase structure, and to the cirrus clouds of the tropical troposphere.

Ice Formation in the Atmosphere

The factors controlling ice formation in clouds remain poorly understood despite the importance of ice for almost all aspects of cloud physics and despite much effort over the past decades directed toward understanding this problem. Neither ice nucleation nor secondary ice formation is understood as well as needed to predict ice concentrations observed in clouds from the controlling variables. Many current studies, including efforts to understand radiative effects on clouds and precipitation formation, require improved understanding of these basic processes.

Improved Understanding of Precipitation Formation

Despite good understanding of the basic processes involved in precipitation formation, we do not yet understand them in detail. Significant gaps remain in the chain of events leading to precipitation by all the major mechanisms for precipitation formation (e.g., coalescence of water drops; ice formation followed by accretion of supercooled cloud water; and the classical Bergeron process). Understanding of ice-phase precipitation is incomplete, not only because of the problem of understanding how ice originates, but also because the detailed roles of aggregation, accretion, melting, breakup, and evaporation remain incompletely understood. Although it is thought that the basic processes involved in warm-rain formation are probably understood, it is uncertain if the current knowledge of collection efficiencies and the drop breakup process is adequate. The roles of large and giant particles in warm-rain precipitation formation need resolution. Definitive tests of all aspects of precipitation formation are lacking, but data are now available to test many of these processes.

Predict Size Distributions of Hydrometeors and the Aerosols That Affect Radiative Transfer in the Atmosphere

The important links among cloud structure, aerosols, and trace gases have been the subject of intensive study in connection with acid rain, and attention has been devoted to some aspects of the connection between cloud condensation nuclei and cloud microstructure. However, with a few exceptions, this area of research suffers from the absence of a comprehensive approach that addresses all components of the problem (trace gases, aerosol, cloud structure, and radiation) as opposed to the coupling between subsets.

Aerosols have many controlling but poorly understood roles in the atmosphere. They influence the Earth's radiation balance, sustain heterogeneous chemical transformations, control most precipitation formation, and determine the microscale structure of clouds. Their influence is evident in many current studies, including those of global warming, stratospheric ozone depletion, aircraft icing, radiative effects of cirrus, and possible effects of aircraft emissions on climate. To understand the influences of aerosols on these processes, we need a better understanding of the role of aerosols in heterogeneous chemical transformation in the atmosphere, improved definition of the role of particles in the ice formation in clouds, and an ability to predict the formation and life cycle of atmospheric particles. Aerosols thus are an important focus for future research because their study involves fundamental scientific problems that have immediate practical relevance.

The radiative effects of clouds are governed by their hydrometeor size distributions. Questions needing resolution include the following: Are there high concentrations of small ice crystals in cirrus clouds that dominate their radiative characteristics, as some remote sensing observations indicate? How important is oceanic production of dimethyl sulfide (DMS) in stratocumulus albedo? What is the source of the discrepancy between observations and theoretical calculations of cloud absorption? What are the factors controlling the concentrations of ice crystals in cirrus clouds? Thus, many of the questions involving radiative effects also involve fundamental questions of cloud-physics. The radiative effects of clouds cannot be understood, in a predictive sense, until the factors determining the size distributions are understood. In the case of cloud droplets, this is initially the CCN population; in the case of ice, it is initially the ice nucleus (IN) population. To understand how either might change as the climate changes, we need to know the sensitivity of the CCN and IN populations to other climate parameters such as global temperature, solar radiation, surface humidity, and soil characteristics. We are currently far from this understanding; there is no good explanation even for the order-of-magnitude concentration of these particles in the atmosphere, much less a predictive capability for how they might change.

Parameterizations of the Subgrid-Scale Influences of Clouds and of Microphysical Processes on Cloud Models

With increased attention directed to climate-related studies, new efforts have been devoted to developing ways to represent or parameterize the effects of clouds in global-scale climate models. These include ways of representing the radiative effects of cloud droplets or ice crystals; parameterizations of the effects of clouds on the momentum, heat, and water budgets; documentation of the connections between cloud coverage and large-scale conditions such as relative humidity; satellite observations to provide global coverage; and the development of new modeling techniques to represent clouds and liquid water in mesoscale and global models.

Global climate models must represent the effects of clouds in terms of large-scale variables, whereas the scales at which important processes occur in clouds are often many orders of magnitude smaller. The effects of clouds on large-scale transport processes and on radiation must be represented in terms of variables that apply over distances of the order of 100 km. It is not yet clear if this gap can be bridged in a way that represents the essence of the relationships needed for climate prediction. However, there are many promising avenues of approach to this problem. Cloud-resolving models are being used to determine the influences of clouds on scales smaller than those that can be resolved in GCMs. Radiative characteristics of cloud droplets have been determined from observations in ways that provide improved representations of the present climate situation, although probably not adequate for climate prediction or understanding how CCN populations vary. Some parameterizations of ice size distributions have been attempted, but these seem particularly weak at present and are unlikely to be improved until ice-forming processes are better understood. Satellite remote sensing studies have provided global documentation of the radiative effects of clouds, cloud populations, and cloud coverage. They promise to provide still more information on hydrometeor spectra in clouds.

Atmospheric Electricity

As the population, urbanization, technological sophistication, and economic wealth of the country and world increase, the impact of severe weather events and lightning is likely to be increasingly felt (see Part I, Table I.2.1 for data on the impact of severe weather events). Lightning is a leading cause of weather-related fatalities, and it results in huge economic losses through forest fires, power outages, and damage to computers, communications, and other electronic equipment. Determining the total electrical activity shows promise as a future observational tool for monitoring severe weather such as tornadoes, hail, flash floods, winter storms, and hurricanes.

The field of atmospheric electricity traditionally encompasses six areas of research: (1) lightning, (2) cloud electrification, (3) the global electrical circuit, (4) ion physics and chemistry, (5) ionospheric and magnetospheric currents, and (6) telluric (Earth and oceanic) currents. This assessment focuses primarily on the first three areas of research because they are the ones normally pursued in the domain of atmospheric sciences. The topic of ion physics and chemistry is discussed in the report of the Committee on Atmospheric Chemistry (NRC, 1996a). An earlier summary of research in all of these areas is found in *The Earth's Electrical Environment* (NRC, 1986). Although the topics have been separated here, there is considerable interdependence among them. For example, knowledge of the global lightning frequency is important for a better understanding of global electrical activity, NO_x production by lightning, and perhaps global temperature.

At present and in the near future, the following actions are likely to be important aspects of the study of atmospheric electricity.

Mechanisms of Charge Separation in Clouds

Although there is now considerable laboratory, observational, and modeling evidence showing the importance of ice-graupel collisions for the electrification of clouds, there are important gaps in our understanding. We lack fundamental understanding of the physical mechanisms(s) responsible for the charge transfer, which seems to be intimately linked to the nature of the ice surface, which in turn is determined by temperature, liquid water content, particle size, and other microphysical parameters. Improved understanding of ice particle formation and growth is important both in cloud electrification and in radiation-climate feedback and should be a high priority (as discussed further in the "Cloud Physics" section in Part II). Other charge separation mechanisms may also be acting. Continued observational, laboratory, and model efforts are needed on the electrification of clouds containing ice, as well as warm clouds, over geographically different parts of the globe.

Investigate the Global Electrical Circuit and Lightning as Measures of Stability and Temperature in Climate Change Studies

The global electrical circuit may prove useful in monitoring climate change. Lightning production is known to be sensitive to convective updraft speeds, which are influenced by atmospheric stability; thus, lightning rates may increase with increasing instability. It has therefore been hypothesized that global warming could be manifested by increased instability and lightning frequency. Continued monitoring of ionospheric potential, air-Earth current, and Schumann resonances may detect global trends in tropospheric stability and perhaps surface temperature and moisture as well. Global monitoring of lightning is now also technologically feasible. However, since electrical conductivity and electrification are connected to particulate concentrations, these quantities may require simultaneous measurement in order to understand the relationship between electrical circuit properties and climate.

Nature and Sources of Middle-Atmosphere Discharges

Electrical discharges above storms into the middle atmosphere (stratosphere and mesosphere) have only recently been recognized as occurring quite frequently. At least two different types of events take place, extending into the stratosphere and mesosphere. Observations are needed to understand the nature and mechanism(s) responsible for the discharges and their possible effects on radio propagation and stratospheric chemistry.

Production of NO_x by Lightning

The production or loss of ozone in the upper troposphere is strongly dependent on the distribution and strength of NO_x ($NO + NO_2$) (i.e., nitric oxide + nitrogen dioxide) sources, and there is ample evidence that lightning is an important and perhaps dominant source. An immediate challenge is to assess the importance of lightning to global and regional NO_x concentrations. Before the effects of emissions from present and future fleets of commercial aircraft or anthropogenic surface sources can be assessed, quantification of this natural source of NO_x is essential. Proper investigation of this topic requires observational and modeling expertise in lightning physics and morphology, atmospheric chemistry, cloud dynamics, and mesoscale and global dynamics. These studies are interdisciplinary; but instruments, techniques, and models are now becoming available that can address the problem. This constitutes an important challenge with high priority in terms of understanding and protecting our atmospheric environment.

Boundary Layer Meteorology

As the part of the atmosphere in which all of humanity lives, the atmospheric boundary layer has a special relevance to our lives. The boundary layer is defined as the part of the atmosphere that is turbulently coupled to the Earth's surface and includes fields of nonprecipitating shallow cumulus or stratocumulus clouds. The boundary layer plays a central role in weather and climate because it couples processes at the Earth's surface such as evaporation and sensible heat flux with the rest of the free troposphere. This transfer of energy and momentum from the Earth's surface to the atmosphere, although complicated, is crucially important as a determinant of the behavior of the atmosphere.

Structure of Cloudy Boundary Layers

Boundary layer cloud has a climatically crucial radiative effect. It is also inextricably coupled to the turbulent and convective dynamics of the boundary layer in which it is embedded. We understand the vertical thermodynamic structure of boundary layers capped by solid stratocumulus clouds fairly well, and those consisting of shallow cumulus clouds to a lesser degree. However, we are only beginning to understand the dynamics of boundary layers that are transitional between these two types, although transitional conditions cover a substantial fraction of the subtropical and middle-latitude oceans. Over land, there are few integrated studies of boundary layer cloudiness, turbulence, and surface fluxes to compare with models. In the arctic summertime, persistent clouds form in a stable or multilayered boundary layer. One challenge is to model and develop parameterizations that realistically represent the tight coupling among clouds, microphysics, radiation, and turbulence in these diverse boundary layers. A second challenge is to provide integrated data sets over land, over arctic sea ice, and over the middle-latitude oceans to test these models.

Turbulence and Entrainment

For unsaturated boundary layers driven by convective heating, we have a good understanding of turbulence statistics, turbulent fluxes, and large eddy structure, with good agreement between large eddy simulation (LES) models, one-dimensional models, and observations. Fundamental challenges arise when we consider turbulence and entrainment in other ubiquitous types of boundary layers. The relation between entrainment rate, turbulence characteristics, and cloud-top profiles of temperature and moisture in radiatively driven stratocumulus-capped boundary layers is still controversial. The stably stratified boundary layer is especially challenging because of intermittent turbulence, small length scales, and sensitivity to surface variability. It is also of great importance because of its propensity for accumulating trace constituents released at or near the surface.

Effects of Inhomogeneity and Baroclinicity on the Boundary Layer

We have a good understanding of how the boundary layer behaves over horizontally homogeneous surfaces, but in the real world, surfaces are almost inevitably inhomogeneous. A major challenge is dealing with this heterogeneity, which occurs due to both complex terrain and varying surface characteristics. An important practical application is developing techniques for scaling up flux estimates obtained over nonuniform surfaces to scales applicable to meso- and large-scale models. Boundary layer heterogeneity is also tightly coupled to convection. Localized downdrafts from deep convection interacting with surface fluxes create heterogeneity in the subcloud temperature and moisture field that is important in determining when and where future convection will occur. In middle-latitude cyclonic storm systems, the boundary layer often is highly baroclinic. Large-scale models do not accurately represent the vertical wind profile or surface momentum fluxes in a baroclinic boundary layer. Most boundary layer modeling studies purposely exclude consideration of baroclinicity. The challenge of the next few years is to tightly couple observational strategies, numerical modeling experiments, and parameterizations to deal with inhomogeneous and baroclinic boundary layers.

Measurements of the Exchange of Water, Heat, and Trace Atmospheric Constituents at the Earth's Surface

Surface heat and moisture fluxes are fundamental to the atmospheric heat engine on all length and time scales. Many trace atmospheric constituents are of climatic importance: ozone, carbon dioxide, methane, and particles. Their atmospheric distribution cannot be understood without knowledge of their surface fluxes. Modeling of these fluxes is empirically based and requires accurate flux measurements. This in turn requires continued development of sensitive and fast-response sensors for direct eddy flux measurements, the development of alternative techniques for measuring fluxes when fast-response sensors are not available, and investigation of techniques for measuring turbulent fluxes remotely.

Interactions of the Planetary Boundary Layer, Surface Characteristics, and Clouds

The boundary layer over land is particularly important as the environment for most human endeavors. Boundary layer processes and boundary layer clouds have important regional impacts on the climate over land and must be considered in tandem with land surface processes that exchange heat and moisture with the atmosphere. These include vegetation and soil moisture models that are currently in rapid evolution. Turbulent processes in the boundary layer redistribute the heat and moisture and help determine the surface temperature and moisture. Clouds also affect surface temperature and evaporation, and produce rainfall. One current forecast problem is to predict the daily cycle of temperature correctly over land directly from a numerical model. This has proved to be surprisingly difficult. Flooding and hydrology are also involved, because storms both replenish soil moisture and feed off evaporated water from the soil. In the Arctic, the surface is a jumble of sea ice, meltpools, leads, seasonal snow, and tundra. The complexity of the surface is matched by a complex boundary layer microphysics, including clouds and suspended ice crystals.

Models of these interactions are interdisciplinary and require regional field experiments for their verification. Past studies in boundary layer meteorology have focused on determining the vertical profiles of wind and turbulence, surface stress, surface heat, and moisture fluxes in the boundary layer and have made substantial progress toward understanding these features. The interaction with clouds and with various land surfaces is the frontier on which boundary layer meteorology should focus in the next decade.

Small-Scale Atmospheric Dynamics

It would be rash to conclude that we have learned everything we need to know about small-scale circulations in isolation from other influences. Nevertheless, it seems clear that some of the most interesting research opportunities relating to small-scale circulations lie in questions of the interactions of such circulations with other processes. The following are some examples: How do the large scales control such processes as tropical cyclogenesis and mesoscale convective system formation? How can gravity wave drag be realistically parameterized into global circulation models? How do convection and the resulting precipitation at a given location affect the prospects for future convection in this area? The last question seems particularly pertinent to the development of the floods in the upper midwestern United States during the summer of 1993.

Most current research on small- and middle-scale atmospheric dynamics is related to (1) moist convection and mesoscale convective systems, (2) fronts and middle-latitude cyclones, (3) tropical cyclones, and (4) topographic and other surface-induced flows. At present and in the near future, the following actions are likely to be important aspects of the study of small-scale atmospheric dynamics.

Effects of Moist Convection in Large-Scale Models

We have witnessed a long series of field programs focused on the challenge of determining the structure and evolution of convection and the mesoscale systems in which convection is embedded. These have occurred in a wide variety of geographic settings (e.g., the continental high plains, the western mountains, southern U.S. coastal regions, the tropical oceans), and we have a reasonably good idea of the morphology of convection around the world. Numerical modeling of convection saw its first success in helping to understand the supercell thunderstorm. Subsequent work has elucidated the crucial role of the cold pool and gust front in multicell storms. Models are now reaching the stage where agglomerates of thunderstorms, known as mesoscale convective systems (MCSs), can be simulated.

Large-scale models must correctly incorporate the effects of convection using convective parameterizations. One of the major challenges is to incorporate our current understanding of convection into the development of improved parameterizations. Attempts to do so reveal limitations in our understanding that must be filled by a combination of observations and high-resolution cloud modeling. Fundamental questions about issues such as the development, distribution, and evaporation of precipitation; entrainment and detrainment under various environmental conditions; the transport of heat and momentum by convection; and the control of convective initiation and amount by environmental factors have to be answered.

Dynamical Representation of Small-Scale Features in Midlatitude Cyclones

Quasi-geostrophic theory and semigeostrophic theory have been very successful in explaining the gross features of middle-latitude cyclones and associated fronts. Analyses using the operational network and special field programs have verified these ideas. The current focus is on smaller-scale features of these disturbances. The three-dimensional structure of fronts, including the study of frontal waves and small-scale unbalanced effects, is a topic of current interest. The relative contributions of adiabatic dynamics, surface fluxes of energy and momentum, and latent heat release in fronts and cyclones are also being investigated. The truly intense parts of cyclones are often small in scale and therefore cannot be described adequately by conventional balance models. There is some interest in using more accurate balance schemes, such as nonlinear balance, in describing these systems. The breakdown of balance and the production of gravity waves are topics that are beginning to be addressed. These studies could lead to better understanding of local severe weather, large-amplitude gravity waves, clear air turbulence, and stratospheric-tropospheric exchange in tropopause-fold areas.

Incorporation of Surface-Induced Flows into Large-Scale Models

The challenge of replicating the flow over and around various mountain ranges has been addressed by several observational programs. Theoretical and modeling studies have progressed from highly idealized two-dimensional calculations to three-dimensional numerical simulations taking into account nonlinearities and surface fluxes. Some agreement is often seen in comparisons with observed mountain-wave and lee-wave structures. Progress is also being made in understanding the process of lee cyclogenesis, which often occurs on a relatively small spatial scale. The challenge in all of these cases is to incorporate our knowledge of these phenomena into large-scale models.

DISCIPLINARY RESEARCH CHALLENGES

The topics discussed here are suggested components in a research program for the next decade. They address some of the challenges in current research and arise from opportunities presented by recent developments in research and technology. After each is presented briefly, some broader aspects considered of highest priority are discussed at the conclusion of this Disciplinary Assessment.

Most of the research topics discussed here are intrinsically cross-disciplinary. Indeed, many of the opportunities involve studies of the interactions between processes that have conventionally been studied in isolation. This is especially true in cloud physics, where the interactions of clouds with radiation, trace chemical species, atmospheric dynamics, and electrification are among the focal points. Nevertheless, it is important to maintain a balance that advances the traditional bases of these disciplines while developing the new themes. There are significant deficiencies in our understanding of the traditional topics that will surely limit our ability to conduct these new cross-disciplinary studies, and such barriers can be removed only if we continue to direct a significant fraction of our attention and resources to the solution of the fundamental problems.

Develop Adequate Representations or Parameterizations for Physical Processes Occurring on Subgrid Scales in Climate Models

General circulation model (GCM) calculations currently use prognostic variables having grid spacings of the order of 100 km. Significant fractions of the fluxes of mass, humidity, energy, and chemical constituents occur at smaller scales, so the effects of these subgrid-scale (SGS) transport processes must be represented parametrically in GCMs. Calculations of radiative transfer must also consider SGS distributions of cloudiness and other inhomogeneities in the atmosphere and must represent the effects of irregular ice crystals in cirrus clouds in terms of the GCM variables. Boundary layer fluxes are dependent on the nature of the land surface, which is often quite variable within the grid box of a GCM. Transformations among the phases of water, with associated heating and cooling effects, also occur primarily on scales smaller than the grid spacing. The parameterization problem is thus to represent these SGS processes as functions of the GCM prognostic variables (and perhaps other available information such as geographic location and season).

Improve, Test, and Verify Models for Radiative Transfer in the Atmosphere Using Observational Data

Over the past 30 years, remarkable progress has been made in our ability to model radiation processes during clear weather. However, there are serious deficiencies in our understanding and modeling of radiative fluxes and heating rates, especially in cloudy atmospheres. This, in turn, affects our understanding and modeling of cloud-radiative interactions and the role of clouds in climate.

The principal reasons for these deficiencies are inadequate representation of the spatial and temporal variabilities of the atmosphere, cloud cover, and cloud microphysical and macrophysical properties. Recently, some innovative work has begun to address aspects of these deficiencies, but a great deal of effort remains. Work must move forward with a combined program of modeling and observation to address these issues simultaneously.

Recent innovations in radiative transfer modeling include a variety of approaches to three-dimensional radiative transfer and improved treatment of combined gaseous absorption and particle scattering. In addition, there is considerable interest in applying new insights gained from fractal mathematics to the study of cloudy environments. These new models suggest that there may be the theoretical tools to treat radiative transfer in actual three-dimensional cloud fields. However, the science is a long way from having adequate models and testing such models in real environments. To test these models, it would be necessary to make simultaneous measurements of both the three-dimensional cloud field and its properties and the resultant radiances and irradiances. A variety of sophisticated observing tools, particularly millimeter-wavelength radar, hold the promise of providing the necessary information.

The highest priority is to establish the validity of both the input quantities and the related three-dimensional model calculations. At the present time, the most pressing problem is a lack of sufficiently sophisticated radiation instrumentation and platforms. Once there is an accurate assessment of the ability of three-dimensional models to compute radiative transfer, these models can be used to assess the performance of the simpler models that must necessarily be employed in weather and climate activities.

Develop an Ability to Predict the Extent, Lifetimes, and Microphysical and Radiative Properties of Stratocumulus and Cirrus Clouds

We are well positioned to improve parameterizations of marine stratocumulus boundary layer cloud from the insights and data gained from past field experiments in the subtropics and a proposed Arctic experiment in 1997. In particular, the connection between the vertical structure of the boundary layer and the type and amount of cloud cover is becoming much better understood. It will be particularly useful to compare new GCM parameterizations against these regional data sets as well as global data sets such as those from the International Satellite Cloud Climatology Project. Large eddy simulations also are beginning to show skill in predicting how cloud and boundary layer properties depend on large-scale variables and may prove useful in parameterization development. Further focused field research is needed on specific issues such as entrainment, where different models are in disagreement and where new instrumentation should allow us to more tightly test model predictions.

Improved boundary layer cloud prediction should help resolve some important climate modeling problems. Major failures of present coupled ocean-atmosphere models can partially be traced to deficiencies in predictions of boundary layer clouds. For instance, "full-physics" coupled models currently cannot maintain realistic El Niño oscillations, partially because they do not predict a large expanse of low cloud off the coast of South America that serves as a "refrigerator" for the eastern subtropical Pacific. Over the Arctic, climate models have predicted a large warming that has not been observed. Interactions between persistent boundary layer cloud and sea ice may be responsible.

The indirect climate effect of aerosol through its impact on the microphysics of boundary layer clouds should also continue to be a particularly fruitful research area. Here, small-scale modeling and more observational work are necessary before any climate model can reliably incorporate this aerosol-cloud-albedo feedback effect.

The properties and radiative feedbacks of boundary layer clouds over land have not been sufficiently studied either in the field or in detailed models, and such studies should provide an important opportunity for progress in the next few years. Here, better observations and more modeling studies on all scales are required.

Cirrus clouds are recognized as important components in the Earth's climate system. Their direct radiative effects can act to either cool or warm the planet, depending on the relative values of solar and infrared optical depths. Recent research has shown that tropical upper-tropospheric cirrus clouds play more subtle, though possibly critically important, roles in determining the vertical distribution of water vapor throughout the tropical atmosphere and the strength of tropical atmospheric circulation systems; these sensitivities are apparently manifest through a modulation of the static stability and resulting increase in convective motions and moistening of the upper troposphere by evaporation of cirrus cloud ice. It is very important to capture the essence of these tropical upper-tropospheric cirrus cloud systems if we are to simulate the climate successfully. We have the opportunity within the next decade to gain significant understanding of the evolution of these systems and their linkages with other processes in the climate system.

Tropical cirrus systems requiring investigation may be grouped into three categories: (1) convective cirrus, which is in close spatial and temporal proximity to the convective systems that produce it; (2) detached anvil cirrus, which can be identified with its convective source but has become spatially detached from the convection and takes on an evolution of its own; and (3) subtropopause cirrus, a spatially pervasive layer of optically thin aerosol detectable a high percentage of the time at tropical and subtropical latitudes. Each of these three cirrus systems has distinct evolutionary cycles that must be defined, described, evaluated in terms of climate sensitivity, and incorporated into climate models where appropriate. Advances in in situ and remote sensing instrumentation from aircraft, ground, satellites, and aircraft platforms scheduled to become available in the late 1990s, coupled with models capable of simulating convective-scale systems, provide the opportunity for progress in describing and understanding these climatically important systems.

Investigate the Interactions Among Aerosols, Trace Chemical Species, and Clouds; Develop and Improve Characterization of Atmospheric Aerosols

There are a number of feasible research objectives to scope the relationships between aerosols and their interactions with trace chemical species and with clouds. One would be the development of a predictive capacity of the concentrations of soluble aerosols that are active as cloud condensation nuclei. This capability would then be applied in an atmospheric global model that contains an atmospheric chemistry module and accounts for the effects of other aerosol processes as well as the aerosol effect on radiative forcing.

Another research objective would be to develop representations of the radiative effects of aerosols, suitable for incorporation into a climate model, that are interactive with the atmospheric chemistry in the model. This would require information on CCN (above) as well as the radiative properties (particularly absorption) of the aerosols.

Additional research objectives would include modeling and documenting the effects of heterogeneous reactions in the major atmospheric chemical cycles and determining how these reactions are influenced by aerosol populations and concentrations.

Increasing international sensitivity to the results of extensive biomass burning suggests that research attention should be directed toward determination of the magnitude and contribution of biomass burning to the global aerosol population and to CCN populations. Additionally, it is necessary to document the characteristics and lifetimes of aerosols in the upper atmosphere.

Other field studies have been outlined by the International Global Atmospheric Chemistry (IGAC) Project and in the document *A Plan for a Research Program on Aerosol Radiative Forcing and Climate Change* (NRC, 1996a).

Determine the Sources of Ice in the Atmosphere

Despite the importance of understanding ice formation in the atmosphere, relatively little recent effort has been devoted to studies of ice nucleation. This is a problem ready for a fresh approach. Nucleation processes responsible for cirrus formation are largely unstudied because of the unavailability of suitable research platforms and instruments, but this situation promises to change with the future availability of new high-altitude research aircraft. Global aerosol models offer the possibility that the contributions of desert dust aerosols to ice nucleation might be assessed. Perhaps the foremost need is for development of suitable instrumentation to support these studies in the laboratory as well as in field experiments. A practical goal would be to document the role of aerosol particles in ice formation in some of the simpler cloud systems, including widespread cirrus and upslope stratiform clouds, and to learn the origins of particles responsible for nucleating the formation of ice in conditions, including both homogeneous and heterogeneous nucleation.

Quantify and Parameterize Surface Effects on Atmospheric Dynamics

The boundary layer transfers heat, moisture, and momentum between the surface and the free troposphere, acting as a valve and a reservoir for these quantities. Energy and momentum fluxes from the surface are crucial to many processes in the atmosphere. For many purposes, existing bulk flux formulations valid at moderate wind speeds yield adequate treatment of surface fluxes for large-scale models over the ocean. Recent work has shown how surface heat, moisture, and momentum fluxes behave at low speeds over the ocean. However, major uncertainties still exist for the high wind speeds experienced in tropical storms. The character of these systems depends critically on the relative magnitude of the moisture and momentum exchange coefficients. Over land, the wetness of the soil, terrain variability, and the nature and distribution of the underlying vegetation play major roles in determining surface fluxes. These fluxes are crucial to large-scale dynamics but may play a role on the mesoscale as well.

Feedback by convection onto the boundary layer in the form of moist downdrafts is a crucial process. These downdrafts temporarily suppress further convection, but also enhance the heat and moisture fluxes from the surface and introduce horizontal variability in the fluxes. In this way it appears that the boundary layer plays an important role in the control of deep convection in the atmosphere.

Exploit New Remote Sensors to Broaden the Scope of Boundary Layer Studies

New developments in lidar and radar technology have given us the prospect of remote measurements of velocity and scalar fields from ground-based and mobile platforms. Doppler lidar can resolve mean and turbulent fluctuations in the radial velocity throughout the clear boundary layer. Doppler radar can provide detailed radial velocity, velocity variance, and reflectivity fields in boundary layers, which can be used to examine turbulence throughout clear and cloudy boundary layers, including the entrainment zone at the top of the boundary layer. Analysis of the velocity spectrum of short-wavelength radars (e.g., 8 mm wavelength) can provide the height profiles of drop size distributions for those drops having substantial settling velocity. This remote sensing technology could be used to provide a more complete description of three-dimensional boundary layer flow, including vertical profiles of momentum flux, velocity variance and higher-order moments, length scales, and vertical coherence of turbulent eddies in both cloudy and clear boundary layers. These data sets can then be used for more direct comparisons with numerical simulations than are possible with in situ measurements. However, this opportunity presents challenges. New strategies for digesting such large and complex data sets have to be developed so that they can be used efficiently to address such problems as comparisons with, and validation of, numerical simulations and visualizing in quantitative ways the three-dimensional morphology of turbulent structures.

The development of lidar techniques for estimating concentrations of trace gases in the boundary layer (e.g., by differential absorption lidar or Raman scattering) offers the opportunity to obtain vertical cross sections of, for example, water vapor or ozone, which could be used to study how these trace species diffuse, especially in cases of horizontal inhomogeneity. When combined with Doppler lidar, it may be possible to measure vertical flux profiles either from the ground or from aircraft, and thus to estimate surface exchange, entrainment rates, and in the case of chemically active species such as ozone, the photochemical source-sink term in the species budget.

Another opportunity is provided by the development of boundary layer wind profilers, which may be used in networks to measure boundary layer height and vertical profiles of wind, temperature, and possibly fluxes of heat and momentum. This may be particularly applicable to addressing problems of horizontally inhomogeneous flow on scales larger than those that can be addressed by a single ground-based scanning lidar or radar. Airborne lidars and millimeter Doppler radars also provide new opportunities for studying heterogeneity in clear and cloudy boundary layers.

Utilization of the global positioning system (GPS) for improved aircraft navigational accuracy can be combined with a Doppler laser to provide a major enhancement in the accuracy of wind measurement from aircraft. This has the potential to provide measurements of flow divergence and of mean flow perturbations associated with mesoscale phenomena such as land or sea breezes and flow over variable terrain, more accurate eddy flux and coherent eddy structure measurements, and measurements of wind shear over lengths of the order of ten meters. GPS also may have other important applications. For example, it can be used as a basis for airborne air motion systems that are smaller and less expensive (although less accurate) than inertial navigation-based systems, so that more systems can be available for multiplatform experiments.

Although the preponderance of development currently seems directed toward remote sensing, there are many possible developments in direct sensing that are important. An ever-increasing number of trace constituents can be measured with sufficient sensitivity and time response that they can be used for direct eddy flux measurement. Advances in this area include radiatively important species such as methane and ozone. Several of the species for which sensor technology has advanced to where eddy flux measurements are becoming possible have application as tracers for atmospheric processes such as entrainment and diffusion. At the same time, alternative flux-measuring techniques that do not require as fast a response as eddy correlation are being developed. These include devices that control the flow or accumulation of air in a reservoir according to the vertical air velocity. In this way, the requirement for fast response is placed on the collection strategy and not on the sensor, which broadens the class of species whose flux can be measured.

Investigate the Interactions of Small-Scale Circulations with Larger-Scale Processes

It would be rash to conclude that we have learned everything we need to know about small-scale circulations in isolation from other influences. Nevertheless, it seems clear that some of the most interesting research opportunities relating to small-scale circulations lie in questions of the interactions of such circulations with other processes. Some examples follow: How do the large scales control such processes as tropical cyclogenesis and the formation of mesoscale convective systems? How do cloud physical processes affect the structure and evolution of convective systems? How can gravity wave drag be realistically parameterized into global circulation models? How do convection and the resulting precipitation at a given location affect the prospects for future convection in the area? The last question seems particularly pertinent to the development of the floods in the upper midwestern United States during the summer of 1993.

Large-scale motions are nearly balanced. Whether such motions can be characterized by quasi-geostrophic theory or whether they require something more complex such as semigeostrophic or nonlinear balance, potential vorticity dynamics and the invertibility principle apply. In this picture the prognostic nature of the theory is encapsulated in the advection and nonadvective changes in the potential vorticity and surface potential temperature fields. A knowledge of these fields is then sufficient to obtain all other fields of dynamical interest by the inversion process. Given this picture, the lasting effects of small-scale circulations on large-scale motion are limited to the changes they induce in the potential vorticity and surface potential temperature fields—all other changes are transient and, therefore, of considerably less interest. Recent theoretical work has shown how diabatic heating, friction, and the turbulent transfers of heat and momentum generate nonadvective fluxes of potential vorticity. This work provides a useful framework for viewing the action of smaller-scale processes on the large scale. In the free atmosphere, two types of small-scale phenomena can cause significant nonadvective transport of potential vorticity—gravity waves (and possibly associated shear instability) and moist convection. We now discuss these processes and other effects of convection, such as moisture transport, that have indirect dynamical significance.

Gravity waves are produced in the atmosphere by flow over terrain, by convection, by shear instability, and possibly by geostrophic adjustment. These processes are reasonably well understood with the exception of geostrophic adjustment. Once produced, gravity waves transport momentum through the atmosphere without depositing it until they dissipate. Thus, gravity waves are agents of "action at a distance" in the atmosphere. The propagation of gravity waves is far from simple; multiple reflection and refraction take place. Wave breaking and dissipation are complicated, so the momentum transports are complex and subtle. Unfortunately, although very challenging, this area does not seem to offer much opportunity for great progress in parameterizing the effects of gravity waves on large-scale circulations. Although we should continue to look for fundamental advances in this area, it is perhaps most realistic to try to set bounds on the effects of gravity waves on the large scale. Potential vorticity dynamics should be useful in this respect, because it shows how isolated gravity wave breaking and the resulting deposition of momentum affect the large-scale flow.

The opportunities for producing a reasonably accurate moist convective parameterization look somewhat brighter than they do for gravity waves, mainly because convection acts locally. (This, of course, ignores the generation of gravity waves by the convection itself.) The convective parameterization problem splits naturally into two parts—the control of convection by the large-scale flow and the action of this convection back on the large scale. Many schemes have been proposed for resolving the control aspect of the problem. This issue remains controversial, but it is one with many opportunities for progress in the next decade. An opportunity to resolve the lively debate between the "convergence causes convection" school and the "instability causes convection" school may lie in examination of the results of existing field experiments such as the Tropical Ocean Global Atmosphere (TOGA) Coupled Ocean-Atmosphere Response Experiment (COARE). Also important to this effort is increased understanding of the types of convection that result from particular kinds of large-scale situations. For instance, what effects do varying degrees of environmental shear and midlevel relative humidity have on the character of the convection? The compilation of individual case studies, as well as composite analyses and cloud models, are all required to fill in this picture.

Convective ensemble simulations of convection should also be helpful. Such simulations differ from the usual type of calculation in that the convection is allowed to develop naturally from large-scale forcing, rather than being initiated by a buoyant bubble or some other imposed feature. They are computationally expensive, but the increasing power of computers over the next decade may make them more feasible. Understanding the action of convection on the large scale requires a reasonably accurate model of how convection works. The traditional picture of convection as an ensemble of entraining plumes has been challenged by models in which air moves both up and down under the influence of condensation and evaporation of cloud particles. More work is needed to clarify this picture, particularly since the study of convection has been hampered by inadequate observations.

A particularly difficult aspect of convective action on large scales is in the realm of convective momentum transfer. It has been shown that convective systems transfer momentum up the gradient and sometimes down. The net effect is uncertain. A great deal of work has been done to document the stratiform rain areas that are often associated with agglomerations of convection. The updrafts and downdrafts in these systems are very different from those of the convection itself. The relative strengths of the convective and stratiform parts of MCSs are known in the average sense for particular geographical areas, but variations in this ratio with environmental conditions have to be better understood.

Moisture has indirect dynamical significance due to the energy transformations associated with phase transitions between vapor, liquid, and ice. However, in other respects it acts like other trace constituents of the atmosphere in that convective motions distribute it vertically. (It also differs from other constituents in that precipitation processes remove it from the atmosphere.) Convective transport and precipitation of water are particularly important to large-scale dynamics because of the diabatic processes with which they are associated. The convective fluxes of water substance are poorly understood. This situation occurs partly because the dynamics of clouds are still somewhat uncertain, but it also stems from our lack of understanding of cloud microphysical processes in the complicated context of convection.

Conduct Tests of Current Understanding of Major Precipitation Mechanisms and Their Dynamic Consequences; Improve Ways of Representing These Processes in Parameterized Form

Cloud modeling and observational capabilities have progressed to the point that critical comparisons of predictions with theory are possible. The sensitivity and availability of polarization diversity radars, when combined with trajectory calculations based on measured, modeled, and retrieved fields of cloud properties and on current and emerging capabilities for detailed microphysical modeling, coupled with realistic cloud dynamics, make such a critical comparison possible. Some attempts at determining if the speed of warm-rain formation is consistent with current theoretical predictions are now planned, and further investigations directed at similar comparisons in other locations and for other precipitation mechanisms are now possible. Such comparisons are necessary to develop confidence in current theory and modeling or to learn where improvements are needed. They also contribute to developing improved parameterizations for the precipitation process suitable for use in situations where complete simulation of the microphysical details of this process is not practical.

The speed with which precipitation forms not only determines the precipitation efficiency of many cloud systems but also influences dynamical and radiative processes by affecting the distribution of condensate, the lifetime of the systems, and the size distribution of hydrometeors. The two principal mechanisms for rain formation, the warm-rain and ice-phase precipitation processes, are both understood only partially, and hence difficult to parameterize from current knowledge of the fundamental processes. There is a need for critical observational data against which to test key aspects of currently accepted theory. For example, the collision efficiency for collisions among water droplets determines the speed of precipitation formation via the warm-rain process, and rates of secondary ice production influence the concentration and sizes of thunderstorm ice crystals that enter anvil regions. One realization of this goal would be to verify the predictions from cloud models of the warm-rain process or of anvil generation.

The production and evaporation of precipitation influence the dynamic evolution of precipitation storms. The factors that control the partitioning of precipitation into convective and stratiform components are not well understood. This partitioning is important because convective and stratiform precipitation are subject to very different fates. The former generally falls out with less evaporation, whereas the latter is often carried tens or hundreds of kilometers from the generating cloud and is subject to a great deal of evaporation from the freezing level downward. In addition, a fraction of the stratiform condensed water is in the form of small ice crystals that are important to the radiation balance of the atmosphere. The evaporation of rain is of extreme dynamical importance, and its rate depends on the size distribution of the raindrops, which is quite variable.

Developing a verified ability to predict rain production is also an important step toward successful weather modification and toward understanding the inadvertent weather modification that results from anthropogenic emissions. Weather modification research appears to have stagnated. Demonstration of a better understanding of the natural precipitation process, in a verified quantitative model, could revive this area of research by providing a basis for developing and testing hypotheses and for assessing the likely effects of modification programs. One additional consequence of this effort would be an updating of parameterizations of precipitation formation, which in most cases continue to be based on the Kessler parameterization or the parameterization determined by Berry and Reinhardt by integration of the stochastic coalescence equation. Neither has been verified by comparison to experimental data, and both are based on theoretical formulations of the coalescence process that are now outdated. Because various forms of these parameterizations enter many climate calculations as well, verifying or improving them would have widespread applicability in the modeling of precipitation-producing systems.

Determine the Utility of Lightning Observations and Measurements of the Global Electrical Circuit as Proxy Atmospheric Data

The lightning location and detection systems that have become operational in the past decade have been an invaluable aid to forecasters in tracking the motion and intensification or decay of storms, particularly in the western United States where radar coverage is incomplete. However, case study research suggests that much additional information about storm behavior and the likelihood of severe weather events, such as tornadoes, flash floods, winter storms, and hail, might be obtained by combining radar and meteorological data with continuous measurements of both intracloud and cloud-to-ground lightning flash rates, the former being more difficult to measure. To be of most benefit for "nowcasting" and warnings, the combined meteorological-electrical behavior of a statistically significant number of different types of storms should be determined. The needs for forest fire forecasting in the West are very different from the needs for tornado forecasting in the Midwest, or for protection of the power network in the East. A significant number of agencies and industries could benefit from increased use and availability of lightning data.

Another recent development suggests that lightning systems operating at very low frequencies can provide lightning detection over a much longer distance than the present National Lightning Detection Network. Such systems may provide information over much of the globe from a few remote sites. For example, the intense deep convection associated with tropical cyclogenesis over the eastern Atlantic Ocean could be detected by a couple of widely spaced stations in North America. This detection could precede by several days the arrival of a full-fledged hurricane at the Florida coast.

Lightning frequency and the electrical circuit of the globe may also prove valuable as indicators of climate change. There has been great concern and debate over the possibility of global warming due to increases in greenhouse gases, but it is very difficult to obtain reliable measures of changes in global temperature. The global electric circuit, an old and well-established concept, affords a new approach to the issue of global change by virtue of its natural integration of the electrical contributions of weather worldwide. The empirical argument for global circuit sensitivity is that increases in surface air temperature create changes in buoyancy [Convective Available Potential Energy (CAPE)], that enhance cloud updrafts, leading to increased electrical activity. The global electric circuit thus provides a mechanism for monitoring the entire planet from a single, or a few, measurement sites. Available measurements suggest that the global circuit is indeed positively correlated with temperature. However, additional investigations are needed to demonstrate the utility of using global circuit measurements. Another potential link between global climate and electrical activity is the possibility that if ice production and charge separation are associated, cirrus cloudiness and electrical activity may be correlated. Clearly, opportunities for progress exist and further studies are warranted.

Determine the Mechanisms of Charge Generation in Clouds, Middle-Atmosphere Discharges, and the Propagation of Lightning

Important progress toward understanding cloud electrification and lightning propagation has been made in the past decade through observational, laboratory, and modeling studies. In regard to charge separation in clouds, these studies continue to show the importance of the development of ice and particularly of graupel, which requires stronger updrafts for growth, to the electrification process. Laboratory studies have also demonstrated the importance of the collisions of smaller ice particles with simulated graupel in separating significant charge. Despite this progress, fundamental questions remain with regard to the basic physics of the charge transfer process, and progress has been limited by the need for more and better observations of the electrical (particularly particle charge) and simultaneous microphysical and dynamical structure in a wide variety of clouds.

In situ measurements of electric fields and particles, lightning location systems, networks of electric field meters to identify lightning discharge locations and characteristics, and polarization Doppler radar measurements—all provide valuable opportunities to obtain information on cloud electrification and microphysics. Even some simultaneous measurements of charge, size, and particle-type characteristics are beginning to become available, which is an important parameter in testing different theories of electrification. Although each of these observations requires care and scrutiny, they are now possible. It is time to coordinate these different measurements and to make the measurements in a larger variety of clouds in different geographical regions of the world. In view of the importance of tropical thunderstorms to the global lightning budget and climate change, it is of particular importance to give added emphasis to clouds in the tropics.

In the past decade, there has been considerable progress toward characterizing the peak current, rise time, electric fields, maximum voltages, and other physical properties of the lightning discharge itself. The results of this research show that the rise times and currents are even shorter than expected and that they create quite large electromagnetic disturbances. One of the difficulties encountered in the course of this research has been in simulating the lightning stroke. The combination of high-current and high-voltage discharges over long distances has not been adequately duplicated.

Although there have been significant advances in understanding the properties of lightning, there has been relatively little work on the fundamental physics. The reasons for the initiation point of discharges in clouds are at best poorly understood. The geometrical development of discharges is largely unknown. The radiation patterns, microscale processes in the channel, chemistry in and near the channel, and attachment processes that are responsible for damage are all areas in which opportunities exist for important advances. New initiatives in the development of physical models are needed to help interpret the observations that are available.

Ordinary lightning is a large-scale electrical discharge commonly found within and beneath thunderclouds and confined to the troposphere. In recent years, at least two exceptionally large-scale discharges occurring above the top of large thunderstorms have received considerable attention. One of them, which has been called a stratospheric discharge, is initiated within intense thunderstorms and propagates upward into the stratosphere. The other phenomenon—a weak, luminous discharge called a sprite—has a reddish hue with the greatest intensity in the mesosphere in the vicinity of 50 to 80 km height. This discharge seems to be associated with the mature phase of large intense convective complexes when positive cloud-to-ground lightning is present, as in the storms responsible for the flooding of the Mississippi during 1993. The behavior and morphology of both of these phenomena are poorly understood. Since the impact on the atmosphere is virtually unknown, they offer both a challenge and a singular opportunity to increase our knowledge of a newly discovered fundamental phenomenon. Although the expected temperature increases are not large, the affected volumes of the sprites are large, and the potential impact on atmospheric chemistry via NO production or alteration of the mixing ratios of other species requires investigation. Likewise, investigation of the spectral characteristics of these discharges is needed.

A better understanding of cloud electrification could lead to a much better ability to predict when and where clouds are likely to become electrified. One example of the potential economic benefit of better understanding of cloud electrification is in the launch of space vehicles, which is often postponed because of lightning, at the expense of about a million dollars per day. Lightning threats to aircraft also result in significant course deviations and costs for commercial airlines. Low-voltage devices are sensitive to low-level transients, and there is concern about the use of composite materials in commercial airplanes because of their susceptibility to lightning damage. Thus, better understanding of electrical phenomena in the atmosphere would have widespread benefits, not only through reduced hazards to humans but also through direct economic payoffs.

Determine the Rate of Generation of NO_x by Lightning

The concentration of NO_x ($NO + NO_2$) is one of the major factors in determining whether there is a net in situ loss or production of ozone in the troposphere. Ample field and laboratory evidence shows that lightning is an important source of NO_x, but the impact on a global scale remains uncertain. Current estimates of the global production rate of NO_x by lightning range between 2 and 200 Tg N/yr. However, models have not accepted the higher estimates and usually restrict the source strength to 3 to 5 Tg N/yr. Even these lower estimates are comparable to, or larger than, source strengths in the middle and upper troposphere from stratospheric exchange and subsonic aircraft, and the upper estimates are comparable to anthropogenic sources in the boundary layer. Likewise, lightning is reported to produce other chemical species in lesser concentrations, and further studies are needed here as well. The newly recognized phenomena of middle-atmosphere discharges, discussed above, may also have an impact on the chemistry of the middle atmosphere.

Determining the importance of lightning as a source of NO_x from observations is difficult, because it requires not only measurements of NO_x within or near storm systems but also characterization of the number and type of lightning flashes within a given area and, for some regions, discharges. It also requires knowledge of the transport and evolving chemistry of the species from the cloud to the regional and global scales. Although this may be a difficult problem, it is one that is tractable and provides an opportunity for real advance. Most of the technical and modeling capabilities are in place, and relatively minor improvements to others would be necessary. Chemistry instruments to measure the proper species, airborne platforms from which measurements can be made, Doppler radars for cloud motion measurements, lightning interferometers to identify location and type of lightning in conjunction with surface networks of electric field sensors, and chemistry and dynamical models on different scales to examine the transport, transformation, and redistribution of the species are now available. One area in which measurement capability is lacking involves the global frequency and location of lightning, which has been discussed above. The main challenge of this problem is to combine and focus the necessary expertise in several interdisciplinary areas.

Improve Observational Capabilities in Support of Studies in Atmospheric Physics

Although some studies of the atmosphere use advanced, modern, state-of-the-art instrumentation to make reliable measurements, there are many glaring weaknesses in our overall ability to measure atmospheric characteristics in support of the studies outlined above. We should learn from the great amount of time wasted in trying to interpret inadequate measurements that it would be more efficient to devote a larger fraction of the community's resources to instrument development, characterization, and improvement. There are numerous examples where current measurement capabilities are inadequate but technical solutions are possible. Yet the atmospheric science community devotes only a very small fraction of its effort to instrument development. Universities are hampered by lack of the long-term funding commitments needed to conduct such research. Private companies do not fill the gap because of the small market. The National Center for Atmospheric Research and other national facilities are always pressed to support a high level of deployment, but have limited personnel and resources to devote to instrument development. The result is seriously inadequate attention to this problem by the atmospheric physics community, perhaps contributing to a community shift toward computer simulations in which technological progress is impressive and new frontiers open every few years. This problem strikes at the infrastructure of the science—the ability to test theoretical understanding with observations. A substantial increase in the overall level of effort devoted to instrument development is needed, especially in the National Science Foundation-supported community. This is a clear "imperative" and is discussed further below.

Develop New Analysis Techniques for Meteorological Data

There is an opportunity to make an initial assault on many of the research challenges identified in this report by using the large data sets previously collected by satellite sensors and in-process experiments. We must seek new ways to express and visualize these data. Examples include multispectral algorithms to infer optical depth, cloud liquid water content, and trace gas concentration from infrared spectrometers. A second opportunity stems from advances made in research applications of remote sensing from the ground. These research tools could be configured to provide routine data on variables important to the parameterization and validation of various physical processes simulated in regional and climate models. Candidate instrumentation includes Doppler infrared lidar, polarized lidar, millimeter radar, microwave radiometers, infrared and solar interferometers, Doppler wind profilers, and 3 cm and 10 cm polarimetric Doppler radars.

At the same time that we seek to optimize the utility of existing data and instrumentation, we must continue an ambitious program of development of new instrumentation if we are to meet the challenges of the next century. The advent of the Department of Energy's (DOE's) Atmospheric Radiation Measurement (ARM) program, the planned launching of the National Aeronautics and Space Administration's (NASA's) Earth Observing System (EOS) platform in 1998, and the new interest in unmanned aerospace vehicles (UAVs) hold the promise of a wealth of new data on radiation and the hydrologic cycle. Some of these data streams will be similar to those being acquired by current systems, but many will be substantially different. There is and will be a tremendous opportunity to bring new and different analysis techniques to bear on these data and associated atmospheric problems. This field can benefit by importing and adapting techniques from areas such as signal processing, pattern recognition, intelligent systems and artificial intelligence, chaos theory, and computer visualization. Such techniques may also have considerable applicability to analyses of model results. As the complexity and resolution of atmospheric models continue to grow, we have to improve our ability to extract meaningful information from them. A number of atmospheric scientists are working in these areas, but more support, particularly for innovative and perhaps risky research, is needed. Unfortunately, in times of static or diminishing support, it is precisely this type of research that is most likely not to be funded.

Exploit Rapidly Increasing Computational Power

As the price of computer cycles continues to fall, new modeling approaches can be used both to circumvent the need for parameterizations and to tackle new issues. In the next decade we may expect numerical modeling of time-dependent problems spanning three orders of magnitude in spatial scale to become feasible. For example, one fundamental problem for which this may provide important new understanding is the organization of deep tropical convection into clusters, superclusters, and synoptic-scale waves. It should be feasible to simulate the motions explicitly in individual cells within a synoptic-scale region of the tropical ocean or over the midwestern United States, avoiding the need for cumulus parameterization. Increasingly sophisticated techniques for placing areas of increased grid resolution adaptively within a flow, so as to cover only the convective regions and not the large expanses of stably stratified nonturbulent flow in between, are a promising way to improve the efficiency of such large computations. These calculations can contribute to fundamental insights into self-organizing mechanisms for moist convection and help in the development of convective parameterizations. As a second example, higher-resolution large eddy simulations of the inversion-capped cloudy atmospheric boundary layer should be able to properly resolve turbulent eddies near the entrainment interface and the surface layer, thereby providing a valuable tool for better understanding and parameterizing entrainment and surface-layer structure.

Increased computer power also allows new and often more fundamental modeling approaches to be used. In radiation, the exploration of three-dimensional radiative transfer in realistic cloud fields and nonspherical scattering problems are becoming feasible and should be fostered. Large eddy simulation models, including explicit drop and aerosol size distributions and even simple chemistry, are a promising tool for exploring climatically important feedbacks such as those between clouds, aerosols, and radiation, as well as for understanding drizzle processes in boundary layer clouds. The effects of orography in climate and forecast models can be represented much more faithfully with increased spatial resolution, while parameterizations of the boundary layer benefit from increased vertical resolution. These are only a few examples of problems in which increased computer power can make a fundamental contribution. However, experience has taught us that increased computer power and model resolution are not substitutes for improvements in the underlying basic physics of the model.

All of the preceding topics are opportunities that should be pursued for their scientific and societal value. However, some stand out because of their particular scientific importance or their likely societal impact. We suggest that the following three topics are of particular importance and should be special foci for research in the coming decade.

Focus 1: Develop and Verify an Ability to Predict the Influences of Small-Scale Physical Processes on Large-Scale Atmospheric Phenomena

The atmosphere is an interactive system with mutually dependent processes that span scales from microscopic to global. In most cases, the important influences among processes arise from the collective effects of an ensemble of interactions. A central impediment to our ability to predict weather and climate is the fact that even when the fundamental physical laws governing individual processes are understood, their collective influences on other phenomena cannot be predicted because of the complexity and number of interactions involved. This problem is particularly evident in climate, weather, and cloud models, where many small-scale processes have collective influences on the largest-scale phenomena. Examples are the interaction of solar radiation with cloud droplets, the effects of cumulus convection and turbulence on the transport of mass and momentum, chemical interactions that influence cloud microstructure and hence radiative properties on a global scale, and the generation of lightning and the global electric field from many small transfers of electrical charge during collisions of hydrometeors. To model all of the processes involved in the atmosphere would require representing processes that occur with scale sizes from at least 0.0001 cm to 10,000 km, a range of 10^{13}. Most modern computer models can only handle scales that cover a range of about 10^3, so we are far from circumventing this problem even with the most optimistic future increases in computing power.

The solution lies in developing organizing principles that make it possible to understand the collective influence of these small-scale processes on larger-scale phenomena in the atmosphere. We need to develop a better understanding not only of the interactions between phenomena that occur with different scales but also of the interactions among processes historically studied by different groups of scientists. In many cases, the linkage between these processes is the essence of the outstanding problem. For example, cloud physicists understand how soluble particles in the atmosphere influence the sizes and number of cloud droplets, and chemists understand the fundamental chemical reactions involved in producing soluble particles. Nevertheless, this understanding has not yet led to a predictive capability for the number of soluble particles in the atmosphere and hence for cloud droplet concentrations and sizes, because these processes interact with each other and with radiation, chemical cycles in the atmosphere, global circulation patterns, and the hydrological cycle in ways that are beyond our current understanding.

We suggest that increased effort directed to this class of problems will provide substantial dividends in the next decade. Our optimism is based on the following considerations:

1. In many cases the treatment of physical processes in models is inconsistent with current understanding of the phenomena. This has resulted partly from a failure to bridge the communications gap between those involved in fundamental studies and those involved in modeling the effects of these studies. An initial basis for improvement in the representation of processes such as turbulent transport, entrainment, boundary layer influences, and precipitation formation already exists.

2. Modeling and computing capabilities are improving rapidly. Model simulations of phenomena on small scales can be used to develop representations of these processes on larger scales, and the techniques and capabilities for this approach are now available but are only beginning to be exploited.

3. New observational capabilities are appearing or expected that will aid in these studies. New satellites for global monitoring will provide extensive new data sources; new weather observing networks are now available; improved atmospheric soundings may become available from new uses of commercial aircraft soundings and remote sensors; new long-range and high-altitude aircraft are

available to support this research; and new fixed sites for global monitoring that are being established or are in operation will provide crucial data.

4. There have been recent satisfying advances suggesting that we are on the threshold of significant progress toward improved understanding of these interactions. Success in predicting radiative transfer for clear skies, new understanding of the structure and lifetime of stratocumulus and cirrus clouds, new appreciation of the roles of aerosols in the atmosphere, the success of cloud-resolving models in representing ensemble influences of clouds, increasingly realistic models of the atmospheric boundary layer, and the development of techniques for inferring hydrometeor and cloud characteristics from satellite observations are all important steps that can support continued and enhanced focus on these and related problems.

5. There has been a significant shift in perspective among scientists capable of addressing these problems—for example, cloud physicists, who traditionally were oriented more toward microscale studies, have developed increasingly global perspectives. This shift has to continue, for example, through field programs that seek to make observations with characteristic scales similar to the grid scale in climate models. Planned programs are formulating better ways of studying the interactive and large-scale consequences of the phenomena being studied.

A key consequence of an improved ability to predict the consequences of small-scale processes will be the development of improved representations or parameterizations for physical processes that occur on scales smaller than the grid scale in models. This will permit the effects of subgrid-scale processes to be represented in simulations of climate, weather, and clouds. Two complementary approaches to parameterization are needed. First, the fundamental interactive processes (e.g., linking radiative properties to hydrometeor spectra and hence to aerosols and cloud dynamics) require clarification. Second, existing knowledge and existing or new data can be used to develop representations that are consistent with current understanding. Such parameterizations, although sometimes unsatisfying because they fall short of representing true physical relationships, are still useful and necessary to support studies of the complex interactions in the atmosphere.

Currently, a wide gulf exists between the parameterization approaches used in climate models and data, and process model studies associated with field programs. The parameterization issue cannot be resolved without bringing to bear the accumulated knowledge from process studies. At the same time, narrowly focused process studies that exhaustively study single cases cannot supply the information needed to improve model parameterizations. Until a much better representation of reality is introduced into parameterizations of processes such as the effects of clouds on radiation, there is little hope of improving confidence in climate and climatic change simulations. Emphasis should be placed on the acquisition of data sets in which cloud and hydrometeor properties, fluxes of radiation, aerosol size distributions, and concentrations of significant trace gases are measured simultaneously. Such data sets will also have to characterize different cloud regimes and geographical areas.

This is an extremely difficult problem, justifying a multifaceted approach. This approach should include increased attention to the acquisition of long-term, temporally and spatially consistent data sets characterizing the properties of clouds, radiation, water vapor, and trace gases. It must also include continued efforts to develop models that explicitly link these quantities. In addition, a concerted effort must be made to link the observational, data analysis, and modeling communities to focus on this problem. Promising approaches include the following:

- using cloud-resolving models and nested models to determine interactions with large-scale variables;
- using cloud models with explicit microphysics to develop parameterizations for microphysical processes suitable for simulations of weather and climate;
- developing parameterizations exclusively from process study observations, then using satellite observations to generalize and extrapolate to the global scale;
- isolating a limited number of empirical parameters that have a physical basis, then fitting to a set of observations to determine the best values of these parameters; or
- using operational models as sources of finer-scale "data" from which to develop parameterizations.

The inability to represent interactions among phenomena, especially those that occur on small scales, is the primary weakness in current models of the climate and weather, so we regard the need for these results as "imperative" and offer this as a key challenge for the next decade. Although a central justification for such studies is the need to improve representations of these processes in models, we nevertheless argue that the challenge—and opportunity—is *to understand* the linkages among various physical and chemical processes. A short-range approach is inappropriate because we are still at an early stage in understanding these interactions. Significant effort must be directed toward improving the foundations that will lead to adequate representations of these processes.

Focus 2: Develop a Quantitative Description of the Processes and Interactions That Determine the Observed Distributions of Water Substance in the Atmosphere

Precipitation is the source of essentially all fresh water on Earth, so the hydrological cycle is truly "vital" to humans and to most plant and animal life on land. Although precipitation and cloudiness are the most evident results of the cycling of water through the atmosphere, water also has many other effects on weather and climate. Water vapor is the most important greenhouse gas, and variations in cloudiness and ice cover are the primary sources of variability in the albedo of the Earth. Clouds scavenge particles and trace gases from the atmosphere, and thunderstorms maintain the Earth's electric field. The latent heat released or absorbed as water changes phase is the source of energy that drives hurricanes and other severe weather systems. Thus, we cannot understand weather and climate without a good understanding of the distribution of water substance in the atmosphere.

Weaknesses in current specifications of the atmospheric water cycle include poor characterization of upper-tropospheric water vapor, uncertainties in surface fluxes, poor understanding of the factors controlling precipitation efficiency, inability to represent the ensemble effects of cumulus convection on the transport of water, poor characterization of rainfall over the oceans, and the absence of a comprehensive understanding of the links between the atmospheric cycle and other components of the hydrological cycle. Emerging technologies and recent developments now can support a comprehensive new approach to these problems. Improved characterization of water vapor in the atmosphere may become possible from sensors on commercial aircraft, remote sensors employing radiometric or GPS technology, and improved research instruments capable of accurate measurements at low humidity. Satellite characterization of precipitation over the oceans will be possible in the coming decade. Existing and new modeling capabilities and data sets can be used to characterize the effects of cumulus convection, and improved models and understanding of boundary layer fluxes are emerging. Comprehensive international programs to study the hydrological cycle over regional scales appear feasible and are planned.

These expected new results should provide an opportunity to characterize the distribution of water in the atmosphere with new confidence and to relate this distribution to the underlying interactions with the global hydrological cycle. Such an improved characterization is needed for accurate determination of radiative transfer in the atmosphere, for climate predictions of precipitation amounts and global temperature, and for improved weather forecasts. These studies thus provide a good match between the opportunities presented by new research capabilities and the needs of current research. However, a comprehensive and systematic approach is required if the many factors and processes entering the atmospheric hydrological cycle are to be understood.

Focus 3: Improve Capabilities to Make Critical Measurements in Support of Studies in Atmospheric Physics

Although impressive advances in instrumentation have been made in some government laboratories, especially in the development of high-technology remote sensing instruments, other areas have fallen far behind what is technologically feasible. For many measurements, research aircraft still employ inadequate, decades-old instruments for lack of better alternatives. Critical measurements of humidity in the upper atmosphere cannot be made because of deficiencies in the humidity sensors used for routine soundings. Our ability to make critical measurements is also lagging far behind current needs in studies of hydrometeor size distributions or radiation. Numerous examples exist in all areas of atmospheric physics.[2]

In most cases there are good candidate techniques for making the needed measurements, but they await implementation. The instrumentation in current use is seriously out of date and is not taking advantage of modern knowledge and modern technology. There is an opportunity for a concerted effort in instrumentation to have a large impact. New experimental instruments to measure temperature and wind from aircraft are beginning to answer crucial questions about cumulus convection that have been posed for 50 years or more. Similar advances can be expected from other improvements in the basic instruments used in atmospheric science research.

There are special needs for new equipment and instrumentation to study the issues raised in the first two imperatives. Foremost is the need for platforms suited to the studies that must be conducted, including new observing satellites, high-altitude research aircraft, and aircraft with long range and flight duration. Remotely piloted vehicles will provide new opportunities to collect measurements at high altitude and over long periods. Examples of missing observations that would make critical contributions include ice mass (especially in anvils), ice crystal radiative characterization (especially near the tops of cirrus and other clouds), wind fields with improved coverage and resolution, upper-troposphere humidity, and cloud fractional coverage as a function of cloud area and altitude. Instrumentation presently available is particularly weak in the following areas:

---

[2]For example, measuring the temperature inside a cloud from aircraft is a problem that has apparently been solved only recently by new radiometric thermometers, and these are not yet thoroughly evaluated or in widespread use. Measurements of equivalent potential temperature in rain are not reliable, so important dynamical influences of precipitation cannot be evaluated. Measurements of cloud-condensed water still do not provide the accuracy required for most studies. Measurement of low concentrations of large precipitation particles, and the sizes and shapes of small ice particles, have not been possible until quite recently, and the new instruments that can make these measurements are still experimental. Unless corrected by GPS measurements, the accuracy of measured horizontal winds from research aircraft is only a few meters per second due to the character of inertial navigation systems, and the degradation of GPS accuracy in nonmilitary applications still prevents removal of these errors in many cases. Broadband radiometers used on research aircraft are not sufficiently accurate and do not respond fast enough to provide the needed measurements of flux divergence in clouds.

- the characterization of hydrometeor size distributions and shapes;
- measurements of total water content and ice water content;
- water vapor mixing ratio, especially in the upper troposphere;
- cloud condensation nuclei, where instruments have to be more widely available and suited to use at high altitude;
- ice nucleus measurements;
- nephelometers;
- other radiation instrumentation for the characterization of irradiance, radiance, and net flux as a function of wavelength, optical depth, and volume absorption coefficient, and for determining the optical properties of ensembles of ice crystals;
- chemistry instrumentation, including trace gas detectors;
- equipment for the detection of organic constituents, especially aerosol instrumentation;
- airborne remote sensing equipment such as lidars, short-wavelength radars, and near-field profilers for temperature and humidity that can provide some extension away from an airborne platform and thus increase the representativeness of its sample; and
- measurement of mean vertical motions to better than 10 cm/s, averaged over areas ranging from cloud-scale to mesoscale.

There are some particular needs for instrumentation to support studies of atmospheric radiation:

- Produce an inexpensive observing system for radiative fluxes that is capable of making continuous, accurate, and reliable observations. Such a system is crucially needed to increase the data base of surface radiative fluxes.
- Make a concerted effort to upgrade all radiometers, particularly broadband flux radiometers used on aircraft. The typical thermopile devices in use are not sufficiently accurate, and their response times are too slow to supply the required in situ data.
- Develop simultaneous multiwavelength, active and passive systems to probe the atmosphere from both space and the surface and to retrieve hydrologic parameters. These systems must be robust and well calibrated. Examples are combinations of passive microwave radiometers and radar, or Doppler lidar and radar.
- Develop observational techniques to sample routinely the regional-scale hydrological cycle. Required observations beyond our current capabilities include water vapor profiles in the upper troposphere and lower stratosphere, ice water path and content, and precipitation.

Although major additional resources are needed for improvements in instrumentation, a modest program, perhaps combining the efforts of university scientists, government laboratories, and research support facilities, could lead to significant improvements in our ability to make critical measurements in the atmosphere.

CONTRIBUTIONS TO NATIONAL GOALS

Human activity is having documented effects on the weather and climate. For example, carbon dioxide and sulfate concentrations in the atmosphere have been changed dramatically by human activity, and the effects of these changes are beginning to appear in the climate of the Earth. In some major cities, pollution advisories warning of hazards to health and leading to restrictions on activities are becoming increasingly common. Biomass burning by humans is the predominant source of particulates and of some chemicals in some regions, and perhaps worldwide. As the global population continues to increase and to industrialize, these problems will become more urgent.

Many of the components of the research program recommended here address sources of uncertainty in climate prediction. Other benefits will result from improved abilities to predict regional climate and weather. When policies to mitigate anthropogenic effects on climate and weather or to manage water and other natural resources are debated, the results of the research proposed here will provide the critical basis for making sound decisions. The difficult choices are likely to involve accepting damage to the economy or to health, penalizing developing or industrialized countries, and burdening current or future generations. In this context, the importance to society of the scientific information that accrues from the research recommended here is indisputable.

PART II
===

2

Atmospheric Chemistry Research Entering the Twenty-First Century[1]

SUMMARY

Atmospheric chemistry came of age during the latter half of the twentieth century. Through the application of modern analytical and computational techniques, scientists were able to elucidate the critical role the atmosphere plays as the "connective tissue" for life on Earth. In the process, another, more disturbing insight was uncovered: the activities of an increasingly populous and technological human society are changing the composition of the atmosphere on local to regional to global scales. Experience has shown that air pollution on local and regional scales can be environmentally and economically destructive. The consequences of chemical change on a global scale have yet to be fully assessed, but the potential for catastrophic effects exists.

---

[1]Report of the Committee on Atmospheric Chemistry: W.L. Chameides (Chair), Georgia Institute of Technology; J.G. Anderson, Harvard University; M.A. Carroll, University of Michigan, Ann Arbor; J.M. Hales, ENVAIR; D.J. Hofmann, NOAA Aeronomy Laboratory; B.J. Huebert, University of Hawaii; J.A. Logan, Harvard University; A.R. Ravishankara, NOAA Aeronomy Laboratory; D. Schimel, University Corporation for Atmospheric Research; and M.A. Tolbert, University of Colorado, Boulder. The group gratefully acknowledges contributions from C. Ennis, NOAA Aeronomy Laboratory; D. Fahey, NOAA Aeronomy Laboratory; F. Fehsenfeld, NOAA Aeronomy Laboratory; I. Fung, University of Victoria, British Columbia; E.A. Holland, National Center for Atmospheric Research; D. Jacob, Harvard University; C.E. Kolb, Aerodyne Research, Inc.; H. Levy, II, NOAA Goddard Fluid Dynamics Laboratory; S. Liu, Georgia Institute of Technology; P. Reich, University of Minnesota; P. Samson, University of Michigan; and P. Tans, NOAA Aeronomy Laboratory.

> **Box II.2.1**
> **Environmentally Important Atmospheric Species**
>
> These species are scientifically interesting and important to human health and welfare because of their radiative (e.g., climate changing) and/or chemical properties. They include the following:
>
> - Stratospheric ozone
> - Greenhouse gases
> - Photochemical oxidants
> - Atmospheric aerosols
> - Toxics and nutrients
>
> Documenting the changing concentrations and distribution of these species, elucidating the processes that control their concentrations, and assessing their impacts on important environmental and ecological parameters will define the principal challenges for atmospheric chemistry in the coming decades.

The scientific questions facing atmospheric chemistry entering the twenty-first century are intellectually profound but are also of vital social and economic importance. They relate to atmospheric constituents that are fundamentally important to our environment: stratospheric ozone, greenhouse gases, ozone and photochemical oxidants in the lower atmosphere, atmospheric aerosols or particulate matter, and toxics and nutrients (see Box II.2.1). It is perhaps a measure of the strides made in recent decades, that the issues of atmospheric chemistry are familiar to the general public, policy makers, and scientists alike. Continued progress in the twenty-first century will require an ambitious, but judicious, commitment of financial, technological, and human resources to document the changing composition of the atmosphere and elucidate the causes and potential consequences of these changes.

Major Scientific Questions and Challenge

The principal focus for atmospheric chemistry research entering the twenty-first century will be the "Environmentally Important Atmospheric Species"—species that, by virtue of their radiative and/or chemical properties, affect climate, key ecosystems, and living organisms (including humans). From an intellectual point of view, these species are interesting because they influence the life support system of our planet. From a societal point of view, they are also of central importance because they directly impact human health and welfare.

The challenge for atmospheric chemistry research in the coming decades follows:

Development and application of the tools and scientific infrastructure necessary to document and predict the concentrations and effects of Environmentally Important Atmospheric Species on a wide variety of spatial and temporal scales.

To meet this challenge, atmospheric chemistry research should be formulated around three fundamental questions:

1. What are the shorter-term periodic and longer-term secular trends in the concentrations of Environmentally Important Atmospheric Species on local to global scales? What are the causes of these trends?
2. How will the concentrations of these species change in the future? What are the most effective and feasible policy options for managing these changes?
3. What will be the totality of environmental effects of present and future trends in the concentrations of these species?

Overarching Research Challenges

The scientific strategy for atmospheric chemistry emerges logically from the application of these fundamental scientific questions to each of the Environmentally Important Atmospheric Species. It is a strategy that endeavors to continuously improve our understanding of the underlying chemical, physical, and ecological processes that control the concentrations of these species, while providing timely and relevant input to decision makers. Toward these ends, the scientific research strategy in atmospheric chemistry must include the following:

- Document the chemical climatology and meteorology of the atmosphere, particularly their variability and long-term trends, through the development and maintenance of diverse and interrelated arrays of monitoring networks.
- Develop and evaluate predictive tools and models of atmospheric chemistry through a synthesis of information gathered from process-oriented field studies, laboratory experiments, and other observational efforts; their representation in mathematical/numerical algorithms; and the testing of these algorithms in well-posed model-evaluation field experiments.
- Provide assessments of the efficacy of environmental management activities through the gathering and interpretation of relevant air quality data.
- Be holistic and integrated in the study of the Environmentally Important Atmospheric Species and of the chemical, physical, and ecological interactions that couple them together.

Disciplinary Research Challenges

The disciplinary challenges listed below focus on the specific, key scientific issues facing the atmospheric chemistry community in the twenty-first century:

- *Stratospheric Ozone Challenges:* Document the distributions, variability, and trends of stratospheric ozone and the key species that control its catalytic destruction; elucidate the coupling between chemistry, dynamics, and radiation in the stratosphere and upper troposphere.
- *Greenhouse Gas Challenges:* Elucidate the processes that control the abundances, variabilities, and long-term trends of atmospheric CO_2 (carbon dioxide), CH_4 (methane), N_2O (nitrous oxide), and upper-tropospheric and lower-stratospheric O_3 (ozone) and water vapor; and expand global monitoring networks to include upper-tropospheric and lower-stratospheric O_3 and water vapor.
- *Photochemical Oxidant Challenges:* Develop the observational and computational tools and strategies needed by decision makers to effectively manage ozone pollution; elucidate the processes that control, and the interrelations that exist between, the ozone precursor species, tropospheric ozone, and the oxidizing capacity of the atmosphere.
- *Atmospheric Aerosol Challenges:* Document the chemical, physical, and radiative properties of atmospheric aerosols, their spatial extent, and long-term trends; elucidate the chemical and physical processes responsible for determining the size, concentration, and chemical characteristics of atmospheric aerosols.
- *Toxics and Nutrients Challenges:* Document the rates of chemical exchange between the atmosphere and key ecosystems of economic and environmental import; elucidate the extent to which interactions between the atmosphere and biosphere are influenced by changing concentrations and deposition of harmful and beneficial compounds.

Infrastructural Initiatives

The following infrastructural initiatives provide the resources and capabilities recommended to accomplish the disciplinary challenges:

- *Global Observing System*: deployment of an observing system for moderately lived species to complement ongoing networks and measurement platforms focusing on long-lived species and stratospheric ozone.
- *Ecosystem Exposure Systems*: deployment of monitoring networks capable of assessing ecosystem exposure to primary and secondary toxics and nutrients.
- *Surface Exchange Measurement Systems*: development and deployment of measurement systems capable of quantifying chemical exchange between the atmosphere and key biological or ecosystems.

- *Environmental Management Systems*: demonstration and assessment of the feasibility of operational "chemical meteorology" as a prognostic tool for environmental managers and regulators.
- *Instrument Development and Technology Transfer*: development of programs and facilities to support the evaluation of new atmospheric chemical instruments and their transfer to the scientific, regulatory, and private sector communities.
- *Fundamental Condensed Phase and Heterogeneous Chemistry*: development and maintenance of laboratory facilities focused on condensed phase and heterogeneous chemical processes relevant to the atmosphere.

Expected Benefits and Contribution to National Well-Being

The scientific questions to be addressed by the atmospheric chemistry research community entering the twenty-first century are central to our understanding of the chemical and physical environment in which we human beings must reside. For this reason, the science of atmospheric chemistry is highly relevant to the future development and economic vitality of our society. Today, the changing chemistry of our atmosphere on local, regional, and global scales is an observational fact. These changes are impacting human health and placing economically and environmentally important resources and ecosystems at risk. At the same time, air quality management activities in the United States cost tens of billions of dollars annually. Research in atmospheric chemistry and the resulting improvements in our predictive capabilities will help us to maximize the environmental and economic benefits gained from these sizable investments in air quality management, while also teaching us how to minimize the deleterious effects of human activity on the chemical and physical environment.

INTRODUCTION AND OVERVIEW

As the world stands on the threshold of a new millennium, the atmospheric chemistry community stands at the portal of a new era of scientific research. During the latter half of the twentieth century, the discipline of atmospheric chemistry came of age. Scientific study revealed the crucial role that the chemistry of the atmosphere plays in the life support system of the planet, acting as a "connective tissue" by which organisms of the biosphere interact and exchange materials and energy. It also uncovered a more disturbing insight: the activities of an increasingly populous and technological human society are changing the composition of the atmosphere on local, regional, and even global scales. Experience has shown that air pollution on local and regional scales can be environmentally and economically destructive. The consequences of chemical change on a global scale could be even more damaging. Thus, the scientific questions facing atmospheric chemistry are not only intellectually challenging but also of vital social and economic importance.

The challenge for atmospheric chemistry as it enters the twenty-first century will be to build on the discoveries of the twentieth century by maintaining its scientific vitality and rigor while also making the results of its scientific and technological advances available to influence the nation's and the world's social and economic development. In this Disciplinary Assessment, we discuss the strategy that will be necessary to address the major scientific issues of the discipline while also providing decision makers with the information and tools they require to manage and maintain environmental and economic vitality. We begin our discussion with a statement of the mission for atmospheric chemistry research entering the twenty-first century.

The Mission

Development and application of the tools and scientific infrastructure necessary to document and predict the concentrations and effects of Environmentally Important Atmospheric Species on a wide variety of spatial and temporal scales.

In identifying the mission for atmospheric chemistry research entering the twenty-first century, we have adopted three basic premises:

1. The financial and human resources available for research and development in the coming decades will be limited.
2. The activities of an increasingly populous and technological society have and will continue to perturb critical environmental factors that affect the natural resources on which our society relies.
3. Unraveling the mechanisms that couple the chemistry of the atmosphere to the life support system of the planet represents one of the major intellectual and technological challenges of the coming decades.

Premises 1 and 2 relate to the resource- and policy-relevant issues that must be considered in defining the mission for atmospheric chemistry research, whereas premise 3 focuses on the intellectual or curiosity-based raison d'etre for the discipline. The prospect of limited resources for research and development indicated in premise 1 demands that a rigorous prioritization be applied to any contemporary research program, so that the most pressing scientific issues can be addressed in the allocation of public resources to the scientific community. Premise 2 suggests that priority should be placed on developing a scientifically robust, predictive, and systematic understanding of the Earth system, its chemical environment, and the relationships between the economic and technological growth of the world's nations and the environmental vitality and natural resources on which they depend.

The development of a research program often requires compromises between

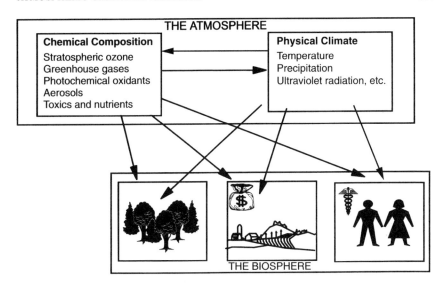

FIGURE II.2.1 Environmentally Important Atmospheric Species are atmospheric constituents that affect human health and welfare and thus are central to a policy-relevant research program in atmospheric chemistry. Because these species drive the interaction between the atmosphere and the life support system of the planet, they are also central to a curiosity-based research program in atmospheric chemistry. Environmentally Important Atmospheric Species include greenhouse gases (e.g., CO_2, CH_4, N_2O), aerosols, and stratospheric ozone—species that, because of their radiative properties, affect the climate and other physical characteristics of our environment. They also include the photochemical oxidants and tropospheric ozone, acid aerosols, and a wide variety of toxic and nutritive substances that, because of their chemical properties, affect humans and ecosystems of economic and environmental importance when they come in direct contact with them. Although the radiatively important species' effects are more commonly felt on a global scale, the effects of the ecologically important species are most often experienced on local-to-regional scales. Nevertheless, despite the varying scales of their radiative and chemical effects, research has revealed that the atmospheric cycles of these species are coupled together through complex photochemical and dynamical interactions. Unraveling these complex interactions represents one of the major challenges of atmospheric chemistry research in the coming decades.

the priorities dictated by policy-relevant issues and those dictated by more theoretical interests. However, in the case of atmospheric chemistry research we find a strong resonance between the two. Atmospheric chemistry research in the coming decades should be focused on documenting and predicting the concentrations and effects of the chemical constituents that most directly affect the physical and biological environment and, by extension, human health and welfare. We refer to these species, here, in the most generic sense, as the Environmentally Important Atmospheric Species that, by virtue of their radiative and/or chemical properties, directly affect living systems and key environmental parameters (see Figure II.2.1).

For atmospheric chemistry to make significant scientific advances in the coming decades, however, its research focus on these Environmentally Important Atmospheric Species must go well beyond simple observation and documentation of chemical content and change, to a rigorous investigation of the underlying chemical, physical, and ecological processes that determine the atmospheric concentrations of these species. It is, after all, only through understanding these processes that a genuine appreciation of the atmosphere and its relationship to the Earth system can be fostered, and a reliable predictive capability can be achieved and made available for the development of effective public policy. Additionally, because the atmosphere and the stresses placed on it are continually changing, a significant portion of the resources made available for atmospheric chemistry research in the coming decades must be used to develop an enduring research infrastructure that can inform decision makers in an open, effective, and responsive fashion.

The mission for atmospheric chemistry research in the coming decades must therefore combine a focus on the Environmentally Important Atmospheric Species with a commitment to the development of a comprehensive, long-term research capability and technological infrastructure. Hence, the mission of atmospheric chemistry in the coming decades:

Development and application of the tools and scientific infrastructure necessary to document and predict the concentrations and effects of Environmentally Important Atmospheric Species on a wide variety of spatial and temporal scales.

In the following sections, we consider how to accomplish this mission most effectively by first considering what is now known about the atmosphere and then identifying the key unresolved scientific questions surrounding a number of Environmentally Important Atmospheric Species and the research challenges that grow from these questions.

Insights of the Twentieth Century

By grappling with a number of critical, but largely unforeseen environmental problems in recent decades, scientists have gained fundamental new insights about the atmospheric chemical system.

The study of atmospheric chemistry as a quantitative, scientific discipline can be traced to the eighteenth century when world-renowned chemists such as Joseph Priestley, Antoine-Laurent Lavoisier, and Henry Cavendish undertook the investigation of the chemical components of the atmosphere (Farber, 1961; Weeks and Leicester, 1968). It was largely through their efforts, as well as those of a number of prominent chemists and physicists who succeeded them in the nine-

TABLE II.2.1 Important Trace Species of the Atmosphere[a]

Species	Concentration (Mole Fraction)	Principal Sources
Methane (CH_4)	1.6×10^{-6}	Biogenic
Carbon Monoxide (CO)	$(0.5 - 2) \times 10^{-7}$	Photochemical, Anthropogenic
Ozone (O_3)	$10^{-8} - 10^{-6}$	Photochemical
Reactive Nitrogen (NO_y)	$10^{-11} - 10^{-6}$	Lightning, Anthropogenic
Ammonia (NH_3)	$10^{-11} - 10^{-9}$	Biogenic
Particulate Nitrate (NO_3^-)	$10^{-12} - 10^{-8}$	Photochemical, Anthropogenic
Particulate Ammonium (NH_4^-)	$10^{-11} - 10^{-8}$	Photochemical, Anthropogenic
Nitrous Oxide (N_2O)	3×10^{-7}	Biogenic, Anthropogenic
Hydrogen (H_2)	5×10^{-7}	Biogenic, Photochemical
Hydroxyl (OH)	$10^{-13} - 10^{-11}$	Photochemical
Peroxyl (HO_2)	$10^{-13} - 10^{-11}$	Photochemical
Hydrogen Peroxide (H_2O_2)	$10^{-10} - 10^{-8}$	Photochemical
Formaldehyde (H_2CO)	$10^{-10} - 10^{-9}$	Photochemical
Sulfur Dioxide (SO_2)	$10^{-11} - 10^{-9}$	Anthropogenic, Volcanic
Dimethylsulfide (CH_3CCH_3)	$10^{-11} - 10^{-10}$	Biogenic
Carbon Disulfide (CS_2)	$10^{-11} - 10^{-10}$	Anthropogenic, Biogenic
Carbonyl Sulfide (OCS)	10^{-10}	Anthropogenic, Biogenic
Particulate Sulfate (SO_4^-)	$10^{-11} - 10^{-8}$	Anthropogenic, Photochemical

[a]After Chameides and Davis (1982).

teenth century, that the identity and concentration of the major components of the atmosphere (i.e., nitrogen, oxygen, water, carbon dioxide, and the rare gases) were established.

In the late nineteenth and early twentieth centuries, atmospheric chemists shifted their focus from identifying the major atmospheric constituents to consideration of the trace constituents, that is, the gaseous and aerosol atmospheric species having concentrations of less than a few parts per million per volume of air (i.e., ppmv). The application of modern chemical analytical techniques revealed the atmosphere to be a reservoir of a myriad of trace species, whose presence can be attributed to a complex array of geological, biological, chemical, and in many cases, anthropogenic processes (see Table II.2.1). Moreover, these trace species were found to have a disproportionately large impact on our environment. In some instances, they adversely affect plant and animal life because of their toxic properties; in other instances, they benefit these or other organisms because of their nutritive properties; in still other instances, they affect the physical climate because of their radiative properties.

The latter half of the twentieth century has seen another major shift in atmospheric chemistry as scientists attempt to grapple with a number of potentially critical environmental problems, including stratospheric ozone depletion, urban

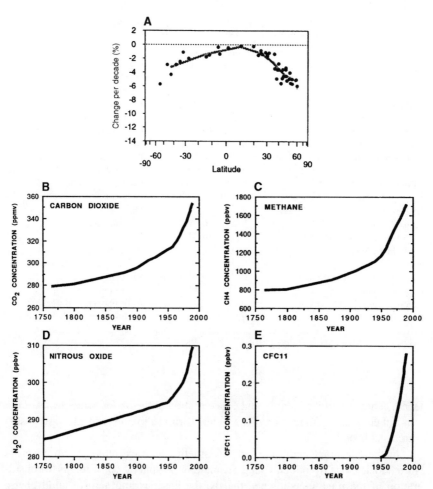

FIGURE II.2.2 From an atmospheric chemistry point of view, global change is an observational fact not a theoretical possibility. (A) Average change per decade in the total atmospheric ozone column as a function of latitude based on recent Dobson station measurements (after WMO, 1995). (B-E) Average global concentrations of CO_2, CH_4, N_2O, and CFC-11 since the mid-1700s (after IPCC, 1990).

photochemical smog, and rising concentrations of "greenhouse gases" (NRC, 1984). In the process, a new, policy-relevant research paradigm for atmospheric chemistry has developed that has profoundly altered its role in society. More importantly, the insights gained from the study of these environmental crises have irrevocably changed our understanding of the atmospheric chemical system in which we as a species must reside. The major aspects of these new insights are outlined below.

The Chemical State of the Atmosphere Has Changed in the Past and Is Continuing to Change

Observations have shown irrefutably that the chemistry of the atmosphere is changing on local, regional, and global scales; indeed, from a chemical point of view, global change is an observational fact not a theoretical possibility. The annual appearance of the Antarctic ozone hole provides striking evidence of the atmosphere's vulnerability to chemical perturbation. Although smaller in magnitude, the depletion of stratospheric ozone in the temperate latitudes over the past decade is perhaps equally disturbing (Figure II.2.2A). Moreover, present-day measurements coupled with analyses of ancient air trapped in ice cores provide a record of dramatic, global increases in the concentrations of a number of long-lived greenhouse gases such as carbon dioxide, methane, nitrous oxide, various chlorofluorocarbons (CFCs), and other halocarbons (Figure II.2.2B-E). Although secular trends in shorter-lived species are more difficult to document, a strong case can be made that the abundances of tropospheric ozone and sulfate and carbonaceous aerosols have also increased significantly in the Northern Hemisphere during the past century (NRC, 1993).

Humans Are a Significant Driving Force in Global Chemical Change

Many of the recent changes in atmospheric composition can be traced to anthropogenic causes. A classic example is that of atmospheric CO_2, whose increasing rates of production from the burning of fossil fuels and biomass closely mimic its rising atmospheric abundance since the Industrial Revolution (see Figure II.2.3). In other examples, the forcing from anthropogenic activities is less obvious, largely arising when the photochemical oxidation and degradation processes of anthropogenic emissions lead to the production of secondary products that perturb important environmental parameters. Examples of these indirect perturbations include the release of chlorofluorocarbons that cause stratospheric ozone depletion, the emission of sulfur oxides that result in increasing concentrations of radiatively important and health-damaging sulfate aerosols, and the emissions of nitrogen oxides and volatile organic compounds that lead to the production of tropospheric ozone and other photochemical oxidants.

Chemical Emissions into the Atmosphere Can Have Long-Term Environmental Consequences That May Be Difficult to Reverse

Because of the long time scales associated with many of the processes that affect atmospheric composition, the chemicals we put into the atmosphere and the environmental effects they engender can persist for decades or even centuries. A prime example is the long-term impact of anthropogenic CFCs on stratospheric ozone.

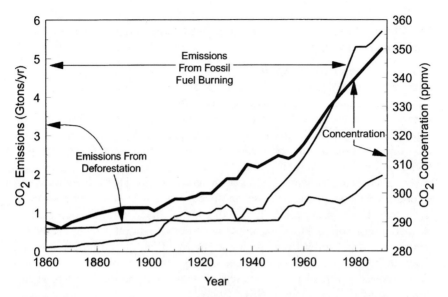

FIGURE II.2.3 Correlation between anthropogenic CO_2 emissions and atmospheric CO_2 concentrations. *Left axis:* Global annual CO_2 emissions from fossil fuel burning and deforestation. *Right axis:* Annual average atmospheric CO_2 concentration, inferred from analyses of air trapped in ice cores and by direct measurements. Increasing concentrations of CO_2 and other greenhouse gases may be triggering a perturbation of global climate that includes increasing surface temperatures and alterations in rainfall patterns and intensities. SOURCE: Figure prepared from data in IPCC, 1995.

During the formation of the ozone hole over Antarctica each austral spring, a major fraction (currently about two-thirds) of the protective column of ozone disappears. Continuous measurements of the ozone column over Haley Bay at 76°S indicate that the ozone hole first began to form in the mid-1970s and has returned each year since, with varying but essentially increasing intensity (see Figure II.2.4A). An international program of research in the 1980s established that the ozone hole is caused by a unique combination of dynamical and anthropogenically driven photochemical processes over the Antarctic. The dynamical processes give rise to a wintertime polar vortex and the concomitant formation of polar stratospheric clouds (PSCs). The photochemical processes, driven largely by chlorine compounds from the photochemical degradation of anthropogenically produced CFCs, give rise to heterogeneous chemical reactions on the PSCs that cause a rapid and dramatic depletion of stratospheric ozone. In large part as a result of these findings, international protocols were implemented in the early 1990s to reduce and ultimately ban CFCs. These protocols should ultimately bring about the demise of this troubling phenomenon. Unfortunately, because of the long lifetime of CFCs in the atmosphere, it is projected that the ozone hole

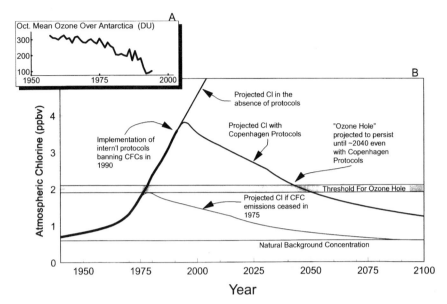

FIGURE II.2.4 History and projected evolution of atmospheric chlorine and the Antarctic ozone hole. (A) October mean ozone column measured over Halley Bay, Antarctica (76°S) from 1957 to 1993 in Dobson Units (DU). (B) Atmospheric concentration of chlorine as a function of year; heavy solid line indicates observed and inferred concentrations from 1940 to the present; thin solid lines indicate projected concentrations under the Copenhagen Protocols if no protocols are implemented and if CFCs had been banned in 1975. Also shown are the estimated natural background chlorine concentration and the estimated threshold chlorine concentration needed to produce an ozone hole. Note that actual chlorine concentrations first attained the threshold concentration in 1975. SOURCE: Figure prepared from data in WMO, 1995.

will continue to appear each austral spring for the next 40 to 50 years (Figure II.2.4B). Interestingly, if CFCs had been banned in 1975, one year after the connection between these compounds and stratospheric ozone depletion was first uncovered (Cicerone et al., 1974; Crutzen, 1974; Molina and Rowland, 1974), it is likely that the occurrence of the ozone hole would have been averted.

The Chemistry and Dynamics of the Atmosphere Interact with Each Other and with the Biosphere Through Complex, Nonlinear Mechanisms

The chemical, dynamical, and biological processes that shape atmospheric composition interact in complex ways and can give rise to unexpected but important phenomena. For this reason, effective management of the environment requires a comprehensive and quantitative understanding of all these processes. A prime example of the importance of understanding the complexities of the

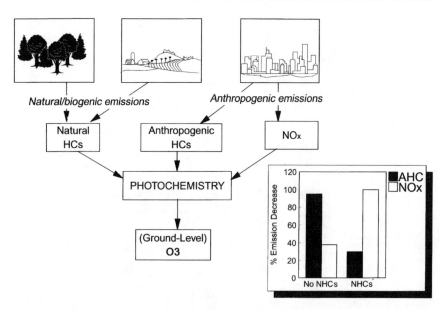

FIGURE II.2.5 Interaction of natural and anthropogenically emitted hydrocarbons (HCs) and nitrogen oxides (NO_x) in the production of ozone (O_3) and photochemical smog. *Inset*: Estimated percentage decrease in anthropogenic HC and NO_x emissions required to bring Atlanta, Georgia, into attainment of the National Ambient Air Quality Standard for ozone for a selected air pollution episode in the summer of 1990 based on model calculations that neglect natural hydrocarbons (labeled No NHCs) and model calculations that include natural hydrocarbons (labeled NHCs). Without the inclusion of natural HCs the calculations indicate that emission decreases in anthropogenic HCs (AHCs) would most effectively reduce ozone pollution. However, inclusion of the effects of natural HC emission leads to the opposite conclusions, namely, that a NO_x-based strategy is most effective. SOURCE: Cardelino and Chameides, 1995.

atmospheric chemical system can be found in the nation's attempts to control photochemical smog.

Photochemical smog, with its high concentrations of ground-level ozone and other oxidants, is most severe during hot, stagnant weather conditions (NRC, 1991). It is generated by a complex series of photochemical reactions involving the oxidation of hydrocarbons in the presence of nitrogen oxides and sunlight (see Figure II.2.5). Thus, the control of photochemical smog can, in principle, be accomplished by decreasing emissions of hydrocarbons or nitrogen oxides or both. In practice, however, the problem has proved much more difficult to solve, as described below.

Despite the implementation of increasingly tighter controls on (mostly hydrocarbon) emissions in the United States since the 1970s, photochemical smog remains a major environmental problem in many of the nation's cities (EPA,

1995). Much of this difficulty can be traced to the complexities of the chemical system. The rate of ozone production is a nonlinear function of the mixture of hydrocarbons and nitrogen oxides in the atmosphere. Depending on the relative concentrations of these species, ozone production rates can be sensitive to hydrocarbons and insensitive to nitrogen oxides (in which case a hydrocarbon control strategy would be most effective) or vice versa. Another difficulty arises from the potential for significant natural hydrocarbon emissions from trees and other vegetation. Because these natural emissions cannot be controlled, a strategy based on decreasing anthropogenic hydrocarbon emissions may not be effective even when ozone production in the area is most sensitive to hydrocarbons (inset in Figure II.2.5). It now appears that the lack of progress in reducing photochemical smog in many of the more sylvan urban areas of the United States can, in part, be traced to a failure to account for the effect of the natural hydrocarbon emissions (Chameides et al., 1988; NRC, 1991).

Linkages Between Atmospheric Species and Their Chemical Cycles Are Complex and Pervasive

The chemical constituents of the atmosphere are not processed independently of each other. They are instead linked through a complex array of chemical and physical processes. As a result of these linkages, a perturbation of one component of the atmospheric chemical system can lead to significant, nonlinear effects that ripple through the other components of the system and, in some cases, to feedbacks that can either amplify or damp the original perturbation. Figure II.2.6 provides a schematic illustration of the kinds of interactions and feedbacks that can occur.

Disciplinary Research Challenges

The successes of atmospheric chemistry research over the past decades have raised a number of intriguing scientific questions about the workings of the Earth system.

The identification and quantification of secular trends in the trace composition of the atmosphere, and the elucidation of complex mechanisms that link atmospheric species to each other and to atmospheric dynamics, are testaments to the vitality and growing technological capabilities of the research programs focusing on atmospheric chemistry and global change. However, while the successes of atmospheric chemistry research over the past decades have answered important scientific questions, they have raised many intriguing new questions that are at the core of understanding the inner workings of the Earth system and must be answered to cope effectively with society's ubiquitous and growing environmental problems. In their most fundamental form, these are the outstanding questions:

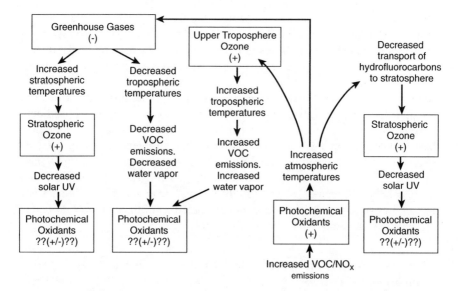

FIGURE II.2.6 Conceptual diagram illustrating potential interactions and feedbacks between photochemical oxidants, greenhouse gases, and stratospheric ozone that might ensue following an initial increase in the emissions of ozone precursor species, that is, volatile organic compounds (VOCs) and nitrogen oxides (NO_x). NOTE: + and - indicate effect of the preceding change on the relevant species' concentration.

1. What are the causes of the variability and secular trends in the concentrations of Environmentally Important Atmospheric Species?
2. How will the concentrations of these species change in the future? What are the most effective and feasible policy options for managing these changes?
3. What are the environmental effects of present and future trends in the concentrations of these species?

These three questions form the basic framework and focus for atmospheric chemistry research in the coming decades. However, because each of the Environmentally Important Atmospheric Species has unique chemical properties and spatial and temporal variabilities, the application of these fundamental questions will require a specific and unique research strategy for each species. The main elements of these strategies are outlined below and then discussed in greater detail.

Stratospheric Ozone

Although great strides have been made in our understanding of stratospheric ozone depletion, major uncertainties remain. Thus, we must remain diligent in our focus on the stratosphere.

Observations have now documented a significant depletion in the levels of ozone in the stratosphere. Epidemiological data showing the role of stratospheric ozone in protecting life from the harmful effects of ultraviolet radiation, along with the scientific evidence linking the depletion of stratospheric ozone to the emissions of anthropogenic chlorofluorocarbons and other ozone-depleting substances, has led to the Montreal Protocol and its subsequent amendments that require a phaseout of CFCs and similar compounds that can deplete stratospheric ozone. Although great strides clearly have been made in our understanding of stratospheric O_3 and its depletion, major uncertainties remain. For example, current models of stratospheric chemistry significantly underpredict the magnitude of ozone depletion observed in recent decades over the midlatitudes during winter and spring (see Figure II.2.7).

If there is one basic lesson to be learned from our experience with the discovery of the ozone hole it is that in the face of scientific uncertainty, we must remain diligent in our focus on the stratosphere. Thus, a continued commitment to stratospheric ozone research in the coming decades is essential. This commitment should focus on two major research challenges: (1) *Continued monitoring of the spatial distribution of stratospheric ozone and the key species responsible for its catalytic destruction* in order to document the evolving nature of the stratosphere; and (2) *elucidation of the coupling between chemistry, dynamics, and radiation in the stratosphere and upper troposphere* in order to develop a more reliable predictive capability for stratospheric ozone.

Trends in Atmospheric Greenhouse Gases

A reliable prediction of future climatic trends demands first and foremost a reliable prediction of future trends in concentrations of the greenhouse gases. However, there are major gaps in our understanding of the processes that control the concentrations of these gases.

Greenhouse gases act as atmospheric thermal insulators; by absorbing infrared radiation from the Earth's surface and reradiating a portion of this radiation back to the surface, they enhance the so-called greenhouse effect that warms the Earth's surface above the warming caused by directly incident solar radiation. These gases include primary greenhouse gases emitted directly into the atmosphere (e.g., CO_2, CH_4, N_2O, CFCs) and secondary greenhouse gases photochemically produced within the atmosphere (e.g., O_3). Water vapor (H_2O), the most important of the greenhouse gases, is both emitted into the atmosphere (via evapotranspiration) and produced photochemically in the stratosphere by the oxidation of CH_4. Ozone is unique in that it not only acts as a greenhouse gas (by absorbing infrared radiation) but also is an important absorber of solar radiation.

As noted earlier, the observed increases in concentrations of CO_2, CH_4, N_2O, and halocarbons provide one of the clearest manifestations of atmospheric global

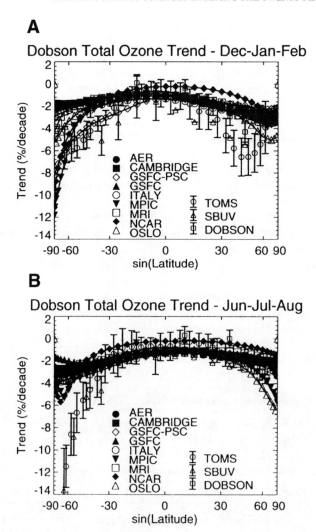

FIGURE II.2.7 Total ozone column change from 1980 to 1990 inferred from TOMS, SBUV, and Dobson measurement systems compared with predicted trends from nine different atmospheric models. (A) Data and model calculations for December-January-February. (B) Data and model calculations for June-July-August. Note the deviations between predictions and measurements in the midlatitudes of each of the hemispheres. The inability of present-day models to fully explain the levels of ozone depletion seen in the atmosphere indicates the need to remain diligent in monitoring the evolution of the stratosphere under declining burdens of CFCs and to continue to investigate the stratosphere to develop a more reliable predictive capability. NOTE: SBUV = Solar Backscatter Ultraviolet Spectrometer; TOMS = Total Ozone Mapping Spectrometer. SOURCE: After WMO, 1995.

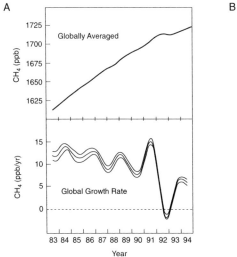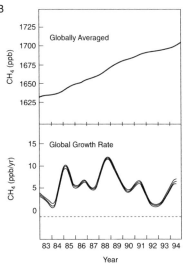

FIGURE II.2.8 Globally averaged concentration (top panel) and annual concentration change or trend (bottom panel) in (A) CH_4 and (B) CO_2 during the 1980s and 1990s. The causes of the large year-to-year fluctuations in the CH_4 and CO_2 trends, especially during the early 1990s, are largely unknown and suggest that major gaps exist in our understanding of the processes that control these species' concentrations on an annual time scale. SOURCE: After P. Tans, private communication, 1996.

change (Figure II.2.2B-E). These increasing concentrations are, in many cases, driven by human activity and related to energy use, industrialization, land use change, and/or agriculture. It is likely that increases in these gases are responsible, at least in part, for the global warming that has occurred in this century, and the warming might have been even greater were it not for the offsetting effects of aerosols such as sulfates (see below).

A reliable prediction of future climatic trends demands first and foremost a reliable prediction of the trend in concentrations of greenhouse gases over time. However, the dearth of data on present-day trends in upper-tropospheric and lower-stratospheric O_3 and H_2O makes projections of future trends for these species highly problematic. Moreover, the seemingly erratic and unexplained variations in the concentrations of many greenhouse gases in the early and mid-1990s (see Figure II.2.8) have exposed major gaps in our understanding of the processes that control the concentrations of these species.

In order to develop more reliable predictions of greenhouse gas trends and concentrations, it is imperative that we (1) *maintain global monitoring for long-lived greenhouse gases, while expanding the monitoring capability for upper-tropospheric and lower-stratospheric O_3 and H_2O* in order to document the changing radiative character of the atmosphere; and (2) *elucidate the processes* that control the concentrations and variability of CO_2, CH_4, N_2O, and upper-tropospheric and lower-stratospheric O_3 and H_2O.

Tropospheric Photochemical Oxidants

The critical scientific issues surrounding photochemical oxidants range from the formation of urban air pollution to the role of tropospheric ozone in global climate change.

Photochemical oxidants are highly reactive compounds produced by photochemical reactions (Haagen-Smit, 1952). These compounds control the oxidizing capacity of the atmosphere and thus the atmospheric residence times and abundances of a host of other Environmentally Important Atmospheric Species (Figure II.2.6). Moreover, because of their reactivity with living tissue, photochemical oxidants can have deleterious effects on human health and ecosystems (LeFohn, 1992).

The most abundant photochemical oxidant is ozone. Paradoxically, whereas ozone in the stratosphere protects living organisms from the harmful effects of solar ultraviolet radiation, ozone in the lower atmosphere can have adverse effects on plants, animals, and human health. As described below, the critical scientific issues surrounding photochemical oxidants range from more fully understanding the processes involved in the development of urban and rural ozone pollution to elucidating the role of changing concentrations of tropospheric ozone in global environmental change.

In the United States, increasingly tighter emission controls have been implemented since the late 1970s to reduce ozone concentrations in the nation's urban centers and thereby protect human health. Measures now being implemented under the Clean Air Act Amendments of 1990 to address this problem are estimated to eventually cost the nation more than $10 billion annually. Despite these measures (see Figure II.2.9), 77 cities are still cited by the U.S. Environmental Protection Agency (U.S. EPA) for failure to comply with the ozone National Ambient Air Quality Standard (NAAQS) designed to protect human health (EPA, 1995). In addition, there is growing concern that enhanced O_3 concentrations in the rural areas of the United States are harming agricultural and forest ecosystems of economic and environmental import (NRC, 1991).

On regional and global scales, data are sparse but suggest that significant changes in tropospheric ozone concentration may also be occurring, at least in some regions, and that these changes are distinctly different from those occurring in cities in the United States. Retrospective analyses of data gathered in the nineteenth century suggest that tropospheric O_3 concentrations over the European continent (and perhaps even the Northern Hemisphere, as a whole) may have increased by a factor of two or more this century (see Figure II.2.10). Contempo-

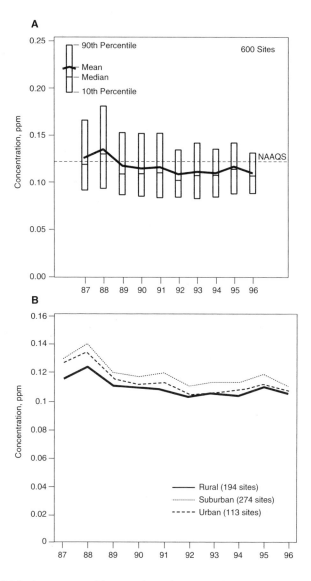

FIGURE II.2.9 Average trend in annual maximum one-hour ozone concentration measured at urban and suburban sites in the United States and recorded in the Aerometric Information Retrieval System. Data indicate a modest (~6 percent) decrease from 1986 to 1995 in both overall air quality (II. 2.9 A) and in its components (II.2.9B). This modest decrease occurred in the face of sizable investments made in pollution control over this period, on the one hand, and significant economic expansion, on the other. SOURCE: After U.S. Environmental Protection Agency, 1995.

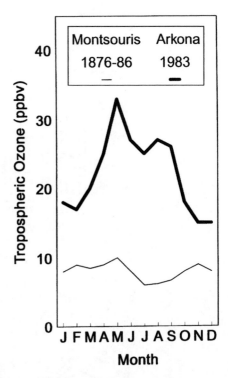

FIGURE II.2.10 Average O_3 concentration as a function of month measured at high altitude sites in Europe. Lighter solid line represents concentration inferred from a retrospective analysis of data collected in the late 1800s. Heavier solid line represents data collected in 1983 using modern instrumentation. SOURCE: After Volz-Thomas and Kley, 1988..

rary data collected at remote surface sites over two decades indicate a more complex pattern, with increasing O_3 levels in some regions of the troposphere, little or no change in other regions, and decreasing O_3 levels in still others (see Figure II.2.11).

The reasons ozone is increasing in some regions of the globe and decreasing in others have yet to be fully explained. Moreover, the climatic and ecological impacts of these changes are poorly understood. Perhaps most importantly from the viewpoint of global change, the long-term trend in the overall oxidizing capacity of the atmosphere implied by these changing ozone concentrations is not known. As delineated in Figure II.2.6, a change in the overall oxidizing capacity of the atmosphere could have effects that ripple throughout the atmosphere's chemical system, ultimately affecting greenhouse gas and stratospheric ozone concentrations, as well as those of the photochemical oxidants.

Two research challenges emerge from this discussion: *It is recommended (1) that the observational and computational tools and strategies needed by decision makers to devise more effective urban- and regional-scale ozone pollution abatement strategies and test their efficacy once implemented be developed; and (2) that the processes that control and the interrelations that exist between the ozone precursor species, tropospheric ozone, and the oxidizing capacity of the atmo-*

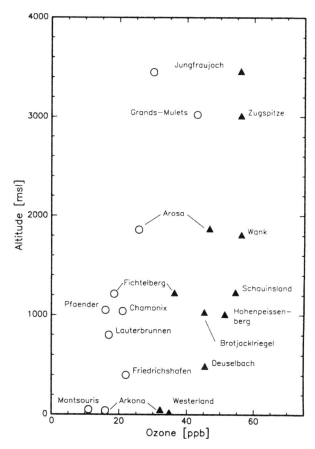

FIGURE II.2.11 Historical (circles) and recent (triangles) surface ozone concentrations from different locations in Europe as a function of altitude. SOURCE: After Staehlin et al., 1994.

sphere be elucidated in order to develop a more robust predictive capability and better understand the consequences of perturbation of the cycle of photochemical oxidants in the atmosphere. The research strategy for addressing these challenges is presented later.

Chemical and Physical Properties of Atmospheric Aerosols

Aerosols play an important role in climate change, stratospheric ozone depletion, and air quality issues. However, large uncertainties in the chemical, physical, and radiative properties of atmospheric aerosols render quantitative assessment of their effects problematic.

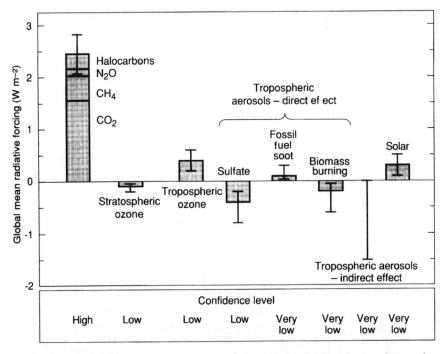

FIGURE II.2.12 Estimates of globally averaged radiative forcing due to changes in greenhouse gases and aerosols from the preindustrial era to the present and changes in solar variability since 1850, with uncertainties in estimates indicated by vertical lines. Although of "low" and "very low" confidence, calculated cooling from the direct and indirect aerosol effect may be large enough to have significantly offset warming from greenhouse gases. SOURCE: After IPCC, 1996.

Atmospheric aerosols consist of solid and liquid particles suspended in the atmosphere. They include primary aerosols that arise from direct particulate emissions and secondary aerosols that arise from chemical reactions involving gaseous precursors. Aerosols have the potential to play a critical role in shaping our environment. They affect climate and stratospheric ozone concentrations and may pose a significant health threat in urban and industrial centers. However, large uncertainties in our knowledge of the chemical and physical properties of atmospheric aerosols, the processes that cause their formation, and the processes by which they influence cloud formation and radiative transfer of energy render quantitative assessments of their effects and society's influence on them problematic.

Atmospheric aerosols can affect climate in two ways: (1) directly, by which aerosols, depending on their properties, either cool the Earth by reflecting a portion of incoming solar radiation back to space or warm the Earth by absorbing radiant energy; and (2) indirectly, by which aerosols cool the Earth's surface by increasing the reflectivity of clouds (Charlson et al., 1992; IPCC, 1995; Ramaswamy et al., 1995). Because of increases in anthropogenic emissions of compounds that lead to the formation of aerosols in the atmosphere (mostly sulfur oxides and the combustion products of biomass burning), atmospheric aerosol loadings have undoubtedly increased since the Industrial Revolution, and this in turn has likely caused a climatic cooling (cf. Charlson et al., 1991). Calculations suggest that this cooling may have been large enough to offset much of the warming from increasing greenhouse gas concentrations over parts of the Northern Hemisphere (see Figure II.2.12; IPCC, 1996). However, the uncertainties in these calculations are significant because the sources, physio-chemical properties, and radiative effects of atmospheric aerosols remain poorly characterized.

Because atmospheric aerosols provide sites for heterogeneous reactions between gases, liquids, and solids, they can have a major impact on the gaseous- as well as the condensed-phase composition of the atmosphere. In the stratosphere, for instance, aerosols in the form of polar stratospheric clouds within the polar vortex, as well as sulfate aerosols from volcanic eruptions, play a pivotal role in the catalytic destruction of ozone by facilitating the conversion of chlorine compounds from relatively unreactive to reactive forms (Solomon et al., 1986; 1993). Similar processes also affect the chemistry of the troposphere. Tropospheric aerosols affect the rate of growth and the properties of clouds. In turn, chemical reactions on the surface and interior of cloud droplets affect the chemistry of the aerosols that result after cloud droplets evaporate. However, large gaps remain in our mechanistic understanding of these processes. As a result, aerosols are currently treated in atmospheric models through relatively crude parameterizations whose application to reliable forecasting of future trends is questionable.

On more local and regional scales, atmospheric aerosols inhibit visibility and can have a deleterious effect on human health (American Thoracic Society, 1996a,b). In fact, there is a growing body of epidemiological data suggesting that aerosols may cause a significant increase in human mortality and morbidity at concentrations significantly below the current National Ambient Air Quality Standard set by the U.S. EPA for particulate matter with diameters less than 10 µm (PM 10). As a result, the EPA has promulgated a new, more stringent NAAQS for fine particles, that is, particulate matter with diameters less than 2.5 µm (PM 2.5). Here again, however, the specific chemical and physical properties of aerosols that give rise to deleterious health effects at such low concentrations remain poorly defined.

The research challenges for atmospheric aerosols are thus to (1) *document the chemical and physical properties of atmospheric aerosols* to provide more accurate assessments of the effects of aerosols on climate, stratospheric ozone, tropospheric oxidation, human health and welfare, and ecosystem functioning; and (2) *elucidate the chemical and physical processes responsible for determining the size, concentration, and chemical characteristics of atmospheric aerosols* in order to develop a more robust predictive capability.

Effects of Toxics and Nutrients on Biospheric Function

We know far too little about the atmospheric transport and deposition of toxics and nutrients, and their interactions with biota, to quantify their present or future effects on natural ecosystems.

The study of toxics and nutrients strives to (1) quantify the rates at which biologically important trace species are transferred from the atmosphere to terrestrial and marine ecosystems through dry and wet deposition; (2) document the net effect on metabolic function within the biosphere of the multiple stresses and benefits brought on by this deposition; and (3) elucidate the biochemical and biophysical mechanisms by which these effects are brought about. The importance of these processes to the long-term viability of our economically and environmentally important ecosystems is significant. The examples are numerous. There is evidence that forests are being damaged by ozone exposure (LeFohn, 1992), as well as the deposition of a variety of atmospheric pollutants (Godbold et al., 1988; Aber et al., 1989; Schulze, 1989; Van Dijk et al., 1990). Important fisheries, such as the Chesapeake Bay and the Great Lakes, are being decimated by the runoff of pesticides and nutrients from agricultural activities. Acid rain has contributed to the eutrophication of waters that once supported fisheries and recreation.

Although we are beginning to understand the more acute effects of atmospheric toxicity and overfertilization on natural ecosystems, we know far too little about the transport and deposition of toxics and nutrients and their interaction with biota to assess the extent of these problems or to predict new ones that might arise. From this brief discussion, two research challenges emerge: (1) *document the rates of chemical exchange between the atmosphere and key ecosystems of economic and environmental import* to provide more quantitative estimates of atmospheric chemical impacts on the biosphere and biospheric emission rates to the atmosphere; and (2) *elucidate the extent to which interactions between the atmosphere and biosphere are influenced by changing concentrations and deposition of harmful and beneficial compounds* to better assess the long-term anthropogenic influences on the coupled atmosphere-biosphere system.

Overarching Research Challenges

Four overarching research challenges emerge from the many disciplinary challenges.

The discussions in the preceding sections have identified a series of disciplinary research challenges aimed at addressing the major scientific issues concerning the Environmentally Important Atmospheric Species. These challenges are summarized in Box II.2.2.

> **Box II.2.2**
> **Summary of the Disciplinary Research Challenges for Atmospheric Chemistry***
>
> • *Stratospheric Ozone Challenges*: Document the concentrations and distribution of stratospheric ozone and the key species that control its catalytic destruction; elucidate the coupling between chemistry, dynamics, and radiation in the stratosphere and upper troposphere.
> • *Greenhouse Gas Challenges*: Elucidate the processes that control the abundances and variabilities of atmospheric CO_2, CH_4, N_2O, and upper-tropospheric and lower-stratospheric O_3 and water vapor; expand global monitoring networks to include upper-tropospheric and lower-stratospheric O_3 and water vapor.
> • *Photochemical Oxidant Challenges*: Develop the observational and computational tools and strategies needed by decision makers to effectively manage ozone pollution; elucidate the processes that control and the interrelations that exist between the ozone precursor species, tropospheric ozone, and the oxidizing capacity of the atmosphere.
> • *Atmospheric Aerosol Challenges*: Document the chemical and physical properties of atmospheric aerosols; elucidate the chemical and physical processes responsible for determining the size, concentration, and chemical characteristics of atmospheric aerosols.
> • *Toxics and Nutrients Challenges*: Document the rates of chemical exchange between the atmosphere and key ecosystems of economic and environmental import; elucidate the extent to which interactions between the atmosphere and biosphere are influenced by changing concentrations and deposition of harmful and beneficial compounds.
>
> ---
>
> *The disciplinary research challenges focus on the efforts and activities needed to address key scientific issues associated with each of the Environmentally Important Atmospheric Species.

Inspection of Box II.2.2 indicates that the individual disciplinary challenges can be organized into three Overarching Research Challenges related to documenting the chemistry of the atmosphere, developing predictive capabilities through the elucidation of mechanisms and processes, and supporting environmental management activities. Below, each of these challenges is discussed, along with a fourth related to the interactions between each of the Environmentally Important Atmospheric Species.

Overarching Research Challenge 1: Document the Chemical Climatology of the Atmosphere

The first priority of atmospheric chemistry research must be to establish the present chemical climatology of the atmosphere by documenting the spatial distributions, temporal trends, and variability of the Environmentally Important Atmospheric Species and of the species that affect their formation and removal. This will require a commitment to long-term and careful measurements of atmospheric species at the appropriate spatial and temporal scales. Networks and observing systems have already been implemented to monitor the global distributions of many of the key long-lived species of the atmosphere and of stratospheric ozone. However, systems for monitoring the shorter-lived species, whose distributions tend to be more variable in time and space, are woefully inadequate. Therefore new initiatives will be required to (1) develop reliable and cost-effective instrumentation capable of monitoring the concentrations of these species; and (2) develop and implement monitoring networks with adequate spatial and temporal resolution to document their concentrations and trends. At the same time, resources must be provided to maintain the current monitoring capability for long-lived species.

Overarching Research Challenge 2: Develop Reliable Predictive Tools and Models

The fact that the chemistry of the atmosphere is changing today and that these changes could have significant societal impacts points to a critical need for predictive models that can (1) forecast future trends and responses to new anthropogenic and natural forcings, and (2) provide insight into the probable consequences of remedial actions and policies to reverse or slow undesirable trends. Unfortunately, our predictive capabilities in atmospheric chemistry are presently inadequate for this task because of major gaps in our knowledge of the chemical, physical, and biological processes and mechanisms that control the concentrations of the Environmentally Important Atmospheric Species. In order to develop sufficiently robust predictive models of the chemistry of the atmosphere and adequately evaluate the performance of these models, it is necessary that we commit adequate resources to relevant field studies designed to elucidate process and mechanisms.

Overarching Research Challenge 3: Support and Assess the Efficacy of Environmental Management

The Clean Air Act Amendments of 1990 (CAAA-90) address a variety of environmental issues related to atmospheric chemistry, including urban ozone pollution, acid rain, air toxics, and stratospheric ozone depletion. Although the environmental problems that the CAAA-90 address pose significant economic as well as environmental costs to the nation, management of these problems will also be quite costly. It is estimated, for example, that the incremental costs of full implementation of the CAAA-90 will amount to about $25 billion per year (J. Bachman, personal communication, 1996). Given the importance of the environmental problems being addressed by the CAAA-90 and the high costs associated with addressing them, it is imperative that atmospheric chemistry research be structured to support the regulatory activities called for in these amendments (and subsequent environmental legislation). This support should be focused on providing the tools and data required to assess the efficacy of the regulatory actions that have been implemented and to refine and improve these actions.

Overarching Research Challenge 4: Develop a Holistic Research Strategy

Thus far, our discussions have focused on the activities needed to address the uncertainties associated with each of the Environmentally Important Atmospheric Species. However, these species and the processes that cause their appearance, transport, transformation, and ultimate removal from the atmosphere operate within a highly coupled and interactive system of chemical and physical processes. Research programs that attempt to address individual environmental problems by focusing on only a single aspect or component of the system have the potential therefore to produce incomplete or misleading results. It is imperative that atmospheric chemistry research recognizes the complexities inherent in the atmospheric chemical system and undertake research aimed at elucidating and quantifying these interconnections. To accomplish this, atmospheric chemists will have to interact increasingly with scientists focused on related disciplines such as climate, cloud physics, terrestrial ecology, ocean sciences, and meteorology, as well as the social sciences and medicine.

Infrastructural Initiatives

The problems addressed by the field of atmospheric chemistry are critically linked to global economic vitality and environmental health. Seen in this light, the price for new initiatives that develop the infrastructure in atmospheric chemistry research, and thus strengthen its ability to address these problems, is relatively modest.

The scientific and technical issues confronting atmospheric chemistry in the twenty-first century are complex and challenging. However, recent and anticipated advances in instrumentation, aeronautical systems, and computational facilities promise to greatly expand the ability of atmospheric chemists to meet the challenges for atmospheric chemistry research outlined in the previous section. However, the field of atmospheric chemistry cannot be expected to undertake these new challenges and absorb these new technologies without significant investments to strengthen its research capabilities and infrastructure. In this section, six Infrastructural Initiatives in atmospheric chemistry are proposed that together would provide firm undergirding for meeting the research challenges outlined above.

Before turning to a discussion of the Infrastructural Initiatives, it is relevant to note that the resources being made available for research in the United States are increasingly limited, and for this reason, the implementation of any new research initiatives cannot be undertaken lightly. However, the problems confronted by atmospheric chemistry are critical to the continuing economic vitality and environmental health of our nation and to the sustainable development of the world. Seen in this light, the costs of the new initiatives in atmospheric chemistry needed to address these problems are relatively modest.

Infrastructural Initiative 1: Deployment of an Observing System for Moderately Lived Species

The need for long-term measurements of atmospheric species has been a recurring theme in our discussions. To meet this critical need, a new observing system must be designed and deployed that is able to establish the regional and global distributions and temporal trends in moderately lived species of environmental import and the chemical and physical properties of atmospheric aerosols.

As illustrated in Figure II.2.13, the residence or lifetimes of the trace species in the atmosphere are quite variable, ranging from centuries or more for long-lived species to seconds or less for short-lived species. Moreover, as Figure II.2.13 illustrates, the spatial and temporal variabilities of species are inversely related to their lifetimes. Long-lived species tend to be relatively uniformly mixed throughout the troposphere; shorter-lived species exhibit much more variability. Because of the uniformity in the distributions of long-lived species, a network consisting of a few dozen surface sampling sites strategically situated at various latitudes and making concentration measurements on the order of once a day has been adequate to characterize the primary distribution features and temporal trend of the long-lived gases. The situation for the shorter-lived gases, however, is quite different. As species' lifetimes become shorter and their distributions become more temporally and spatially variable, the task of designing a monitoring network becomes more complex. Greater numbers of sampling sites are required to ensure adequate spatial coverage, and increasing sampling frequencies are required to ensure proper measurement of short-term temporal variability. Unfortunately, the sparse nature of present observing networks greatly limits our understanding of the trends and distributions of these shorter-lived species.

The development of a network for moderately lived gases represents a significant challenge that will require instrumentation development and careful network design and implementation. Moreover, because of the important role of upper-tropospheric species such as ozone and water vapor, it is especially critical that the global system be configured to monitor the concentrations of key species in the upper troposphere and lower stratosphere as well as at the surface. This can

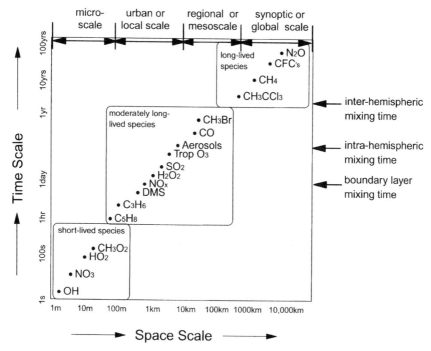

FIGURE II.2.13 Spatial and temporal scales of variability of a number of key constituents in the atmosphere. NOTE: C_3H_6 = propene; C_5H_8 = isoprene; CH_3Br = methyl bromide; CH_3CCl_3 = methyl chloroform; CH_3O_2 = methyl peroxy radical; DMS = dimethyl sulfide; H_2O_2 = hydrogen peroxide; NO_3 = nitrogen trioxide; OH = hydroxyl radical; SO_2 = sulfur dioxide; Trop = tropospheric.

best be accomplished by taking advantage of existing platforms not yet used by the atmospheric chemistry community while also bringing into operation new observing platforms uniquely capable of collecting data at remote sites over extended periods of time.

Infrastructural Initiative 2: Deployment of Exposure Assessment Networks

Current monitoring networks have largely been designed to characterize the chemical climatology of the atmosphere or to monitor compliance with specific environmental regulations; they are not well suited to examining interactions between the biosphere and the atmosphere or to characterizing the exposure of ecosystems to toxics and nutrients. A new infrastructural initiative that is specifically aimed at assessing exposures of targeted populations and biomes through the deployment of exposure assessment networks is therefore critically needed.

Infrastructural Initiative 3: Development of Surface Exchange Measurement Systems

Surface exchange is a fundamental component of the cycling of elements through the atmosphere. Not only does it help determine the concentrations of atmospheric species by controlling their rates of emission into and removal from the atmosphere, it also characterizes the interactions between the biosphere and ocean and the atmosphere. Thus, the accurate characterization and elucidation of surface exchange rates is highly relevant to Overarching Research Challenges 2 and 4, discussed earlier. Unfortunately, the rates of surface exchange and the mechanisms that control these processes are not well understood; as a result, these processes are treated in a highly parameterized fashion in atmospheric models. To aid in the development of predictive models and to better understand biospheric-atmospheric interactions, a new initiative is needed that focuses on the development and evaluation of measurement platforms designed to quantify the surface emission and deposition rates of Environmentally Important Atmospheric Species at a wide variety of spatial scales.

Infrastructural Initiative 4: Demonstration of an Operational Chemical Meteorology System

The standard approach in atmospheric chemistry research generally means that months or even years go by between initial observations and the analysis of these observations and model simulations. Such an approach slows the development and evaluation of predictive models and limits the utility of models to the policy-making community. By contrast the meteorological community has found that the use of real time data to continuously generate short-term and long-term predictions using three- and four-dimensional data assimilation modes has greatly aided scientific understanding and the development and evaluation of modeling systems, while also increasing the relevancy of the science to society and decision makers. The atmospheric chemistry discipline has now reached a level of maturity where the adoption of this so-called meteorological approach would be feasible. It is therefore proposed that a project be initiated to demonstrate the concept of an operational "chemical meteorology" system, involving the daily prediction of selected air quality parameters (e.g., oxidant and acid aerosol concentrations, visibility, ozone column densities) on local and regional scales through the real-time integration of meteorological and chemical observing systems, data analysis schemes, and statistical and/or mechanistic predictive air quality and atmospheric chemical transport models. Such an initiative would support efforts to meet Overarching Research Challenges 2 and 3, discussed earlier.

Infrastructural Initiative 5: Instrument Development and Technology Transfer Program

One of the significant successes of the atmospheric chemistry discipline in recent decades has been the application of modern chemical analytical techniques to the measurement of atmospheric trace species under field conditions with research-grade instrumentation. However, the evaluation of these instrumentations, the development of standards and protocols for them, their dissemination to the general research community, and their ultimate transfer to the private and regulatory sectors have proved more problematic. To facilitate this process and thereby aid in the development of global sampling systems and data bases (Overarching Research Challenges 1 and 2) and in the support of environmental management activities (Overarching Research Challenge 3), it is proposed that a program be initiated to evaluate new instrumentation and aid in its transfer to the scientific, regulatory, and private-sector communities through the development of suitable standards and protocols.

Infrastructural Initiative 6: Condensed-Phase and Heterogeneous Chemical Process Laboratories

The atmosphere is an environment of extreme and varying temperatures, pressures, humidities, and fluxes of solar radiation. These conditions give rise to a myriad of gas-phase, condensed-phase, and heterogeneous chemical processes that ultimately determine the current and future composition of Earth's atmosphere. Historically, characterization of these chemical processes has depended critically on state-of-the-science laboratory measurements. Because the study of homogeneous gas-phase reactions is a relatively mature field with well-defined methodologies, the gas-phase chemistry of many important atmospheric species is reasonably well understood. In contrast, the study of heterogeneous reactions and the methodologies needed to study these processes are in their infancy. As a result, very little is known about the surfaces involved in atmospheric heterogeneous reactions or about how reactions on these surfaces occur. The growing appreciation of the importance of condensed-phase and heterogeneous reactions in the stratosphere and troposphere, along with the realization of the significant climatic and health impacts of atmospheric particulate matter, indicates that we must improve our basic knowledge of multiphase chemical processes. Toward this end, we propose an infrastructural initiative aimed at developing and maintaining laboratory facilities that focus on elucidating fundamental condensed-phase and heterogeneous chemistry.

Conclusion

A carefully designed research strategy will yield significant advances in our understanding of atmospheric chemistry and our ability to manage many of the important environmental problems that confront society.

The challenges facing the atmospheric chemistry community for the next century are imposing. The array of chemical species that must be measured and ultimately understood is large. The concentrations at which these species require quantification often tax, and in some cases exceed, modern analytical chemical techniques. The fact that many of these measurements must be made outside the laboratory, in an uncontrolled environment, makes the task all the more difficult. Equally challenging is the variety of spatial and temporal scales at which the problems must be addressed (see, for example, Figure II.2.13). To make progress in understanding the most pressing and scientifically important issues in atmospheric chemistry, while also providing timely and relevant information for decision makers and regulators, we have proposed a matrix of four Overarching Disciplinary Challenges and six Infrastructural Initiatives. These are summarized in Table II.2.2.

Although the scientific and technical issues of the twenty-first century are complex and challenging, new technological advances in instrumentation, aeronautical systems, and computational facilities promise to greatly expand our ability to address these issues comprehensively. By capitalizing on these advances and working closely with scientists from related disciplines, a carefully designed research strategy should yield significant advances in our understanding of atmospheric chemistry and our ability to manage many of the important environmental problems that confront society.

THE ENVIRONMENTALLY IMPORTANT ATMOSPHERIC SPECIES: SCIENTIFIC QUESTIONS AND RESEARCH STRATEGIES

Stratospheric Ozone

In the past couple of decades, column ozone abundance has decreased substantially over major portions of the globe, and the scientific evidence overwhelmingly points to anthropogenically produced chlorofluorocarbons and other halogenated compounds as the main cause of this ozone loss (WMO, 1995). The consequences of this reduction are serious; hence, steps have been taken by the international community to reverse the loss. As the stratosphere slowly recovers to what will hopefully be conditions similar to those of the pre-CFC era, the atmospheric chemistry community must continue to monitor this change and improve its basic understanding of the stratosphere.

In formulating a strategy for studying the stratosphere, we have identified

TABLE II.2.2 Overarching Disciplinary Challenges and Infrastructural Initiatives

	Overarching Disciplinary Challenge			
Infrastructural Initiative	1. Chemical Climatology	2. Development of Predictive Models	3. Support of Environmental Management	4. Holistic Research Approach
1. Observing system for moderately lived species	X	X	X	X
2. Exposure assessment networks	X	X		X
3. Surface exchange measurement systems		X	X	X
4. Operational chemical meteorology			X	X
5. Instrumentation and technology transfer	X		X	X
6. Condensed-phase and heterogeneous chemistry	X	X	X	X

five basic scientific questions that we believe will motivate research on stratospheric ozone in the coming decades:

1. Will the evolution of the Antarctic stratospheric ozone hole proceed as expected with a short period (~5 years) of continued increasing intensity, followed by a much longer period (~50 years) of recovery to normal conditions?
2. What will be the evolution of midlatitude ozone depletion, and can we develop models to simulate this evolution correctly?
3. What is the role of the tropical region of the stratosphere in global ozone change?
4. What are the interactions between stratospheric ozone depletion and climate change?
5. What are the consequences of current and future perturbations, such as emissions from aircraft and volcanic eruptions, for stratospheric ozone concentrations?

The essential research activities that will be required to address these questions are outlined in Box II.2.3.

Monitoring the Distribution of Stratospheric Ozone

The centerpiece of the strategy for research on stratospheric ozone depletion must be uninterrupted observations of the total concentration and vertical distribution of ozone made with high temporal resolution and accuracy using a combination of intercalibrated instruments on space-based and ground-based platforms as well as small balloon sondes. The critical nature of this research need cannot be overemphasized in light of the recent decline in our capabilities to monitor stratospheric ozone distributions from space.

At the very least, continuous measurements of stratospheric ozone will document the extent of the ozone loss and its expected recovery as stratospheric chlorine concentrations decrease. This will provide a critical gauge of the sufficiency of (or compliance with) international treaties devised to halt ozone depletions.

Perhaps even more importantly, continuous measurements of stratospheric ozone will provide rapid warning of any unanticipated trends in stratospheric ozone. The relationships between stratospheric ozone, anthropogenic activities, and various naturally occurring agents of chemical change are very complex and, as yet, not fully understood. As the stratospheric chemical composition evolves under changing chlorine and bromine loadings, the effects of perturbations—both natural and anthropogenic—may vary in unexpected ways. For example, we now know that volcanic eruptions into a chlorine-rich stratosphere can have a profoundly different effect on stratospheric ozone than they most likely had in the pre-CFC era (Brasseur and Granier, 1992; Solomon et al., 1993). Similar phe-

> **Box II.2.3**
> **Recommended Research Tasks for**
> **Stratospheric Ozone Research**
>
> 1. Monitor stratospheric ozone to
>   - observe unanticipated changes,
>   - confirm that remediation policies are effective, and
>   - provide short-term UV-B radiation forecasts.
> 2. Examine the distribution of radical species to
>   - map the loss rate of ozone, and
>   - refine our ability to predict ozone loss.
> 3. Map the distribution of tracer species to
>   - identify mechanisms for tropospheric and stratospheric transport mechanisms, and
>   - refine our ability to predict winds and temperatures.
> 4. Characterize critical chemical processes to
>   - provide data for chemical models,
>   - predict chemical reactivities, and
>   - examine the effects of heterogeneous reactions.

nomena may very likely apply to anthropogenic perturbations such as those arising from certain aircraft emissions (see for example, Bekki and Pyle, 1993).

Extending Stratospheric Measurements of Critical Gas- and Condensed-Phase Species

Although monitoring the distribution of stratospheric ozone will be the central activity for stratospheric ozone research in the coming decades, the need to understand the evolving nature of stratospheric chemistry in the face of changing trace gas concentrations gives rise to an additional important research activity: mapping the distribution and variability of the species that determine the magnitude of ozone depletion and characterizing the complex of chemical reactions among these species. The total rate of stratospheric ozone destruction is governed by several catalytic loss processes whose individual rates are limited by the abundances of specific free-radical species such as hydroxyl radical (OH), hydroperoxyl radical (HO_2), chlorine monoxide (ClO), bromine monoxide (BrO), nitric oxide (NO), and nitrogen dioxide (NO_2), as well as atomic oxygen. A fundamental scientific question central to our understanding of the chemistry of ozone depletion is, What rate-limiting steps actually dominate ozone loss as a function of altitude, latitude, and season? Of equal importance, How do the rates and relative roles of each of these loss processes change as hydrogen-, chlorine-, bromine-, and nitrogen-containing species are added to or removed from the

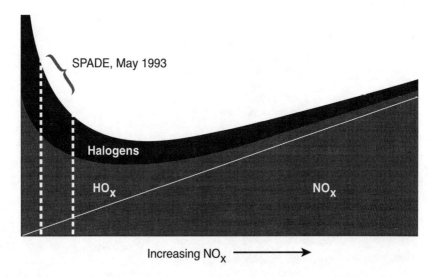

FIGURE II.2.14 Schematic illustrations of ozone loss rate as a function of NO_x concentration. Because of the coupling that exists between radical families, the response of the total ozone loss rate to changes in NO_x is nonlinear. For example, at the low NO_x concentrations observed during the Stratospheric Photochemistry, Aerosols and Dynamics Expedition (SPADE), the ozone loss rate was found to be inversely correlated with NO_x. SOURCE: Wennberg et al., 1994. Reprinted with permission of the American Association for the Advancement of Science.

stratosphere and as coevolving changes in temperature, aerosol surface area, water vapor, and so forth, occur?

Clearly, answering these questions requires mapping the relevant radical concentrations in the upper troposphere and the stratosphere. However, to be able to define the full spectra of stratospheric ozone responses to the myriad of current and possible future perturbations of the stratosphere (e.g., volcanic injections of sulfur; aerosols, nitrogen oxides, and water vapor from subsonic and supersonic aircraft; emissions of bromine and chlorine compounds; temperature changes resulting from ozone depletion and greenhouse gas forcings), it is essential that we map the specific rates of change of rate-limiting reactions with respect to the variables on which the concentrations depend, as well as the radical concentrations themselves. Figure II.2.14 schematically illustrates the diagnostic power of this dual approach. In this case, the loss rate of ozone from the catalytic cycles driven by HO_x, halogen, and NO_x radicals is illustrated as a function of increasing NO_x concentrations, with the total ozone loss given by the sum of the individual contributions. Because of the couplings that exist between NO_x and other radical families, the response of the ozone loss to a change in NO_x concentrations turns out to be a complex function that depends on the individual gradients in each of the rate-limiting steps with respect to NO_x. Because of the nature of these gradients, the total ozone loss exhibits a minimum at an intermediate NO_x concentration, as shown in the figure. Similar effects can be expected to determine the dependence of ozone destruction cycles on other key radical species.

Since the relationships between critical chemical species that control ozone loss change over various spatial and temporal scales, it is important to extend measurements of these critical species to the full range of relevant conditions. This will require the development of new instruments and new measurement platforms capable of covering the relevant regions of the stratosphere, with an emphasis on measurements that characterize the variations of each of the rate-limiting radicals and its catalytic ozone loss rate with respect to each of the other critical species.

Elucidating the Coupling Between Chemistry, Dynamics, and Radiation

The stratosphere is a coupled photochemical, dynamical, and radiative system in which ozone is exported from the high-altitude tropics to mid- and high latitudes along surfaces defined by constant mixing ratios of tracers such as N_2O and CH_4. The coherence of these tracer surfaces as a vertical coordinate is revealed by the tight relations between them and other reactive species; this dramatic feature has led to important insights into the dynamics of the stratosphere and the response of stratospheric ozone to chemical and physical perturbations. Air enters the stratosphere primarily in the cold inner tropics through a process that desiccates the air and confines the upwelling flow to the middle stratosphere, as shown in Figure II.2.15.

This exchange of air between the upward-moving tropical air mass and the stratospheric midlatitudes apparently occurs on time scales of a few months and varies with season. However, this exchange is largely uncharacterized and is a potential source of uncertainty in our ability to predict the response of the stratosphere to perturbations.

Polar regimes are characterized by rapid cooling in the fall, with subsidence of many kilometers associated with the establishment of a strong polar jet. This in turn restricts mixing with the midlatitudes, thus tending to isolate the polar stratosphere from the rest of the stratosphere during the winter months, especially in the Southern Hemisphere. This isolation of the wintertime polar stratosphere plays a significant role in the annual, dynamical cycle of the stratosphere and is an essential element in the formation of the Antarctic ozone hole.

To quantify the mechanisms for polar and midlatitude ozone loss, we must understand the patterns of air exchange between the stratosphere and troposphere, and between the tropical, extratropical, and polar stratosphere. This gives rise to another critical research activity that endeavors to elucidate the coupling between chemistry, dynamics, and radiation in the stratosphere. Because of the tight relationships between air mass origin and tracer concentrations, this can best be

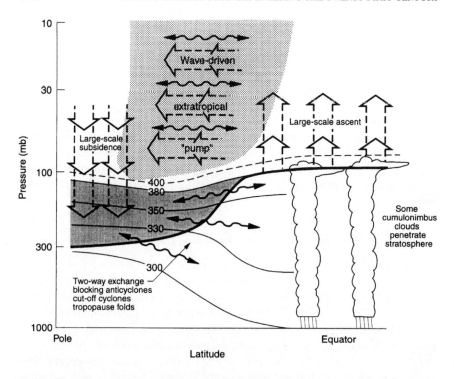

FIGURE II.2.15 Dynamical aspects of stratospheric-tropospheric exchange and intrastratospheric transport. Tropopause is shown by the thick line. Thin lines are isentropic (constant potential temperature) surfaces labeled in kelvin. Heavily shaded region is the lowermost stratosphere, in which isentropic surfaces span the tropopause and exchange occurs by tropopause folding. The region above the 380 K isentropic surface lies entirely in the stratosphere. Light shading area denotes wave-induced forcing (the extratropical "pump"). Wavy double-headed horizontal arrows denote meridional transport by eddy motions, which include tropical upper-tropospheric troughs and their cutoff cyclones, as well as their midlatitude counterparts. Broad vertical arrows show transport by global-scale circulation, which consists of tropical upwelling and extratropical downwelling. This large-scale circulation is the primary contribution to exchange across the isentropic surfaces that lie entirely in the stratosphere. SOURCE: Holton et al., 1995. Reprinted with permission of the American Geophysical Union.

accomplished with spatially resolved (0.1 km), highly accurate, in situ observations of the phase relationships between CO_2 and H_2O; measurements of the seasonal changes in the relationships between N_2O, CH_4, SF_6 (sulfur hexafluoride), CFC-11, CFC-12, O_3, and NO_y; and determination of the age of the air mass in which the measurements are made. Such measurements can be carried out using ground-based instruments, piloted-reconnaissance and robotic aircraft, and small satellites because of the short time and low cost with which they can be deployed. Trajectory and three-dimensional models are required to interpret these observations.

Because of the sharp concentration gradients, the region from the upper troposphere to 30 km must rely heavily on in situ observations. At higher altitudes where vertical and horizontal gradients begin to soften, small satellites will be very useful. The altitude overlap of these two observational strategies must be enhanced, because it is essential to obtain these coupled observation sets with seasonal (about one-month) frequency for a number of annual cycles.

Quantification and Characterization of Critical Gas-Phase and Heterogeneous Mechanisms

Predicting future changes in stratospheric ozone concentrations will require a more thorough understanding of the fundamental processes, both chemical and physical, that govern the balances between formation, transport, and destruction of ozone in the stratosphere. The chemical transformations that take place in the stratosphere either in the gas phase or on condensed matter ultimately determine the composition of this region. The quantification and characterization of these processes through laboratory experiments (and in some cases computational techniques) provide the essential building blocks for interpreting, simulating, and predicting stratospheric processes. Continued development and application of laboratory experiments and computational techniques to the study of the relevant chemical processes of the stratosphere is thus an important component of any research strategy. Especially critical will be the investigation of heterogeneous processes, and of the chemical and physical processes of particle nucleation and growth, where our understanding lags far behind that of gas-phase homogeneous reactions.

Atmospheric Greenhouse Gases

Greenhouse gases absorb infrared radiation from the Earth's surface and reradiate a portion back to the surface. In so doing, they act as thermal insulators, warming the Earth's surface through the so-called greenhouse effect. Predicting future climate changes in response to secular trends in greenhouse gases requires not only an accurate knowledge of how the major greenhouse gases evolve in time, but also an understanding of the processes that control the production and removal rates of these gases in the atmosphere. These gases include the primary greenhouse gases emitted directly into the atmosphere (e.g., CO_2, CH_4, N_2O, CFCs) and the secondary greenhouse gases produced within the atmosphere by photochemical processes (i.e., O_3). Water vapor, the most important of the greenhouse gases, is both emitted into the atmosphere (via evapotranspiration) and produced photochemically in the stratosphere by the oxidation of CH_4. Ozone is unique in that it not only acts as a greenhouse gas (by absorbing infrared radiation) but also is an important absorber of solar radiation.

The observed increases in the concentrations of CO_2, CH_4, N_2O, and CFCs (see Figure II.2.2) provide one of the clearest manifestations of global change in the atmosphere. Historic trends in H_2O and O_3 have yet to be quantitatively characterized. However, limited data suggest that background, surface-level ozone concentrations may have increased by a factor of two or more in this century (see Figure II.2.10), whereas stratospheric ozone concentrations have decreased over the past 20 years (see Figure II.2.2). In addition there are preliminary indications that stratospheric H_2O concentrations are currently on the rise.

For the most part, the changing concentrations of atmospheric greenhouse gases appear to be driven by human activities such as energy use, industrialization, land use change, and agriculture (IPCC, 1995). The fact that these anthropogenic drivers of global change are not likely to abate in the coming decades without significant political and economic intervention has given the scientific debate over greenhouse gases and their climatic impact a sense of urgency. Moreover, although significant advances in our understanding of the sources and sinks of greenhouse gases have been attained in recent years, we do not as yet have a reliable predictive capability for evolution of their concentrations in the future. In the absence of such a predictive capability, our ability to predict climatic trends—and hence to formulate effective long-term policy responses—remains limited.

In this section, we outline a basic research strategy for studying atmospheric greenhouse gases. This strategy is motivated by one central question:

What will be the trends in concentrations of greenhouse gases during the twenty-first century?

From this motivating question, we have identified several critical scientific questions for investigation:

1. How do the natural carbon and nitrogen cycles control the amounts of CO_2, CH_4, and N_2O in the atmosphere. How are these cycles perturbed by human activities? More specifically:

- What are the regional sources and sinks of CO_2 other than fossil fuel burning?
- How large are the individual CH_4 and N_2O sources?
- How might future climate changes affect these sources?
- What causes year-to-year changes in the trends of greenhouse gases?

2. Are the Montreal Protocol and its successor agreements effective in mitigating the climatic warming from CFCs and hydrochlorofluorocarbons (HCFCs)? Which new halogenated compounds may affect climate in the future?

3. What are the trends in O_3 in the troposphere and stratosphere, and what are their causes?

> **Box II.2.4**
> **Recommended Research Tasks for Atmospheric Greenhouse Gases**
>
> These must address the following:
>
> 1. The primary greenhouse gases emitted directly into the atmosphere, which include
>   - the biogenic greenhouse gases (CO_2, CH_4, N_2O) that are closely coupled to ecosystem and biospheric processes, and
>   - the halogenated greenhouse gases (CFCs, HCFCs, SF_6) whose sources are dominated by anthropogenic activities.
> 2. Ozone (O_3): this compound is a secondary greenhouse gas produced in the atmosphere by photochemical processes.
> 3. Water vapor (H_2O): this compound is emitted into the atmosphere at the Earth's surface by evapotranspiration, but is also produced photochemically in the stratosphere as a result of the photochemical oxidation of CH_4.

4. What are the trends in water vapor in the upper troposphere and lower stratosphere, and what are the reasons for these trends?

The essential research activities that will be required to address these questions are outlined in Box II.2.4 and below. These activities are organized into three categories: (1) research related to the primary greenhouse gases that are emitted directly into the atmosphere; (2) research related to ozone, whose source is entirely photochemical; and (3) research activities related to H_2O, whose sources are both surface emissions and photochemical. In general, these research activities focus first on enhancements of key strategies currently under way and then progress to new strategies that will require technology development and, in many cases, the investment of new resources. It should be noted that many of the research activities described here are relevant to other key issues in atmospheric chemistry highlighted in this Disciplinary Assessment. For example, investigations of the distributions and surface exchange rates of N_2O, CH_4, and the halogenated compounds, as well as O_3, are clearly relevant to the study of stratospheric ozone and photochemical oxidants. Conversely, the research activities described under the other key atmospheric chemistry issues discussed in this Disciplinary Assessment are relevant to understanding the greenhouse gases discussed here. For example, phytotoxics and nutrients can strongly influence the structure and functioning of ecosystems, changes in which can significantly influence the balance of CO_2, N_2O, O_3, and H_2O in the atmosphere.

Primary Greenhouse Gases

The primary greenhouse gases can be divided roughly into two categories: (1) the greenhouse gases whose sources and sinks are closely linked to biospheric processes (i.e., CO_2, CH_4, and N_2O), and (2) halogenated compounds (e.g., CFCs, HCFC, SF_6) that are entirely anthropogenically produced (NRC, 1993; IPCC, 1995). However, some halogenated compounds such as CH_3Cl (methyl chloride) and CH_3Br (methyl bromide) have important interactions with the biosphere and thus fall into both of the categories listed above. Recommended research tasks are listed in Box II.2.5 and discussed below.

Maintain Current Concentration Monitoring Networks The most robust large-scale signature of sources and sinks of the greenhouse gases and their time dependence will be variations in the mixing ratios of CO_2, CH_4, N_2O; in their isotopic ratios; and in O_2, which aids in distinguishing between terrestrial and

---

Box II.2.5
Recommended Research Tasks for Primary Greenhouse Gases

1. Maintain current monitoring networks to
   - document trends and
   - elucidate source-sink signatures.
2. Monitor vertical profiles to elucidate mesoscale variations in sources and sinks.
3. Conduct flux measurements to
   - elucidate biological mechanisms, and
   - establish relationships between climatic variations and surface exchange rates.
4. Conduct large-scale gas exchange studies to establish methodology for extrapolating small-scale flux information to the meso- and global scales.
5. Monitor gas concentrations near source regions to
   - better characterize regional-scale source strengths, and
   - develop methodology for monitoring compliance with international emission agreements.
6. Improve ocean flux measurements to
   - better characterize air-sea exchange rates, and
   - provide tighter constraints on atmospheric budgets and process models.
7. Devise new measurement systems for space-based, robotic aircraft and other innovative platforms capable of extended duty times on station to
   - provide global flux information,
   - provide long-term data from remote and inaccessible regions, and
   - provide technology for verifying compliance with international emissions agreements.

marine sinks of CO_2. It is therefore essential that the current global networks for monitoring these parameters be maintained. For the data to be useful, high accuracy and precision will be required because the pertinent geochemical information is derived from small spatial and temporal variations.

Expand Monitoring Networks to Include Vertical Profile Measurements over Continents Predicting the future concentrations of greenhouse gas requires, in part, an understanding of how the exchange rates of these gases with the biosphere behave as a function of season, soil moisture, and so forth, and of how these exchange rates will respond to local, regional, and global climate change. The effects of regional-scale climate variations on greenhouse gas emissions are not adequately captured by the present surface network. In addition, measurements from land surface sites are often difficult to interpret because of the effects of local sources and sinks. In contrast, vertical profiles over the continents can help us to infer important flux information for the biogenic greenhouse gases on the scale of 2,000 to 3,000 km. A key advantage of profiles is that they are less sensitive to details of vertical mixing. The sampling should be carried out throughout the year and should be of sufficient spatial density. The appropriate spatial scale for such measurements is suggested by the current distribution of ecosystems and the spatial extent of major climate "anomalies" such as the drought in the United States in 1988. For the North American continent, this would probably require several dozen sites. Consideration of the expected signal-to-noise ratio suggests that a feasible and economical strategy for the near future would be the collection of automated flask air samples aboard light aircraft at twice-weekly intervals (Tans et al., 1996). The advantage of such a system is that samples could be analyzed for multiple species and for isotopic ratios with tightly controlled calibration. Seasonal, and perhaps monthly, mass fluxes over the continent could be determined from such a system; these could be related to variations in climate and analyzed in the context of three-dimensional climate and chemical transport models. In addition, they would provide a large-scale analogue of the smaller-scale flux measurements described below. Larger-scale studies are also required as described below to extrapolate such subcontinental-scale flux studies to larger hemispheric and global scales.

Conduct Multiyear Flux Measurements over Different Ecosystems Predicting future greenhouse gas concentrations requires understanding how the surface exchange rates of these gases will respond to local, regional, and global climate variations. For the biogenic greenhouse gases, this will require surface flux measurements, in concert with hydrological and climatic observations, over a variety of biomes and climate regimes for a multiyear period (see, for example, Baldocchi et al., 1996). These measurements will prove critical to the establishment of empirical relationships between climatic conditions and biospheric emission and uptake rates, as well as the biological mechanisms that are responsible for these relationships.

Conduct Large-Scale Studies of Gas Exchange Although multiyear flux measurements provide insight into the mechanisms responsible for gas exchange on scales of a few to perhaps hundreds of meters, larger-scale studies are needed to establish the methodology and the validity of extrapolating this flux information to regional and global scales. These studies, such as the one planned for Brazil in 1998, represent a natural progression from the mesoscale studies carried out in the 1980s and early 1990s and will require ground experiments, aircraft campaigns, and significant meteorology and transport modeling efforts. These experiments would be greatly enhanced by the availability of advanced flux measurement technology, drone aircraft, and improved instrumentation.

Conduct Surface-Based Measurements near Source Regions In principle, estimates of the continental source of a trace gas can be obtained from measurements at a surface site that is subject to intermittent pollution events from a nearby continent (Prather, 1985, 1988). For example, a pollution event measured on the west coast of Ireland is readily identified from trace gas concentration data; it appears as a temporary enhancement in the local concentrations of long-lived gases of anthropogenic origin. Prather was able to derive emission ratios of various halocarbons and N_2O to $CFCl_3$ (trichlorofluormethane) by measuring the covariance in their atmospheric concentrations. If the magnitude of emissions of a reference gas is known and the sources are co-located, absolute source strengths of the other gases can be derived. Ultimately, with a three-dimensional model that can reproduce synoptic-scale pollution events, it should be possible to infer the continental source strength of a gas without scaling to a reference gas (Prather et al., 1987).

Thus, the implementation of a monitoring network to document changing greenhouse gas concentrations downwind of major source regions would provide critical information about the magnitude of greenhouse gas emissions from these regions. Such information would prove highly valuable for constraining atmospheric budgets for these gases and could ultimately represent a means of verifying compliance with any future international emissions agreements. However, the problem of defining the source strength for biogenic greenhouse gases using this method is somewhat more challenging than for the CFCs whose emissions are essentially constant through the year. Because the sources of biogenic gases vary with season, their background concentrations also vary seasonally. Thus, continuous measurements at both upwind and downwind sites will be required. The challenge will be to develop algorithms that can reliably unravel this information in a quantitative manner, and to define the background concentration and the excess over background caused by the advection of pollutants from a given source region. An atmospheric transport model based on observed winds would play a key role in analysis of the data.

Improve Methods of Measuring Fluxes from the Oceans The oceanic flux of biogenic greenhouse gases represents an important component of the atmospheric global budgets for these gases; thus, we must be able to adequately characterize the process of air-sea gas exchange. Unfortunately, at present most air-sea exchange rates are not known to better than a factor of two and in some cases an order of magnitude. Direct measurements of fluxes over the oceans have not been successful in reducing this uncertainty because of shortcomings in current platforms and instrumentation. Conventional eddy correlation methods can in principle provide data of sufficient accuracy and precision, but they require fast-response and highly precise chemical detectors. For the long-lived trace gases, the concentration gradients between atmosphere and ocean that must be measured are a very small fraction of the total concentration; as result, these differences are often masked by much larger effects caused by the fluxes of water vapor and heat.

With recent technological advances, a variant of the eddy correlation method (the conditional sampling method, which accumulates samples from the upward-moving eddies in one container and samples from the downward eddies in another) is becoming a viable and promising approach (Businger and Oncley, 1990). After careful conditioning, the difference between the two containers can then be measured with conventional slow-response instruments. Application of this atmospheric measurement over the oceans would enable us to determine some of the factors driving the kinetics of air-sea exchange, whereas for other factors, time-resolved full eddy correlation measurements are required.

Devise New Systems to Make Accurate Concentration Measurements Because of their global coverage, satellites represent ideal platforms for measuring concentrations of greenhouse gases on the meso- and global scales. Global data bases gathered from such platforms would undoubtedly revolutionize our grasp of the biogeochemical cycles of greenhouse gases if they were carefully linked to direct observations of flux using strategically deployed ground-based and in situ observations. To serve public policy needs, this approach may ultimately provide an important means for verifying compliance with some forms of international agreements to manage greenhouse gas fluxes. Requirements for such systems, in terms of both accuracy (0.1 percent or better) and spatial resolution (0.5° or better with some vertically resolved information) are beyond the currently available technology. High-precision remote sensing devices thus have to be developed and tested with in situ measurements.

Pilotless or robotic aircraft represent another emerging technology that could transform the study of greenhouse gases and constitute an important component in any systematic, global approach to this problem. Such aircraft could provide a platform for making near-continuous flux measurements over remote and/or inaccessible areas of the globe, such as the oceans, tropical forests, tundra, and icepack. To take advantage of this platform, however, lightweight instrumentation with fast response times and high accuracy must be developed. This instrumentation could be designed for either in situ measurements at low altitude within the boundary layer or remote sensing measurements from high altitude.

Improve and Develop Models The need for model development to address the long-term variability of the biogenic greenhouse gases is severalfold. Improved global biological process models are critically needed in order to extrapolate from small-scale intensive studies to larger temporal and spatial scales. The quality of such models is presently limited by the short duration of the field measurements on which they are based. Once based on longer-term field studies such as those described above, these global models can then be driven by parameters such as temperature and moisture and in some cases by satellite observations of vegetation.

Improved atmospheric chemical transport models are needed to better elucidate the effects of changing surface sources and sinks on atmospheric concentrations. Although atmospheric models require improvement on all scales, particular attention must be paid to subgrid-scale transport processes, such as turbulent mixing of the atmospheric boundary layer, and both shallow and deep convective transport. At present, most models do not incorporate the effects of ecosystems on atmospheric dynamics (e.g., through evapotranspiration). Transport models that use assimilated winds also have to be improved and used in tropospheric applications to trace gas budgets, to incorporate the effects of interannual variability in transport; such models will be essential for regional flux studies. Ocean models are needed for the integration of sparse oceanic data, to attain an understanding of the biogeochemical cycles in the oceans and to provide regional estimates of the trace gas exchanges between the oceans and atmosphere. Ultimately, coupled biospheric, oceanic, and atmospheric models that allow for the proper feedbacks must be developed to reach the goal of predicting trends in biogenic greenhouse gases.

Ozone

In addition to its critical role as an absorber of ultraviolet radiation in the stratosphere and in the oxidant chemistry of the troposphere, ozone's infrared absorption (at 9.6 μm) makes it an effective greenhouse gas, especially in the upper troposphere and lower stratosphere (Lacis et al., 1990). Thus, in addition to its role as a greenhouse gas, we have to document the trends in ozone in the upper troposphere and lower stratosphere and elucidate the reasons for these trends. The research activities necessary to address the causes of the ozone trends are outlined later. Below and in Box II.2.6, we concentrate on the research activities designed to elucidate the ozone trend itself.

> **Box II.2.6**
> **Recommended Research Tasks for Ozone as a Greenhouse Gas**
>
> 1. Expand ground-based vertical ozonesonde program to document trends in the upper troposphere and lower stratosphere with high vertical resolution.
> 2. Maintain and expand space-based, remotely sensed ozone observations to document global-scale trends in the upper troposphere and lower stratosphere.

Implement Expanded Vertical Ozone Sounding Programs In situ measurements of the vertical distribution of O_3 are critical to characterizing its long-term trend in the upper troposphere and lower stratosphere; these measurements provide high vertical resolution and offer an independent check on remote sensing data. Although an international ozonesonde program currently exists, the present collection of ozonesonde stations does not provide a coherent or an adequate program (see, for example, Logan, 1994). The stations do not use the same techniques, they do not all maintain adequate calibration programs, and the frequency of measurements is too low at several stations. Some are located in sufficiently polluted locations that the quality of the tropospheric data may be compromised. Most importantly, the number of sites maintained under the current program is simply too few to provide a reliable global picture. A large number of measurement sites is particularly critical for characterizing upper-tropospheric O_3 because of its relatively short lifetime and heterogeneous distribution.

Thus, to bring the current ozonesonde network to an adequate level, many of the existing sonde sites will have to be upgraded. In addition, new sites are desirable, primarily over the tropical continents and oceans. It should be noted in this regard that the planned Network for Detection of Stratospheric Change (NDSC) will not be adequate for the detection of trends in the troposphere, because the measurements are biased toward obtaining stratospheric data and the number of sites is small (about five). The use of pilotless aircraft may provide an efficient solution for this problem.

Maintain and Expand Space-Based Ozone Observations Although sonde measurements provide data with high vertical resolution, space-based measurements provide true global coverage. For this reason, maintenance and enhancement of satellite measurements of the ozone vertical profile are also critical. Measurements from SAGE II (Stratospheric Aerosol and Gas Experiment II) have provided valuable information on the trends in ozone above 17 km since 1984, with the exception of the period after the Mt. Pinatubo eruption in June 1991, when data could not be obtained in the lower stratosphere (McCormick et al., 1992). Launch of a new instrument while SAGE II is operational would allow for overlap and lead to more reliable trends in the future. Gaps in measurements of ozone have been a serious problem.

Unfortunately, although it is extremely useful for inferring ozone trends in the stratosphere, SAGE II does not yield data on ozone in the upper troposphere. Analysis of the combined data from the Total Ozone Mapping Spectrometer (TOMS) and SAGE II has proven useful for inferring tropospheric ozone column concentrations (Fishman et al., 1990, 1991), but the lack of vertical resolution and gaps in the data record make this technique of limited use for tracking upper-tropospheric O_3 trends. Non-U.S. satellites may help fill this void to some degree. On the other hand, new remote sensing techniques have been developed that could form the basis for space-based measurements of tropospheric O_3 with 2 to 3 km resolution. Deployment of this instrumentation should be a high priority for the near future.

Water Vapor

As in the case of ozone, elucidating the role of changes in water vapor concentration in an enhanced greenhouse effect most critically requires characterization of the species' trend in the upper troposphere and lower stratosphere (IPCC, 1995). Strategies for accomplishing this are described below and in Box II.2.7.

Implement Sonde Program for H_2O in the Lower Stratosphere As for O_3, in situ measurements (using sondes) offer the advantage of high vertical resolution and provide a means to validate satellite data. Measurements over Boulder,

---

**Box II.2.7
Recommended Research Tasks for H_2O**

1. Implement lower-stratospheric sonde program to document trends in the lower stratosphere with high vertical resolution.
2. Maintain space-based observations to document long-term, global-scale trends in the lower stratosphere.
3. Develop and implement upper-tropospheric measurement techniques to
   - close the upper-tropospheric gap in the atmospheric data base for H_2O, and
   - document long-term, global-scale trends in the upper troposphere.

Colorado (Oltmans and Hofmann, 1995) indicate that a carefully executed monitoring program, using relatively simple, inexpensive instrumentation should be capable of documenting the probable changes in the concentration of stratospheric H_2O that may occur in the twenty-first century. Such measurements at about five global locations, such as NDSC sites (i.e., polar and midlatitudes in each hemisphere and in the tropics), would be adequate, provided these measurements are complemented by global-scale data from satellites.

Maintain Satellite Program to Measure Long-Term Trends in Stratospheric H_2O Instruments such as SAGE II (McCormick et al., 1992), and the Halogen Occultation Experiment and the Microwave Limb Sounder on the Upper Atmosphere Research Satellite (UARS) have demonstrated that stratospheric water vapor can be measured from satellites with adequate precision to characterize temporal trends at some levels. Because of the global coverage that space-based platforms provide, continuous measurements using these techniques are critical to properly tracking the potential causes of global climate change.

Develop and Implement Techniques for Determining H_2O Trends in the Upper Troposphere Compared to those typically found in the lower and midtroposphere, H_2O concentrations in the upper troposphere are extremely low (H_2O concentrations from the surface to the tropopause typically decrease by about three orders of magnitude or more). As a result, the technologies used for routine weather soundings do not have the sensitivity to reliably monitor H_2O in the upper troposphere. Moreover, current space-based platforms (e.g., SAGE II) are able to quantify upper-tropospheric H_2O only when aerosol loadings are low (McCormick et al., 1993; Rind et al., 1993). For these reasons, new approaches must be developed for measuring H_2O trends in the upper troposphere. Ideally these approaches would be amenable to remote sensing from small satellite platforms, thus affording a strategy for obtaining global coverage at a reasonable cost.

Photochemical Oxidants

Elevated concentrations of oxidants on urban and regional scales in the industrialized countries of the world are proving to be among the most intractable air quality problems (NRC, 1991). Thus, the goals of atmospheric chemistry research in the twenty-first century must include the development of a more complete understanding of the chemical processes occurring in the boundary layer and troposphere that determine the distribution and trends of the photochemical oxidants and their precursors on urban, regional, and global scales.

To achieve this goal, several critical scientific questions outlined below, must be addressed in the coming decades.

1. What determines the ability of the atmosphere to cleanse itself of pollutants via free-radical oxidation, both now and in the coming decades? More specifically:

- To what extent does our current understanding explain simultaneous measured OH concentrations and the principal OH chemical production and loss processes?
- Can the oxidation of compounds or the appearance of their oxidation products be successfully used to infer concentrations of OH?
- To what extent do oxidants other than OH (O_3, NO_3, H_2O_2, halogen atoms, etc.) play significant roles in atmospheric chemistry?
- To what extent do changes in stratospheric ozone, climate, and/or cloud cover affect the oxidizing capacity of the lower atmosphere?

2. What determines the distribution of ozone in the troposphere and how will this distribution change in the coming decades? More specifically:

- What fraction of tropospheric O_3 can be attributed to transport from the stratosphere, and how does this change with meteorology and season?
- What portion of O_3 precursors is emitted from biogenic sources, and how will these emissions change with natural (e.g., meteorological variability) and human-induced (e.g., land use, climate change) perturbations?
- What is the contribution of urban pollution to rural and regional O_3, and conversely, what is the impact of rural or regional O_3 on urban pollution?
- How does meteorological variability affect the trends of O_3 and/or its precursors?
- What are the major sources of the oxides of nitrogen in each region of the atmosphere over various geographic regions? What are the rates of emission of NO_x from these sources?
- Which major reservoir and oxidizing species and which gas-phase and heterogeneous chemical processes are responsible for partitioning within the NO_y family?
- Where and when is the production of O_3 limited by the availability of volatile organic compounds (VOCs) or NO_x?
- What are the trends in regional and local O_3 precursors (NO_x, VOCs, carbon monoxide)?

3. How can atmospheric models be improved to better represent current atmospheric oxidants and better predict the atmosphere's response to future levels of pollutants? More specifically:

- What laboratory research is required to provide sufficient understanding of the fundamental chemical processes (heterogeneous as well as gas phase) involved in tropospheric oxidant formation?

- What atmospheric measurements are required, and with what precision and accuracy, to apply diagnostic and predictive models of tropospheric oxidant chemistry?
- What are the quantitative uncertainties associated with the estimates from diagnostic and predictive models of tropospheric oxidant chemistry?
- How can models of tropospheric oxidant chemistry be improved to incorporate direct and indirect effects of multiple, interacting forcing agents (e.g., climate change, stratospheric ozone depletion, anthropogenic perturbations)?

4. What will enable us to evaluate and improve our air quality management strategies for photochemical oxidants? More specifically:

- What design and implementation strategies will provide monitoring networks capable of determining if control measures for photochemical oxidants are having the intended impact?
- What design and implementation strategies will yield monitoring networks capable of determining, for a particular air quality problem, what portion of the problem is essentially irreducible (i.e., natural emissions of ozone precursors and stratospheric influx of ozone) and what portion of the ozone problem is potentially controllable (i.e., human-made precursor emissions)?

To successfully address these questions in the coming decades, it must be recognized that research on photochemical oxidants is truly "data poor" and "measurement limited." As a result, significant progress in this area will require a commitment to acquire high-quality, observational data sets that, collectively, are global in coverage but, individually, are of high enough spatial and temporal resolution to elucidate the important chemical and physical processes responsible for the production, transport, and removal of photochemical oxidants. To accomplish this, a research strategy that is both evolutionary and revolutionary will be required. The beginning of such a strategy focused on the management of urban- and regional-scale photochemical oxidant pollution in North America has recently been developed [see North American Research Strategy on Troposphere Ozone (NARSTO) and charter available from NARSTO Home Page at URL: http://narsto.owt.com/Narsto/]. The research strategy outlined below and in Box II.2.8 is similar in many respects to this previous work but also addresses longer-term and globally relevant issues.

Continue Development and Validation of Chemical Instrumentation

Instrument development and validation should aim at improving the sensitivity, specificity, and sampling rates of instruments needed to measure the compounds of interest throughout the atmosphere from the measurement platforms of choice (Albritton et al., 1990). The focus should be on (1) the development of

> **Box II.2.8**
> **Recommended Research Tasks for Photochemical Oxidants**
>
> 1. Continue development and validation of chemical instrumentation to
>   - provide techniques for long-term monitoring;
>   - provide continuous, fast-response techniques for flux divergence methods;
>   - provide miniaturized techniques for airborne platforms; and
>   - provide long-path spatially resolved techniques for making multidimensional measurements.
> 2. Continue implementation of integrated field campaigns to
>   - elucidate fundamental processes;
>   - document key species' trends, sources, and sinks; and
>   - evaluate air quality and chemical transport models.
> 3. Carry out observation-based studies to
>   - elucidate trends and distribution in short-lived radical species;
>   - independently infer emission inventories; and
>   - infer ozone precursor relationships.
> 4. Develop and deploy monitoring networks to
>   - document the chemical climatology of photochemical oxidants, and
>   - document the response of ozone to changes in precursor concentrations (e.g., as a result of emission controls).
> 5. Develop analytical models and tools to support integrated assessments.

simpler and more reliable instruments to be used in long-term monitoring; (2) the miniaturization of instruments to accommodate a wide array of measurements on airborne platforms; (3) the development of continuous, fast-response instruments to be used for flux measurements and airborne applications; and (4) the use of spatially resolved, long-path methods (e.g., Lidar) that can be operated from airborne and mobile platforms to determine distributions of compounds of interest over considerable distances.

Continue Implementation of Integrated Field Campaigns

Integrated field campaigns are undertaken to increase our understanding of fundamental atmospheric processes; elucidate the distributions, sources, and sinks of key species; and provide data for the evaluation of air quality and chemical transport models. Scientific guidance is required to carefully define how key uncertainties are going to be reduced and what key science questions will be addressed in a specific field campaign. Atmospheric chemistry and meteorology must be integrated in the planning and deployment of air quality measurements and monitoring. The questions that are presently before us will require multidisciplinary teams that can address chemistry, transport, and ecosystem feedbacks. Modeling tools adequate to depict or simulate these processes must be available to guide the planning of measurements as well as the interpretation of results. Moreover, an adequate fleet of research aircraft must be available to the atmospheric sciences community in order to make these studies feasible.

Carry out Inferential Observation-Based Studies

Carefully designed observations of specific tracer compounds or suites of tracer compounds can be used in conjunction with diagnostic and/or observation-based models to independently infer (1) the long-term trends, seasonal variability, and regional distribution of short-lived free-radical species not amenable to continuous, spatially extensive monitoring; (2) urban-, regional-, and global-scale emission inventories of ozone precursors; and (3) the sensitivity of ozone and other photochemical oxidants to precursor compounds. It should be noted however that interpretation of field measurements will require a solid understanding of the fundamental mechanisms involved in related atmospheric processes.

Develop and Deploy Monitoring Networks

The development and deployment of monitoring networks are necessary to establish the chemical climatology of ozone, other photochemical oxidants, and their precursors. This climatology will help shorten the time required to unequivocally observe a response in ozone to changes in the concentration of its precursor compounds. These networks must include a meteorological component that captures the role of meteorology and dynamics in the redistribution of airborne chemicals. Moreover, a comprehensive chemical climatology for the photochemical oxidants must include data from the free troposphere as well the surface. It is thus likely that these networks will require the use of balloon sondes; robotic, pilotless aircraft; and space-based platforms, in conjunction with newly developed instrumentation based on small, lightweight, low-power technology.

Support Integrated Assessments

Integrated assessments draw from a wide range of scientific information and disciplines in order to provide more comprehensive guidance on scientific and technical matters to the decision-making community. A thorough understanding of the distributions and trends in photochemical oxidants and the processes that determine their production and removal is not yet in hand, and this seriously limits our ability to conduct a rigorous integrated assessment of global change (Logan, 1994; IPCC, 1995). The research strategy in atmospheric chemistry should support these assessments by providing analytical and modeling tools that can be readily applied to these integrated assessments.

Atmospheric Aerosols

Atmospheric aerosols play a critical role in the chemistry and radiative transfer of the atmosphere. Minute amounts of particulate matter in the stratosphere, along with increased levels of anthropogenic chlorine, are responsible for the Antarctic ozone hole and probably for the less dramatic but nevertheless significant global-scale ozone depletion (WMO, 1995). Aerosols emitted by industrial activity and biomass burning are now believed to be responsible for partially masking the expected increase in surface temperature associated with greenhouse gas radiative forcing (IPCC, 1995; NRC, 1996a). Atmospheric aerosols also have important impacts on human health and materials degradation (American Thoracic Society, 1996a,b). Despite our recent advances in appreciating the importance of aerosols, our understanding of this critical class of atmospheric species is in its infancy. Why is this the case? An outstanding reason is the complex nature of aerosols and the forces they exert. Unlike the atmospheric gases, aerosols have an infinite number of sizes and a variable, mixed composition. We are not able to fully comprehend the impacts of aerosols now and are not in a position to make predictions about how these impacts will change in the future due to mankind's activities.

The important questions that must be addressed in the twenty-first century involve the effects of atmospheric aerosols on climate, atmospheric chemistry, and human health and well-being and in a fundamental form can be stated as follows:

1. What is the role of natural and anthropogenic aerosols in climate, and how will future changes in the levels of aerosol precursors affect this role?

2. How will future natural and anthropogenic aerosols impact stratospheric and tropospheric ozone and the oxidizing capacity of the atmosphere?

3. What is the role of atmospheric chemistry in changing the composition of aerosols that impact human health, the environment, visibility, and infrastructural materials?

To answer these questions, we must go far beyond our current state of knowledge of atmospheric aerosols. The essential elements of the research strategy that will be needed are outlined below and in Box II.2.9. A more detailed discussion of many aspects of this strategy can be found in *Aerosol Radiative Forcing and Climate Change* (NRC, 1996a).

Maintain and Expand Stratospheric Aerosol Measurement Capability

Limb scanning of solar extinction from satellites has been very successful in monitoring the global stratospheric sulfate layer and its spatial and temporal response to volcanic perturbations. When validated by in situ measurements of

> **Box II.2.9**
> **Recommended Research Tasks for Atmospheric Aerosols**
>
> 1. Maintain and expand stratospheric aerosol measurements to
>   - document aerosol chemical effects on stratospheric ozone, and
>   - monitor impact of volcanic injections.
> 2. Develop new suite of tropospheric aerosol measurements to
>   - document complex chemical and physical aerosol properties, and
>   - expand remote sensing capability.
> 3. Deploy monitoring networks to document spatial and temporal trends in aerosol characteristics and their impact on climate, human health, and so forth.
> 4. Design and implement intensive field campaigns to better understand processes that control aerosol formation, transformation, transport, and loss.
> 5. Develop and evaluate models to provide predictive capability.

particle size distributions from balloons and stratospheric aircraft for validation, satellite multiwavelength extinction measurements have provided stratospheric aerosol particle surface areas with an accuracy adequate for heterogeneous chemical applications. New instruments with higher wavelength resolution, possibly deployed on small satellites, will be the main monitoring tool for this component in the twenty-first century.

Design and Implement New Suite of Measurement Technologies for Tropospheric Aerosols

The complexity of tropospheric aerosol presents a considerably more difficult problem. Past in situ measurements have focused on determining the size distribution or chemical composition of aerosols at specific locations. Several new techniques under development are probing the chemical composition of single aerosol particles. However, these are essentially point measurements that yield little information about spatial and temporal variability. Moreover, there are few methods for analyzing the composition of organic aerosols, which are emitted from biomass burning and industrial activity. Clearly, a new suite of in situ instrumentation is needed that can quantitatively document the complex chemical composition of tropospheric aerosols in regions of the globe that are of interest for atmospheric chemistry.

Current remote sensing technology allows the measurement of gross tropospheric aerosol parameters over large spatial regions, but features such as composition and a complete size distribution cannot be measured yet. Technologies such as scanning polarimeters in the visible and near infrared appear to hold promise because they are able to retrieve tropospheric aerosol scattering characteristics from measurements of multispectral radiance and polarization by resolving aerosols from clouds. Moreover, surface and airborne lidars can be used to map tropospheric aerosol backscatter and, combined with Raman scattering techniques, can provide limited information on aerosol characteristics. Preliminary measurements with nadir-viewing lidars from the space shuttle show promise for obtaining detailed gross features of the tropospheric aerosol on a global basis. However, adequate opportunities for deployment of such instruments do not presently exist and must be a priority for the twenty-first century.

Design and Deploy Networks to Document Aerosol Climatology

With the development of new instrumentation, monitoring networks can be deployed to document the spatial and temporal trends in key aerosol characteristics. These characteristics include aerosol number, size distribution, chemical composition, and radiative properties. Moreover these networks must be designed in such a way that they can address issues on varying spatial scales. For example, urban-scale monitoring networks are needed to uncover the characteristics of aerosols that lead to pulmonary health effects in humans; regional-scale networks are needed to better establish the relationships between aerosol precursor species and visibility; and global-scale networks are needed to better quantify the role of aerosols in climate change.

Design and Implement Intensive Field Programs

To be able to predict how future anthropogenic activities will affect aerosols, and their consequent impacts on climate, chemistry, the environment, and human health, we must go beyond an aerosol climatology to a deeper understanding of the processes that control aerosol formation, transformation, and removal. This will require the design and implementation of intensive field programs that bring together chemical and physical aerosol measurements and precursor gas studies utilizing surface, aircraft, and ship measurements. It is relevant to note in this regard two novel experimental strategies that have emerged for resolving some of the key questions concerning tropospheric aerosols and their effects [see, for example, the ACE-1 Science and Implementation Plan (IGAC, 1995)]. The first of these is the "closure experiment," in which an overdetermined set of variables is measured. A subset of the observations and the relevant theories are then used to predict the "closure variable," which is also measured independently. The result is a test of both measurements and theory, with an opportunity to evaluate the quality of our understanding in each experiment. With instrumentation now available, it is possible to perform closure experiments on aerosol number concentration (using a variety of sizing instruments), mass (based on measurements of relevant inorganic and organic species), radiative properties (using chemical composition, relative humidity, and Mie theory), and the integrated column effect of aerosols on short- and longwave radiation. Closure experiments on aerosol mass can help answer questions about chemical composition, since missing species will make closure impossible. Theories concerning the impact of aerosols on radiative forcing of climate can also be tested by local and column closure experiments. Most of the aerosol experiments planned for the next decade depend heavily on this strategy, since it offers a rigorous test of both measurements and the process models on which more comprehensive models depend.

The other new strategy is to observe the evolution of aerosols and their precursor gases in a Lagrangian reference frame. The idea of Lagrangian experiments is not new, and variations on this theme have been used from time to time. Recently, however, there has been considerable work on tagging airmasses with balloons and chemical tracers, so that aircraft carrying large suites of instruments can revisit the airmass over a period of days to observe changes with time (Huebert, 1993; Draxler and Hefter, 1989). Although these experiments cannot eliminate the effects of dispersion and vertical mixing on concentrations, with ample dynamical measurements they make it possible to sort out the chemical and physical processes that cause changes in aerosols. These processes include gas-to-particle conversion, chemical transformations, wet and dry deposition, entrainment of air from other strata, and mixing through the sides of the "airmass" (dispersion). Although these experiments tend to be complex and expensive (at least one ship and one or two aircraft are required), they offer the potential to test the aerosol models that presently exist or will be developed from future laboratory work and other process studies.

Develop Predictive Model Capability

The overall strategic goal for the twenty-first century should be development of a predictive model that can be used to calculate atmospheric temperature and chemical species concentration fields and, from this information, to derive aerosol formation rates, predict the chemical content and size distribution of the aerosol fields, and determine their concomitant influence on atmospheric radiation and the reflectivity and lifetime of clouds. Since current atmospheric models generally impose, rather than predict, aerosol distributions, it will be necessary to achieve significantly more sophistication in representing precursor gas and gas-particle kinetics, nucleation and agglomeration kinetics, and vapor-particle interactions in future models. One way to naturally stimulate the necessary improvements in aerosol modeling capabilities is to encourage the modeling community to participate directly in the planning, execution, and data analysis portions of the strategic field measurements programs described above.

Furthermore, predictive aerosol models will require currently unavailable quantitative mechanistic and kinetic input data describing a large number of heterogeneous growth, nucleation, agglomeration, and accommodation or evaporation processes. These quantitative input data will have to come from a vigorous laboratory program in heterogeneous kinetics and aerosol microphysics.

Toxics and Nutrients

The atmosphere and biosphere are fundamentally coupled through the exchange of gases and aerosols. Ecological systems, including economically important ones such as those dedicated to agriculture and forestry, can be profoundly impacted by the wet and dry deposition of both toxic and nutritive atmospheric substances (e.g., Ridley et al., 1977; Duce, 1986; Aber et al., 1989; Schulze, 1989; Van Dijk et al., 1990; Lindquist et al., 1991; Benjamin and Honeyman, 1992; Vitousek et al., 1993; Shannon and Voldner, 1995). Although many of the atmosphere's naturally occurring components can have toxic and/or nutritive effects on the biosphere, there are a myriad of toxic and nutritive substances in the atmosphere that are significantly influenced by anthropogenic activities. These include nutrients such as sulfur and nitrogen compounds; heavy metals such as mercury, cadmium, and lead; and toxic organic compounds such as pesticides, polychlorinated biphenyls (PCBs), plasticizers, dioxins, and furans.

Although we are beginning to be able to identify the more acute effects of atmospheric toxicity and overfertilization on key ecosystems, our understanding is far too limited for us to assess the current extent of these problems or to predict future ones. Overall, the motivating scientific questions for the study of toxics and nutrients are as follows:

1. How are interactions between the atmosphere and biosphere influenced by changing atmospheric concentrations and by the deposition of harmful and beneficial compounds?
2. What are the rates at which biologically important atmospheric trace species are transferred from the atmosphere to terrestrial and marine ecosystems through dry and wet deposition?

The essential elements of a research strategy to address these questions are outlined below and in Box II.2.10.

Develop and Evaluate Techniques for Measuring Deposition Fluxes

Many of the key questions about toxics and nutrients cannot yet be answered comprehensively because we lack the necessary methods for measuring deposition fluxes on the appropriate spatial and temporal scales. This problem is most severe in the case of dry deposition, where technologies for reliably measuring many of the most biologically important fluxes do not yet exist. Adequate support for development of the necessary techniques in this area is thus critical; relaxed eddy accumulation, eddy correlation, and gradient methods offer particular promise.

> **Box II.2.10**
> **Recommended Research Tasks for Toxics and Nutrients**
>
> 1. Develop and evaluate deposition flux measurement techniques to
>   - provide new methods for measuring dry deposition rates, and
>   - provide methods for obtaining more spatially comprehensive deposition data.
> 2. Design and implement ecosystem exposure monitoring networks to develop long-term record of stresses and benefits to ecosystems of economic and/or environmental import.
> 3. Carry out process-oriented field studies to
>   - develop and evaluate deposition flux algorithms,
>   - contribute to the development of coupled ecosystem-atmospheric chemistry models, and
>   - provide tools for integrated assessments.

In the case of wet deposition, reliable techniques have in principle been developed, but serious questions exist about sampling representativeness and contamination problems. The problem is most severe for measuring wet deposition fluxes over the ocean, where it is virtually impossible to collect uncontaminated rain samples from a buoy in midocean and samples from shipboard platforms are necessarily intermittent. Present marine deposition estimates, often the result of comparing model calculations with a very small suite of shipboard and island observations, are typically subject to uncertainties of factors of three or more (Duce et al., 1991). The development of new techniques that will allow for more representative determination of wet as well as dry deposition fluxes, perhaps from a low-flying airborne platform, must therefore also be considered a high priority.

In some cases, such as high-altitude forests and foggy regions, the deposition of cloud droplets may be the primary avenue by which toxics and nutrients are delivered to the Earth's surface (Vong et al., 1991). It is extremely difficult to measure such fluxes, because the droplets are so transient that their flux is easily altered by the presence of measuring devices. Thus, new methodologies should be developed to assess the importance of droplet deposition and allow reliable flux measurements.

Design and Implement Ecosystem Exposure Monitoring Networks

In the recent past, deposition monitoring networks have proved useful for assessing the ecological impacts of atmospheric deposition (e.g., Cooperative

Programme for the Monitoring and Evaluation of Long Range Air Pollutants in Europe, National Crop Loss Assessment Network). However, these networks have been largely limited to monitoring the deposition of a specific chemical or class of compounds (e.g., acid deposition, ozone). For this reason, they have provided very limited information on the full suite of stresses and benefits experienced by an ecosystem from atmospheric deposition and, thus, on the long-term effects of this deposition. With the development of new deposition measurement techniques, it should be possible to design more comprehensive atmospheric deposition and exposure monitoring networks. Implementation of these networks for key ecosystems and biomes (e.g., at Long-Term Ecological Research sites) would provide a long-term record of atmospheric deposition; with co-located ecological monitoring, this record would no doubt prove useful in establishing causal relationships between atmospheric deposition and ecosystem vitality and succession.

Carry out Process-Oriented Field Studies for Algorithm Development and Evaluation

Even with reliable and fully evaluated deposition measurement techniques, it will never be possible to measure dry and wet fluxes for all species of interest over all ecosystems of interest, over all time. For this reason, process-oriented field studies, involving observations of fluxes under a carefully selected range of conditions, have to be undertaken in order to identify the factors that control such fluxes. With these factors identified, algorithms and parameterizations describing deposition fluxes can be developed, tested by further observations, and incorporated into regional and global atmospheric chemistry models, as well as integrated atmospheric-biospheric response models.

PART II

3

Atmospheric Dynamics and Weather Forecasting Research Entering the Twenty-First Century[1]

SUMMARY

Progress in understanding and predicting weather is one of the great success stories of twentieth century science. Advances in basic understanding of weather dynamics and physics, the establishment of a global observing system, and the advent of numerical weather prediction put weather forecasting on a solid scientific foundation, and the deployment of weather radar and satellites together with emergency preparedness programs led to dramatic declines in deaths from severe weather phenomena such as hurricanes and tornadoes.

Basic research in atmospheric science has been one of the most cost-effective investments that society has made in science. Progress in the basic understanding of phenomena such as severe thunderstorms has led directly to improved warnings and the reduction of loss of life, while technical advances in numerical weather prediction, application of statistics to model output, and advanced satellite and radar technology have contributed to much improved forecasts of all kinds.

---

[1]Report of the Ad Hoc Group on Weather Dynamics and Storm Systems: K. Emanuel (Chair), Massachusetts Institute of Technology; K.C. Crawford, Oklahoma Climatological Survey; R. Rotunno, National Center for Atmospheric Research; L. Shapiro, NOAA/AOML/Hurricane Research Division; J. Smith, Princeton University; R. Smith, Yale University; L. Uccellini, NOAA/National Meteorological Center; M. Wolfson, MIT/Lincoln Laboratories. The group gratefully acknowledges contributions from A. Betts, L. Bosart, C. Bretherton, J. Derber, K. Droegemeier, B. Farrell, R. Fleming, J.M. Fritsch, R. Houze, M. LeMone, D. Lilly, M. Shapiro, A. Thorpe, S. Tracton, and E. Zipser.

Society chooses to invest in basic research not only because of perceived tangible benefits but also because of the intrinsic value of pushing back the frontiers of knowledge. Few would deny the largely intangible but very real value of intellectual achievements such as the formulation of quantum mechanics, the discovery of DNA, or the characterization of the physics of deterministic but nonperiodic systems. In the United States, the intellectual appeal of progress in the atmospheric sciences rivals that of such fields as cosmology and molecular biology.

Atmospheric science is poised to make another series of major advances, many of which will lead directly to improved weather warnings and predictions. Great strides in the basic understanding of the dynamics of weather systems and the development of new techniques such as ensemble forecasting combine with the deployment of new measurement systems and advanced means of communicating information to offer the promise of much improved forecasts to the American public.

To realize these potential improvements, new means of measuring the atmosphere, oceans, and land surface must be developed and implemented, and existing measurement systems such as rawinsondes, mobile radars, and research aircraft must be maintained and upgraded. We cannot stress enough the continued need for in situ and ground-based remote sensing capabilities and are alarmed at the deterioration of fundamental observing systems such as the global rawinsonde network. In surveying the state of basic research in weather dynamics, time after time we came to the conclusion that further progress was limited by the lack of appropriate measurement capabilities. For this reason, many of our recommendations focus on the need for better measurement systems. However, it must be recognized that we have the ability to predict, with some accuracy, how improvements in observing systems or techniques might actually improve forecasts. This capability is largely unexploited. One of our most important conclusions is that far more must be done to exploit known techniques, such as observing system simulation experiments, to make a priori estimates of optimal combinations of observing systems and forecasting techniques for application to specific forecast-related problems. Further, we feel that atmospheric scientists must work much more closely with other disciplines, particularly economists, to determine the potential costs and benefits of new observing systems and forecasting methods.

The major body of this Disciplinary Assessment was completed just as the U.S. Weather Research Program (USWRP) was being defined. Much of what is contained here is strongly consonant with the objectives of the USWRP as outlined in Emanuel et al. (1995).

Emerging Research Opportunities

We have identified a number of emerging basic research, technique, and technological developments that, on the basis of their intrinsic intellectual value and/or potential economic or societal payoff, should be given high priority in the coming decades. Here, these key developments are summarized, and specific recommendations based on them are offered. The developmental foundations behind the identification of these opportunities are delineated later.

1. *The fundamental physics of land-air interaction:* Basic understanding of the nature of the interaction between atmospheric and land surface processes is at the threshold of major advances and has the potential, when coupled with greatly improved routine measurements of land surface properties, to lead to substantial improvements in understanding and forecasting convection, boundary layer cloud cover, and regional climate anomalies. The link between soil moisture and precipitation may be the key to improved quantitative precipitation forecasts.

2. *Seasonal climate variations and their dependence on the stochastic, internal variability of the atmosphere as well as variations linked to longer time-scale phenomena in the oceans, atmosphere, and land surface:* Research on blocking and on land-atmosphere interactions has the potential to yield significant improvements in seasonal forecasts. The seasonal prediction problem is highly dependent on proper representation of sources and sinks of heat, moisture, and momentum, whereas short-range prediction depends more on advection of these quantities.

3. *The continued development of ensemble forecasting and data assimilation techniques:* These offer great promise for improved numerical weather forecasts and the quantification of forecast uncertainty.

4. *Adaptive observation strategies:* Budding research suggests that ensemble forecasting techniques, including the use of model adjoints and breeding methods, may provide real-time estimates of optimal observation location and timing, given the availability of programmable observation platforms. Observing system simulation experiments could be used to help determine optimal combinations of observing systems. This may lead to large gains in the skill of numerical weather forecasts for a relatively small investment in additional observations.

5. *Improved understanding of the hydrological cycle and much better measurements of atmospheric water:* Ongoing advances in understanding the control of atmospheric water (in all phases) will lead to much improved understanding of and ability to predict a variety of dynamical systems. Critical physical processes include the control of water vapor by convection and cloud microphysics, and the coupling of the atmospheric boundary layer with the underlying surface. Improved understanding of these processes, together with the advent of much improved techniques for measuring soil properties, atmospheric water vapor, and condensed water, is essential for solving the difficult problem of quantitative precipitation forecasting and will be necessary for adequate modeling of climate as well.

6. *Coupling the atmospheric boundary layer with deep convection and merging the understanding of cell-scale dynamics and prediction with the understand-*

ing of convective ensemble dynamics: There have been enormous advances in understanding the cell-scale dynamics of moist convection, and in understanding and representing the interaction between ensembles of convective cells and larger-scale circulations; the time appears ripe for a productive synthesis of these developments.

7. *The dynamics of deep convective downdrafts:* These play a major role in the dynamics of at least some mesoscale convective systems and in the overall heat balance of the tropical boundary layer but have received comparatively little attention in formulating representations of cumulus convection.

8. *The fundamental role of the tropopause in atmospheric dynamics and the possible benefits of better observations at and near the tropopause:* Recent advances in the dynamics of synoptic-scale systems and better analyses of potential vorticity (a measure of atmospheric rotational motion) have pointed to the tropopause as a locus of important dynamical processes and the exchange of chemical constituents. This suggests that future models and observing systems may profit from much improved resolution near the tropopause. One especially promising candidate observing system for improving resolution of the tropopause is the global positioning system (GPS), which can be used to deduce profiles of temperature in the upper atmosphere.

9. *Tropical cyclone genesis and intensity change, including the role of the upper-ocean response and interactions with dynamical systems in the upper troposphere and lower stratosphere:* Tropical cyclones have been implicated in costly weather-related catastrophes in the United States, but there is little skill in forecasting open-ocean intensity change or genesis. Moreover, the modernization of the National Weather Service has done little to improve our ability to observe and forecast these storms.

10. *The dynamics of landfallen tropical cyclones, particularly as they relate to flash floods:* Some of the worst disasters in U.S. history were caused by tropical cyclone-associated flooding, but relatively little research effort has been expended in understanding the dynamics of landfallen tropical cyclones.

11. *The dynamics and cloud physics of mesoscale convective systems and of other convective systems that produce heavy rainfall:* Mesoscale convective systems are responsible for much of the summertime rainfall over the central United States, and research aimed at understanding the underlying dynamics and cloud physics appears to be at the threshold of major advances.

12. *Orographic and other influences on sources and sinks of atmospheric potential vorticity:* The understanding and numerical modeling of synoptic-scale dynamical processes that center on advection is comparatively well developed, but we are only beginning to understand the nature of diabatic and frictional processes. Forecasts beyond a few days rely heavily on an accurate account of such processes.

13. *The interaction of quasi-balanced and unbalanced circulation systems:* This interaction is responsible for, among other things, the generation of internal

waves in synoptic-scale cyclones and the creation of unbalanced flows such as gap winds or Kelvin waves by orography. These mesoscale events are major impediments to improvements in forecasts and warnings.

14. *The development and evolution of mesoscale frontal cyclones:* These are often missed by models, and their dynamics are not well understood.

15. *The development of mesoscale models for forecasting "fire weather" conditions and interactive models for prediction of actual fire development and movement:* Very recent research and modeling results suggest that such developments may aid considerably in the prediction and control of forest and wildfires.

16. *Research on advanced statistical techniques and on optimal blends of numerical and statistical approaches:* The best forecasts available today are based on combinations of deterministic model output and statistical guidance that depends mostly on model output. Further improvements should result from the production of higher-order moments related to the probability of events and from application of model output statistics to nonlocal quantities such as drainage basin-integrated precipitation.

Key Recommendations

We make the following recommendations, based in part on recognition of the value of the research opportunities summarized above and in part on further deliberations:

1. Fundamental improvements in forecasting in the two- to seven-day range have enormous potential economic benefits but require far better collection and utilization of data over the oceans and other data-sparse areas. *We strongly encourage the support of research seeking to determine optimal combinations of satellite and ground-based remote sensing, and aircraft, balloon, and surface observations, as well as the support of key technological developments such as satellite-borne active sensing techniques, near-field remote sensing of atmospheric water vapor, and observations from commercial and pilotless aircraft.* Such research should include comprehensive, well-posed observing system simulation experiments (OSSEs) and data denial experiments. Cost-benefit analysis should play a key role in the definition of "optimal" as it is used above, and *the cost to the nation as a whole, rather than the cost to individual agencies, should be the criterion.*

2. Recent research strongly suggests that adjoint techniques or breeding methods can be used to target specific regions of the atmosphere for observational scrutiny during the subsequent data assimilation cycle, resulting in greatly reduced forecast error. *We advocate enhanced research on adaptive observations and their potential for substantial reduction in forecast error.*

3. *The deterioration of the global rawinsonde network must be reversed or a better substitute developed if progress is to be made on a variety of operational*

and basic research problems. The reduction of in situ measurements in general, in favor of remote sensing measurements, is at best premature, and we reemphasize the desirability of performing research that seeks to determine an optimal mix of observation techniques and placement.

4. Much-improved understanding of land-atmosphere interaction and far better measurements of land surface properties, especially soil moisture, would constitute a major intellectual advance and may hold the key to dramatic improvements in a number of forecasting problems, including the location and timing of the onset of deep convection over land, quantitative precipitation forecasting in general, and seasonal climate prediction. *We see a major opportunity that may be exploited by encouraging interactions between hydrologists and atmospheric scientists and by developing new means of routine and comprehensive measurement of soil properties.*

5. Improvement in understanding the dynamics of atmospheric circulations affected by phase change of water, as well as in numerical weather prediction, especially quantitative precipitation forecasting, is severely impeded by poorly resolved and inaccurate measurements of atmospheric water vapor. *High priority must be given to new water vapor measurement systems and to research that seeks to delineate the water vapor observations necessary to address specific research and forecast problems.*

6. At present, the extent to which seasonal climate variations represent stochastic, internal variability of the atmosphere versus variations linked to longer time-scale phenomena in the oceans, atmosphere, and land surface is poorly understood. Research on blocking and on land-atmosphere interactions presents an exciting opportunity for fundamental advances in understanding and has the potential to lead to significant improvements in seasonal forecasts. The seasonal prediction problem is highly dependent on the correct specification of sources and sinks of heat, moisture, and momentum, whereas short-range prediction depends more on advection. *We encourage enhanced research efforts on blocking, land-atmosphere interactions, and frictional and diabatic effects on atmospheric dynamics.*

7. The worst natural catastrophes in U.S. history were caused by tropical cyclones. Although research on the dynamics of tropical cyclone genesis, intensity and structure change, and motion is ongoing, it has received little emphasis in recent national programs or in the modernization of the National Weather Service. Little is known about the dynamics of landfallen tropical cyclones, and this has limited our ability to forecast related flooding. Detection of hurricanes has been greatly facilitated by satellite-based observations, but much of the current state of understanding as well as the quantitative prediction of storm motion and structure and intensity change has relied on in situ measurements. *We strongly recommend the support of research on the physics of tropical cyclone motion and intensity change, and of research seeking to delineate optimal combinations of measurement systems in aid of hurricane forecasting.*

8. Tropical cyclones and some classes of extratropical marine cyclones are sensitive to local sea surface temperature and are known to influence ocean temperature through wind-induced stirring and upwelling. Modeling studies show that this feedback has an important effect on hurricane intensity, but observations of this interaction are lacking. *We strongly encourage enhanced observations of the upper ocean during the passage of tropical and some extratropical cyclones.*

9. *The resources to maintain a balanced, national, basic research observing infrastructure must be restored, enhanced, and maintained.* Satellites do not provide the spatial resolution or three-dimensional coverage required to diagnose many basic physical processes such as those involving clouds and precipitation. Next Generation Weather Radars (NEXRADs) have operational constraints that compromise their use in basic research, even if combined with other technologies. Mobile and transportable radars, research aircraft, and surface observations used to make high-precision, high-resolution observations with sufficient time continuity are required as research tools in the study of many atmospheric processes.

10. Many of the exciting and potentially beneficial developments identified earlier are of a nature that cuts across traditional disciplinary boundaries, involving much more intimate ties among atmospheric science, oceanography, atmospheric chemistry, hydrology, computational science, economics, communications, and operational forecasting. Yet the breakdown of these barriers is not well reflected in the organizational structures of the principal government funding and oversight agencies, and this is impeding progress on a number of fronts. *We recommend that consideration be given to streamlining federal funding and oversight channels with a view to facilitating interdisciplinary research.*

INTRODUCTION

This section summarizes what we regard as the important elements of current basic research efforts, as well as key developments in measurement and forecast techniques and technology.

Basic Research Foci

Extratropical Cyclones and Associated Mesoscale Processes

Most significant weather events in the midlatitudes occur in association with extratropical cyclones. Mesoscale features characterized by strong upward motion and moderate to heavy precipitation are often embedded in the synoptic-scale region of ascending air in cyclones. These smaller-scale features include fronts, rain bands, and squall lines. Being unable to forecast the formation and ground-relative motion of these mesoscale features is a significant impediment to having accurate precipitation forecasts. Broadly speaking, a front forms as a natural consequence of the deformation field of a large-scale cyclone acting on existing temperature gradients. We need to know more about the diabatic processes (convection, radiation, boundary layer) that modulate this frontogenesis. Mature fronts are known to have characteristic precipitation features such as rain bands; the origin and nature of these have to be understood better. Finally, circulations associated with fronts can instigate significant weather many tens of kilometers away from the front itself (e.g., prefrontal squall lines); the precise nature of these influences is still poorly understood.

Mesoscale processes often exert a significant influence on the larger-scale cyclone behavior. It is still a matter of uncertainty whether the radiosonde network (with station spacing of approximately 400 km) contains enough of the mesoscale information needed to forecast accurately many cases of the large-scale cyclogenesis itself. Sensitivity studies using adjoint models indicate that in many cases, the forecast location and intensity of the cyclones depend on upwind flow features of mesoscale size; in particular, finer-scale information concerning perturbations of the tropopause is needed. Cyclone-scale waves growing on preexisting fronts are still poorly observed and understood. Forecast models still have difficulty with lee cyclogenesis; we need to know more about how, precisely, mesoscale terrain features contribute to the cyclogenesis and how to incorporate this knowledge into a forecast model. Similar comments apply to cyclogenesis in the presence of mesoscale physiographic features such as ice-edge boundaries or coastlines. A major uncertainty is the cumulative effect of moist convection on large-scale cyclone behavior.

In general, diabatic processes and their influence on synoptic-scale dynamics must be better understood. We note that much of the total latent heating and frictional dissipation that occurs in the atmosphere is associated with mesoscale and convective-scale processes. Thus, better understanding of mesoscale and larger-scale processes is inextricably bound.

Observations of the middle-latitude atmosphere show that the gradients of potential vorticity that are so fundamental to large-scale dynamics are usually concentrated at the surface and the tropopause. The mixing and other irreversible processes that lead to nearly uniform potential vorticity distributions in the interior of the troposphere and to the concentration of potential vorticity gradients near the tropopause have to be better understood. We must further explore the consequences of the observed potential vorticity distributions for synoptic and planetary wave propagation and instability.

Better forecasts out to three days are critical for a number of important forecasting problems, such as snowfall, precipitation type, and high winds; these will depend largely on better upstream observations and improvements in understanding and capturing mesoscale phenomena. However, current numerical weather prediction techniques may not be uniformly applicable at the mesoscale. A large issue that must be faced is the initialization problem. Most current techniques are designed to filter out phenomena such as gravity waves and upright and slantwise convection—the very phenomena we wish to forecast on the mesoscale. *We believe that better weather forecasts on time scales less than one day will require much improved understanding of mesoscale phenomena such as gravity waves, slantwise convection, and frontal cyclones, together with advanced numerical weather prediction techniques such as dynamic and diabatic initialization that preserve real internal waves and condensational heating.*

Forecasting at all time scales on the U.S. West Coast and beyond a day or two on the East Coast is seriously impaired by lack of usable data over the Pacific, but there is a paucity of research on the effects of these data voids. *We strongly recommend that OSSEs and data denial experiments be undertaken to estimate the effect of oceanic data voids on medium-range numerical weather prediction and, similarly, to estimate the influence of potential new data sources on numerical forecasts.* Another intriguing technique that should be explored is the use of ensemble forecasting methods and adjoint techniques to make a priori estimates of the distribution, magnitude, and sensitivity of forecast skill measures to upcoming analysis error, so that programmable observation platforms, such as unmanned aerial vehicles or programmed deployment of dropsondes from commercial aircraft, can be directed to focus on sensitive regions. Adaptive observational strategies may serve to help optimize observations in aid of numerical weather prediction.

Tropical Cyclones

Landfalling hurricanes can have catastrophic societal impacts in terms of loss of life and property near the U.S. coastline. Hurricanes have accounted for more than $40 billion damage and costs to the U.S. economy since 1980 and more than 200 deaths. In 1992, Hurricane Andrew alone caused about $25 billion in damage and costs, with 58 lives lost, and was the single costliest natural disaster in the history of the United States.

Detection of an incipient tropical cyclone is generally made by geostationary weather satellite. Although satellites are also used to monitor the evolution of a cyclone, errors from these remote sensors can involve as much as tens of miles in position and tens of knots in wind speed. Measurements from reconnaissance aircraft, coastal radars, ships, buoys, and land stations provide additional sources of data.

The track of a tropical cyclone is determined primarily by the environmental flow in which it is embedded. The internal structure of the cyclone and its interaction with the environment are also important for track and intensity prediction. For accurate forecasts, detailed measurements are required on scales ranging from those of the large-scale environment to the cyclone's small inner-core structure. As noted in a recent American Meteorological Society policy statement (AMS, 1993), however, "The present reconnaissance aircraft fleet and weather satellite information cannot provide the full three-dimensional data required for hurricane track forecasting. Omega dropwinsondes deployed from the aircraft can provide wind, temperature, and moisture information from flight level to the surface, and have been shown to have a positive impact on track forecast models. The aircraft are relatively slow, however, and the information derived from the sondes does not cover the important region above flight level. The remote-sensing satellite data are limited in accuracy and coverage, particularly at the critical middle-tropospheric levels."

More accurate tropical cyclone forecasts and warnings require that improved understanding of basic physical processes and improved depictions of the hurricane and its environment be incorporated into forecast models. Skillful forecasts of hurricane track and intensity require simultaneous, accurate prediction of multiple scales of motion ranging from several thousand kilometers (which determine motion) to several kilometers (which represent intensity).

Research with barotropic models, representing the depth-averaged flow that steers storms, has improved understanding of the mechanisms that influence motion, including effects due to interactions with the environment. Skillful operational track forecasts have been achieved using a barotropic model. The effects of vertical shear on motion have been investigated with baroclinic models, representing the full three-dimensional structure of the hurricane and its environment. The application of initialization schemes that include a synthetic representation of a hurricane has demonstrated the potential for substantial improvements in track prediction. It has also been demonstrated that track forecast improvements of 20 percent or more result from the addition of supplemental environmental observations, including Omega dropwinsondes. Field experiments in the western Pacific have studied the environmental factors, including interactions with mesoscale convective complexes, that influence tropical cyclone motion. We are now in a position to use advanced numerical models to make a priori estimates of the potential benefits of new observing systems for hurricane track forecasts, and the application of objective adaptive observation strategies may be particularly beneficial in the case of hurricane forecasting.

Here, perhaps more than anywhere, the use of hurricane forecast models in suitably designed observing system simulation experiments could delineate a superior mix of observations necessary for accurate forecasts of tropical cyclone movement. There can be little doubt, however, that *improved measurements of the synoptic environment of hurricanes offers perhaps the best opportunity for improved forecasts that would lead to reduced loss of life and property damage.* Platforms that should be considered in estimating an optimal mix of data sources include satellite-borne sea surface scatterometers, Special Sensor Micorowave/Imager (SSM/I), passive water vapor measurements, GPS-based temperature and water vapor profiles, and active radar and Doppler lidar systems as well as in situ and dropwinsonde measurements from manned and unmanned aircraft.

At present, forecasters show little if any skill in hurricane intensity prediction. Current research on intensity prediction indicates that physical processes in the hurricane boundary layer and in the upper-tropospheric outflow layer have a strong controlling influence on intensity changes. High-altitude regions are, unfortunately, the area in which observations and understanding have been lacking. *High-altitude (15-20 km) research aircraft are essential for making measurements that will allow understanding and forecasting of hurricane intensity and structure change by environmental interactions.* Modeling studies demonstrate that hurricane intensity is very sensitive to the ratio of the coefficients governing the exchanges of heat and the momentum at the sea surface, but almost nothing is known about the nature of these exchanges at high wind speeds. *Understanding the nature of the exchange of heat and momentum between the air and sea at high wind speed is important for understanding and predicting hurricane intensity.*

The mesoscale and convective characteristics of a hurricane, including eyewall and spiral rainbands, are being studied. The importance of concentric eyewalls and their associated secondary wind maxima in influencing the short-term evolution of some intense hurricanes has been established, but the basic physical mechanism of this phenomenon has not been conclusively identified. The role of air-sea interactions, including the controlling influence of sea surface temperature, on cyclone intensification is being elucidated. Intensity prediction using a statistical regression model has highlighted the importance of sea surface temperature as a cap on hurricane intensity. Cooling of the ocean surface owing to the passing of a hurricane has been shown to moderate the intensification of the hurricane, but great uncertainty remains about the physics of entrainment of cold water into the ocean mixed layer. Research on this aspect of tropical cyclone physics would profit from better measurements of the ocean response during and after passage of tropical cyclones.

Although the axisymmetric dynamics of hurricane evolution are reasonably well understood, the asymmetric interactions with the environment that influence a storm's intensity are just beginning to be established. Current research emphasizes the importance of upper-tropospheric interactions in modulating storm development. Intensity forecasts with dynamical prediction models show considerable promise but are still at an early stage of development. Data deficiencies, in both the hurricane's inner core and its environment, particularly at upper levels, limit the skill of these models. Innovative numerical techniques are being applied to the development of more accurate prediction in the context of a multinested model. Doppler radar measurements from aircraft are being used to deduce inner-core structure, and satellite imagery is being used to infer storm-associated rain rates.

Earlier stages of tropical cyclone development are also being investigated, from both a dynamical and a statistical perspective. The importance of upper-level influences on tropical storm genesis is being studied from the potential vorticity perspective. The factors that determine the evolution of an incipient easterly wave disturbance to a tropical storm have been the subject of a field experiment in the eastern Pacific. Forecasts of the number of Atlantic hurricanes that will form in a given hurricane season are being made on an operational basis. Especially strong relationships have been found between seasonal hurricane frequency and the El Niño/Southern Oscillation as well as western African Sahelian rainfall. Studies of environmental factors that determine the character of the hurricane season have established the prominent role of vertical shear in modulating the numbers of storms that develop.

Finally, landfallen tropical cyclones often cause major inland flooding and associated damage and loss of life. A recent example of this was the extensive flooding in central Georgia resulting from the stalled remnants of tropical storm Alberto in the summer of 1994. Too little is known about the dynamics of landfallen tropical storms, so prediction remains problematic. *We strongly encourage enhanced research on the dynamics of landfallen tropical cyclones.* Perhaps research aircraft could be directed toward an investigation of landfallen tropical cyclones.

Atmospheric Convection

There has been considerable progress in the basic understanding of moist convection in the atmosphere over the past two decades. This research is beginning to result in improved forecasts ranging from "nowcasts" of severe weather to short-term climate variability.

The late 1970s saw the first numerical simulations of severe, "supercell" convection that bears a strong resemblance to observed severe storms. Advances in supercomputing permitted spatial resolutions of 1 km in the horizontal and 500 m in the vertical, not fine enough to resolve the outer inertial range in ordinary cumulus cells, but evidently large enough to resolve the exceptionally strong, cloud-scale drafts in supercells. Simulations of violent convection have advanced recently to the point that circulations resembling tornadoes are resolved.

By the early 1980s, it became apparent that supercell convection might be predictable on a time scale of many hours. Simulations of a well-observed supercell cluster in Oklahoma captured the trajectory and new formation of many elements of this cluster. Although explicit prediction of actual storm location may not always be possible using initial conditions that precede storm development, real pre-storm thermodynamic and wind soundings may be used to predict the *form* of convection and its general movement. Such information is already being used by the Center for the Analysis and Prediction of Storms (CAPS) and Cooperative Program for Operational Meteorology, Education, and Training (COMET) to help operational weather forecasters predict severe weather.

Numerical simulations and theoretical developments have also helped us understand the dynamics of squall lines and identify the environmental conditions conducive to this form of convection. Here it has been recognized that the dynamical interaction between the surface cold pool formed by cold, evaporation-driven downdrafts and the large-scale shear flow, as well as the convection itself, are key parts of the dynamics of precipitating convection.

There have been several comparatively successful simulations of mesoscale convective systems (MCSs), using parameterized convection. Some of the synoptic-scale circumstances conducive to MCSs have been identified, and theories for their formation and maintenance have been proposed, but only recently have these begun to be rigorously tested against observations or numerical simulations using explicit convection. We feel that a basic understanding of MCSs has not yet been achieved.

Although the dynamics of tornadoes are just being revealed through numerical experiments and observations (Doppler radar and high-quality videos) by storm chasers, there has been some progress in warning thanks to a combination of trained storm spotters and Doppler radar.

Although real progress had been achieved in detection and improved warnings of severe thunderstorms, microbursts, and tornadoes, additional problems remain. Data provided by the demonstration wind profiler network, local mesonetworks, and the Geostationary Operational Environmental Satellite (GOES) 8 and 9; a plethora of numerical guidance produced by more frequent update cycles of mesoscale numerical weather prediction (NWP) models; and a "firehose" of observed and derived information from a network of Weather Service Radar 1988 Doppler Weather Radar Systems (WSR-88Ds) are combining to swamp the process of ingesting storm-scale data, extracting relevant information, and producing more skillful warning decisions. Moreover, there is insufficient knowledge about what is actually happening (or is likely to happen) at the Earth's surface where people live. For example, mesocyclones indicated by radar do not necessarily indicate tornadoes on the ground. Even when the mesocyclones are real, less than half are associated with tornadic thunderstorms.

Of continued practical concern and great research interest is the phenomenon of intense, short-lived downdrafts in convective storms; these are usually referred to as downbursts or microbursts and pose a great threat to aircraft as they approach or leave an airport. Most recently, in the summer of 1994, a U.S. Air jet crashed on approach to Charlotte, North Carolina, probably because of such an event. Owing to their very short duration and to the fact that strong divergence may be limited to a few hundred meters above the surface, microbursts are more difficult to observe with Doppler radar than are tornadoes. Although the advent of NEXRAD and in situ wind shear detection systems at airports will no doubt improve warnings of microbursts, the phenomenon remains an outstanding research challenge and important focus of the warning system.

Better understanding of cloud microphysical processes offers improved understanding and perhaps prediction of convection. In all cases this will require better in situ measurements of cloud microphysical properties, particularly in ice clouds, and the application of remote sensing techniques such as polarimetric radar. Cloud microphysical processes cannot be fully understood without high-resolution wind data having adequate time continuity. Only by mapping the three components of air velocity over the full extent of the storm (including the weakest echo regions and clear air) can the microphysical processes be placed in dynamical context, entrainment processes be better documented, and the interactions between downdrafts and microphysics be determined.

Another scientific issue, which arises in many contexts besides this one, is the dependence of storm initiation and evolution on distributions of atmospheric water vapor. We stress that water vapor, unlike temperature, pressure, and wind, is not constrained by dynamics to vary slowly on the scale of the deformation radius; what evidence exists from aircraft and satellite observations shows significant small-scale structure, even in clear air. Many areas of meteorology will benefit from improved strategies for measuring atmospheric water vapor.

In the present context, the initiation and evolution of convection depend strongly on distributions of water vapor in the subcloud layer, and the dynamics of convective storms is sensitive to evaporation associated with the turbulent entrainment and evaporation of falling precipitation, both of which are sensitive to environmental humidity. Storms often undergo strong transition when they experience changing environmental moisture. Yet existing means of characterizing the distribution of atmospheric water vapor are greatly inadequate if not totally absent. *First-order improvements in both the quality and the quantity of atmospheric water vapor measurements will be necessary.* Some water vapor information can be obtained from satellites, but integral measures, such as precipitable water, are of limited utility.

Another area of opportunity involves boundary layer and land surface properties, particularly soil moisture. There is increasing evidence that the evolution of the planetary boundary layer over land is strongly influenced by the distribution of soil moisture, through its effect on the temperature and moisture of overlying air, but routine measurements of soil properties are seriously inadequate. *We believe that an enhanced research effort on land-atmosphere interaction involving increased collaboration between atmospheric scientists and hydrologists, together with first-order improvements in our ability to routinely characterize soil properties, may lead to dramatic improvements in the prediction of convective storm initiation.*

On small time and space scales, especially over continents, convection can be regarded as a local release of accumulated conditional instability. On larger time and space scales, particularly over oceans, convection can be usefully viewed as a rather special form of turbulence, which can be considered to be in a form of statistical equilibrium with its environment. This equilibrium seems to be characterized by an approximate balance between the rate of creation of potential energy by large-scale processes such as upward motion, radiative cooling, and surface fluxes, and the dissipation of kinetic energy within convective cells. Modeling of convection has historically focused on local release of existing instability, but recent work has taken advantage of increased computational power to explicitly simulate ensembles of convective clouds in statistical equilibrium with applied forcing. Although it is too soon to be certain where the ensemble approach will lead, it is vital for obtaining a quantitative understanding of the interaction of convection with large-scale flows and for improving our ability to represent convection in large-scale models. It has become clear that climate models are sensitive to the way convection is parameterized and, in particular, the way its effect on atmospheric water vapor is represented. More effort has to be directed at the problem of obtaining a high-quality data set for evaluating representations of cumulus convection. Much may be gained from a synthesis of the understanding of cloud-scale dynamics with ensemble behavior of atmospheric convection.

Seasonal Variability

Seasonal forecasting occurs at the boundary between mostly deterministic forecasts at a time scale of 1 to 10 days and climate prediction at time scales of more than several years. Although notable success has been achieved in seasonal predictions of weather associated with specific phenomena such as El Niño, it is not yet clear to what extent it might be possible to make seasonal predictions that show some skill or what mix of deterministic and statistical methods should be brought to bear on the problem.

Several outstanding issues must be addressed in connection with seasonal prediction:

1. How far forward can ensemble techniques be pushed? Traditional application of predictability theory has been based on exponentially growing instabilities in the atmosphere, but recent ideas on algebraic growth of atmospheric disturbances give some hope that quasi-deterministic techniques might be pushed further. Scientists at the European Centre for Medium Range Weather Forecasts recently showed that the interval of validity of deterministic forecasts can be extended in practice to equal that of the best forecast in a series. There is also some hope that deterministic prediction of longwaves might be successful even after the decay of shortwave predictability.

2. Better understanding of low-frequency modes of the coupled atmosphere-ocean and atmosphere-land systems, along with better measurements of the land surface component of the system, might lead to much improved seasonal predictions. Examples showing definite seasonal forecast skill include weather anomalies associated with El Niño and seasonal forecasts of Atlantic hurricane activity based on long-period fluctuations such as El Niño and the quasi-biennial oscillation as well as land surface conditions in sub-Saharan Africa. Sea ice and snow cover on land may also prove to be significant components of the coupled system on seasonal time scales.

3. The influence of high-frequency but extreme events on low-frequency coupled atmosphere-ocean and atmosphere-land surface phenomena must be addressed. For example, a hurricane moving into a region experiencing drought may end the drought by changing the soil moisture distribution.

4. The degree of seasonal predictability is likely to depend on the initial conditions. Some seasonal forecasts may be susceptible to small perturbations in initial conditions, whereas others are not. The degree of fragility must be quantified so that confidence bounds can be placed on seasonal forecasts.

5. The sensitivity of the nonlinear global ocean-land-atmosphere system to small perturbations in boundary conditions is likely to be linear in perturbations, and this sensitivity can be probed by observing the system's response to naturally occurring fluctuations. One way of proceeding is to use the fluctuation-dissipation relation to find the equivalent transfer function. A seasonal prediction model should be consistent with the observed fluctuation-dissipation relation.

6. External influences on short-term climate change have to be better understood. These include small fluctuations in solar output and volcanic eruptions.

7. Observing systems that support seasonal forecasting must be global. The extent to which soil properties can be observed adequately by satellite is uncertain at this time.

Land-Atmosphere Interaction

Contrasts in surface features often lead to geographic contrasts in weather patterns. A prominent and extreme example is associated with the land-sea boundary of Florida. Unlike many regions exhibiting land surface heterogeneities, the physical mechanisms responsible for contrasting weather patterns are relatively well understood in Florida. New observing capabilities open the possibility of improving significantly the understanding of physical processes that control regional climate. Advances in understanding these processes will be of utility both for weather forecasting and for assessing the potential effects of changing climate.

Soil moisture has played a prominent role in research concerning land surface effects on weather because it is the most dynamic component of the land surface and because of its central role in flash-flood hydrology. The link between soil moisture and flash-flood hydrology arises principally in determining the partitioning of rainfall between infiltration of the soil and surface runoff. This partitioning is the major land surface control of flash floods. Links between soil moisture and precipitation processes are an important area of future research and one for which significant advances in weather forecasting are possible. Preliminary studies using the forecast model of the European Centre for Medium Range Weather Forecasts indicate the importance of soil moisture representations for heavy-rainfall forecasting.

Properties of the land surface, especially soil moisture content, exercise an important control on the thermodynamics and water vapor content of the overlying atmosphere. This may have great importance for a variety of issues, ranging from quantitative precipitation forecasting to seasonal climate anomalies. Progress in understanding these issues will require far better measurements of soil properties.

Topographic features warrant special consideration in the context of heavy-rainfall and flash-flood forecasting. Virtually all of the record and near-record rainfall and flash-flood events in the United States have been linked to distinctive topographic features. Common themes that emerge from diagnostic studies of heavy-rain events concern the role of topography in maintaining quasi-stationary storm systems and in sustaining anomalously large moisture flux to storm systems.

Orographic Effects on Weather

The Earth's terrain is known to cause or modify many types of atmospheric phenomena, including the following:

- topographically enhanced rain and rain shadows,
- torrential rain and flash floods,
- forest fire storms,
- shear lines controlling tornado formation,
- sheltering of lee-side locations from strong winds,
- severe downslope and channel winds,
- gravity waves that remotely interact with larger-scale flows,
- cold air damming,
- modification of fronts and cyclones,
- diurnal control of thunderstorms,
- valley pollution and long-range pollution transport, and
- clear-air turbulence.

There are three fundamental difficulties facing researchers and practitioners dealing with meteorology in mountainous areas:

1. *The Continuous Scales of the Earth's Topography*: Atmospheric scientists have traditionally divided the Earth's terrain into two categories: large-scale mountains and small-scale roughness. The airflow disturbance generated by large-scale mountains has been analyzed explicitly, whereas the small-scale roughness has been parameterized. This division is physically inappropriate. The Earth's orography actually has a continuum of scales with no natural dividing scale. Even as the resolution of numerical models has improved from 400 km to 100 km to 25 km or less, an artificial division between resolved and unresolved orographically generated phenomena has remained. Furthermore, there is a par-

tially resolved range of scales near the grid size of the numerical model. These partially resolved scales cannot be treated accurately by parameterization or by direct computation. Among other things, the internal waves excited by flow over topography often break in the troposphere and lower stratosphere, providing a net drag on the large-scale flow. Weather prediction models prove sensitive to the way in which this is formulated, and it is clear that progress in basic research on flow over topography with a continuum of scales is necessary before internal wave breaking can be adequately represented in models.

Over the next two decades, the issue of terrain scale will provide challenges for the theoretician and the numerical modeler. The improved models will begin to capture the horizontal topographic scales of 100 km down to 1 km that contain the gravity wave spectrum of the Earth's atmosphere. The interaction of gravity waves with the larger scales of flow—scales that are already resolved—will bring new physical and numerical problems into our research and applications.

2. *Predictability and Triggering*: The question of whether atmospheric phenomena are more or less predictable in mountainous terrain is now thought to have a double answer. On the one hand, terrain can anchor flow systems in both space and time. On the other hand, mountain airflow patterns exhibit their own instability and triggering characteristics. Slight changes in ambient wind speed, wind direction, or wind shear can lead to sudden reorganization of the airflow and precipitation patterns. For this reason, ensemble forecasting and probabilistic methods will be useful in problems related to orographic influence.

3. *Model Development and Verification*: The numerical simulation of mountain-induced mesoscale phenomena has advanced enormously over the past two decades. The current interest in this subject and the predicted advances in computer technology suggest that the field will continue to move ahead. There remain, however, fundamental questions about numerical techniques and surface boundary conditions. The choice of vertical coordinate in numerical models will continue to be discussed, especially in relation to the diffusion of moisture and the applicability of small-scale parameterization schemes. The degree to which surface roughness and evapotranspiration should be included in mountain-flow models will require further examination.

Although there is growing confidence that high-resolution numerical models can accurately describe mesoscale orographic phenomena, there is less confidence in our ability to verify model output against real data. The problem has to do with the wide spectrum of topographic scales and the full four dimensionality (space and time) of orographic airflow fields. The application of existing measurement technology and new observational tools will be required for evaluation of models in mountainous regions.

Fire Weather Prediction

Forest fires create large losses of timber, property, and sometimes life throughout the United States, particularly in the West. Weather information and forecasts on a variety of time scales are a vital component of fire forecasting and management, both before and after fires have begun. Current forecast models provide reasonable guidance for predicting areas at risk for fires. If the forest is dry, the synoptic meteorological conditions favoring fire formation are low relative humidity, high temperature and winds, and thunderstorms creating cloud-to-ground lightning with little precipitation. Overall improvements in short-range forecast accuracy will also better identify areas of fire risk.

However, *the greatest opportunity for scientific progress in the next ten years may lie in developing and refining computer models of fire spread in mountainous terrain, which can be used to develop an optimal control strategy for a given fire.* The meteorological component of such a model is a non-hydrostatic airflow model with a relatively fine horizontal resolution of less than 1 km, which is used to simulate the immediate vicinity of the fire. This model could be nested in a mesoscale model that provides appropriate larger-scale boundary conditions for the fire-affected area itself. Such models are already well developed. The scientific challenge is to integrate this with a fire-spread model that uses meteorological conditions to predict local fire spread and fuel burn rate, the heat from which modifies the airflow around the fire. There have been promising pilot demonstrations of this idea in which many features of an observed fire were qualitatively simulated, but current fire-spread models are very primitive and must be improved to take full advantage of this modeling approach. With current computer technology, it is becoming possible to run such a model in near real time on a portable computer. This could aid in deciding where to deploy fire fighters or drop fire retardant and could minimize the risk to personnel. Even in control burns that are set deliberately to ameliorate later fire hazards, lives are sometimes lost when a burn goes awry. Prior modeling can be a valuable adjunct to the heuristic rules and human judgment generally used in these cases, owing to the complexity of the airflow that often develops.

Larger-scale mesoscale models can also be used to predict the distribution of smoke plumes from fires. Occasionally, such smoke may be thick enough to significantly affect the temperatures over a large area, mainly by reducing incoming solar radiation. Conceivably, this effect could be included in a forecast model.

Technique Developments

Ensemble Forecasting

Ensemble prediction involves generating multiple forecasts with a forecast model from a set of perturbed initial conditions. Perturbations can be produced by a variety of methods, including time lagging and "breeding," a method in which the most rapidly growing modes are naturally selected in the forecast-data assimilation cycle, and the use of model adjoints to generate particularly rapidly growing perturbations. Two principal benefits result from ensemble forecasting: the spread between members of the ensemble gives a quantitative estimate of uncertainty in the numerical forecast, and as it turns out, the average of all members of the ensemble is statistically a better forecast than any single member. Although many additional conceptual and practical questions must be considered, ensemble prediction is applicable also to shorter-range forecasting with regional models.

Data Assimilation and Adaptive Observations

Data assimilation combines the information in observations with an atmospheric prediction model to provide the best possible estimate of atmospheric state. In the past few years, there have been substantial advances in the theory and practice of data assimilation. These advances can be attributed to improvements in four basic components: the forecast model, the data base, quality control techniques, and analysis or assimilation techniques. The improvement in the forecast models and data base is outside the area of this Disciplinary Assessment, but it should be noted that any improvement in these components immediately results in improvement of the data assimilation system.

Since instruments do not work perfectly and data are collected through a number of different paths (some still using manual means of transmission), data can contain errors. Bad data can cause problems with data assimilation systems. For this reason, it is necessary to perform some type of quality control to eliminate or correct large errors in the data. In most assimilation systems, the observational differences from the model prediction are compared to nearby observational differences interpolated to the observation location. Quality control decisions are based on these differences. At the National Centers for Environmental Prediction (NCEP), for example, a complex quality control system has been developed that, in addition to accepting or rejecting data, corrects some of the observations for common types of errors.

Most operational data assimilation schemes use an intermittent data assimilation technique in which the model is integrated forward for some period of time and then, based on available data, is adjusted using a three-dimensional objective analysis technique. The technique of three-dimensional objective analysis is most commonly some form of optimal (or statistical) interpolation. With the development of variational techniques to solve the analysis problem, many of the approximations contained in optimal interpolation can be eliminated. The advantages of these schemes include the elimination of data selection, the inclusion of more physically realistic constraints, and the easy inclusion of additional data types. As a result of these changes, the independent initialization step can be eliminated and observations such as radiances, refractivities, and scatterometer measurements can be directly incorporated in the analysis system. An even more promising approach is four-dimensional variational assimilation, in which optimization is performed in the temporal as well as spatial domains. With the increased understanding of the theoretical aspects of data assimilation over the past few years, many aspects of the future of data assimilation have become clear.

Quality control will become even more important with the introduction of many new observation platforms and the observation and assimilation of new quantities. Complex quality control, which uses several independent quality checks in order to make a more robust decision, will continue to be improved, and an attempt will be made to salvage information from miscommunicated or improperly coded data.

The observing network will provide many new platforms to assimilate data from such devices as Doppler radar and new satellite sensors. To obtain the maximum amount of information from these data, it is desirable to use them in their most original raw form. Thus, the retrieval step that is currently performed with many data sets (e.g., temperatures and moisture retrieved from satellite measured radiances) can be eliminated, and the observed radiances used directly. To do this, it is necessary to have a high-quality forward model that transforms the model fields into the same form as the observations. This step of incorporating observed quantities directly in the analysis is vital for fully utilizing the data. However, significant effort is required to use properly each new type of data.

Assimilation systems of the future will also be required to include many new quantities (e.g., clouds, soil moisture, skin temperature, precipitation, ozone, other trace gases). To properly assimilate such quantities, it is necessary to incorporate them in the prediction model, develop the proper statistics, and include observations influenced by these quantities. All three of these steps will require substantial effort. In future systems, it is likely that diabatic processes will play a larger role. As the coupling between dynamics and physics becomes more important, the inclusion of more exact constraints will become necessary.

The final configuration of data assimilation systems of the future is not completely decided. It may be based on the extension of a three-dimensional variational system to a four-dimensional system or some approximation of a Kalman filter.

One exciting potential by-product of ensemble forecasting and data assimilation schemes is the concept of *adaptive observations*. Here ensemble techniques are used to identify, in a 12- or 24-hour forecast, regions of the atmosphere that are particularly sensitive to observational error, and/or adjoint techniques are used to estimate the sensitivity of a given forecast error measure to perturbations in these regions. Then programmable platforms (such as high-flying, dropsonde-equipped aircraft) are deployed to the regions. Experiments with low-order models show large potential increases in forecast skill from application of this technique. It may represent an optimal way of deploying limited observational resources and will provide a means of optimizing forecasts with respect to a chosen error measure, which may be local in some cases. Thus, we may be able to choose, in a given meteorological circumstance, to make those observations that minimize errors in, for example, the 72-hour forecast of a violent storm over a populous region.

Applications of Advanced Computer Architectures

Research and operational numerical models are just beginning to be run on massively parallel processors (MPPs; see below). Several problems have to be solved before the application of MPPs becomes routine, however. A major problem is that of software for translating standard code into code that makes efficient use of MPP capability. Experience to date shows that codes written expressly for MPPs are very difficult to understand and to update, making them impractical for many applications.

It will be absolutely necessary to have a stable vendor environment for MPPs before full-scale development can proceed. The recent demise of several MPP vendors underscores the risks involved for operational NWP centers in these early stages of MPP development.

Parameterization of Physical Processes

Accurate prediction of the moisture field, including horizontal and vertical cloud distributions, is one of the most important items for both numerical weather prediction and climate forecasting. Recent studies have shown the importance of predicting the horizontal distribution of shallow clouds for the coupled ocean-atmosphere system. The distribution of upper-tropospheric moisture is an important component in the radiative heat budget, but one that is neither well observed nor well predicted by current models. It is clear that the prediction of moisture requires an accurate formulation of its sources and sinks, as well as extreme care in its advection by the model wind fields. A full treatment of model-parameterized processes is beyond the scope of this Disciplinary Assessment, but the current status of some of the most important components can be outlined briefly: cumulus parameterization, explicit prediction of atmospheric suspended liquid or ice concentration, and surface physics.

Current cumulus parameterization schemes can be classified into three basic types:

1. Adjustment schemes (e.g., Manabe-type moist convective adjustment and the Betts and Miller scheme)
2. Mass flux schemes (e.g., that of Arakawa and Schubert)
3. Schemes based on statistical equilibrium of water (Kuo schemes)

Recently, many operational centers have decided to use one of the mass flux schemes. Overall experience in testing cumulus parameterization schemes in forecast models has indicated the importance of including (1) saturated downdraft effects, (2) the interaction of penetrative convection with boundary- and subcloud-layer mixing processes, and (3) cloud-radiative interactions. There is some indication from coupled atmosphere-ocean experiments that, perhaps owing to inadequacies in cumulus/boundary layer interactions, model-generated convective precipitation is less sensitive to changes in sea surface temperature than in nature. Evaluation of cumulus parameterization schemes is, however, difficult since cumulus convection interacts with many other physical processes that themselves may not necessarily be represented adequately in the model. Although tuning may ameliorate the most obvious problems, questions still remain about the theoretical foundation of cumulus parameterization.

The prediction of cloud liquid and ice water content is now being attempted in mesoscale and global forecast models. The interaction between cloud water and radiation is also being explicitly calculated. Preliminary results indicate improvements in many aspects of the forecasts, including the amount and location of precipitation, and cloud amounts. These improvements are being documented quantitatively using comparisons with satellite data.

More attention should be paid to accounting for cloud microphysical processes both in explicit clouds and in parameterized convective clouds. The water vapor content of the atmosphere is very sensitive to assumed cloud physical processes, and better prediction of water vapor content will be necessary for improvement in quantitative precipitation forecasts, longer-range numerical weather prediction, and climate simulation.

Considerable effort has been made to improve the parameterization of surface physics, particularly ground hydrology. These improvements include two-layer soil thermodynamics and hydrology with explicit evaporation, transpiration, and canopy intercept for latent flux estimates; improved surface exchange coefficients; and parameterizations for drainage, runoff, and snow cover. Results indicate improved forecasts of screen temperature and precipitation over land and, in general, improvements in the diurnal cycle over land.

A comprehensive program covering both numerical upgrades and a review and development of physical parameterizations will be necessary to take advantage of recent research developments and greater computer resources. Upgrades to physical parameterizations should be based on more refined theory, much better evaluation techniques, better model resolution, and affordability. In the near term, one may expect advances to occur in the above-mentioned areas of surface physics, cloud and cloud-radiation parameterizations, and the increasingly realistic simulation of phenomena associated with severe weather such as squall lines, outflow boundaries, and mesocyclones.

Numerical Techniques

The nature of the typical weather forecast problem motivates the continued search for accurate and cost-effective ways of making discrete approximations to the continuous equations. Much effort is currently being spent on developing *two-time-level, semi-Lagrangian techniques,* although the outcome of these efforts may be the development of a two-time-level Eulerian method that avoids its originally encountered unattractive features (e.g., instability, damping) while keeping its natural advantages (e.g., simplicity, conservation easily enforced). Another technique is to increase resolution over only those parts of the computational domain which it is required. The technique of *automatically adjusting selective grid enhancement* presents a promising avenue of exploration since many significant weather phenomena are related to sharp horizontal variations of the meteorological fields (e.g., fronts and airmass contrast across coasts) that form, move, and dissipate. A related recent development, which will continue, is the implementation of a single general model for almost all meteorologically relevant scales (i.e., from the planetary to the cloud scale). These *unified models,* run with selective grid enhancement, may allow simultaneous computation of nonhydrostatic clouds and/or breaking gravity waves on nested domains with enhanced resolution and of large-scale flow on a planetary-scale domain. Other important numerical issues include the following:

Model Vertical Coordinate One of the very recent advances in numerical techniques is the utilization of isentropic coordinates as the vertical coordinate in both global and regional models. Because isentropic coordinates are quasi-Lagrangian, they do not suffer from the problems usually associated with vertical differencing. On the other hand, they cannot be used readily in nonhydrostatic models, in which phenomena such as breaking internal waves may cause isentropic surfaces to overturn. This weighs against the use of isentropic coordinates in a unified model. A hybrid-coordinate approach has been proposed to overcome the lack of resolution near the ground and the problem of isentropic surfaces intersecting the ground. More investigation is needed into the feasibility of using such hybrid coordinates and determining whether there is an advantage to their use in numerical weather prediction and climate modeling. Some operational mesoscale forecast models use the eta coordinate, which is a variation of the sigma coordinate that remains relatively horizontal and uses step mountain topography. This permits a more accurate description of orography with rapidly changing slopes than some spectrally based topographic representations. With recent increases in computer power, mesoscale models can have a horizontal resolution of the order of 20 km with 50 layers in the vertical, which represents a major improvement over present models.

Numerical Techniques for Advection Although many forecast models still employ classical numerical schemes, the semi-Lagrangian approach of treating the dynamics has received much attention in recent years. This approach, when coupled with semi-implicit time integration, allows much longer time steps compared to those allowed by the traditional Courant-Friedrichs-Lewy stability criterion, minimizes phase errors and computational dispersion, and easily allows shape preservation. These considerations have led to the use of semi-Lagrangian formulations in many research and operational models. One major drawback of semi-Lagrangian methods is their lack of an a priori conservation guarantee, which is viewed by some as essential for climate modeling. Although further work is needed on this subject, a novel semi-Lagrangian approach has recently been proposed, in which exact conservation of mass and any other scalar variables can be achieved. The semi-Lagrangian technique is also being applied to nonhydrostatic systems in both regional and global models. One unresolved problem with semi-Lagrangian schemes that will require future attention is the treatment of flow around steep orography when large time steps are being used.

Technological Developments

Ground- and Aircraft-Based Measurement Systems

In situ measurements from surface stations, ships, aircraft, and balloons remain the backbone of the global observing system in aid of weather forecasting. Improvements of this in situ measurement capability and of ground- and aircraft-based remote sensors offer the promise of much improved knowledge of the overall state of the atmosphere, which should lead to improved understanding of weather dynamics and physics and to improved forecasts.

In situ measurements have also proved valuable to the very short-range prediction problem. The State of Oklahoma recently funded a high-density mesonetwork of surface observing stations. Data from these stations have led demonstrably to improvements in short-term local and regional forecasts and to associated economic benefits, as well as improved public awareness and science education. Efforts are now under way to enhance the mesonet by adding profiler measurements and measurements of soil moisture, and to network the data through primary and secondary schools. The panel believes that such mesonets are a cost-effective route to much improved short-range regional and local forecasts, as well as better science education. Such mesonets may be funded ideally through public-private partnerships.

Some effort has been made to improve the technology of the rawinsonde. The next generation of balloon sounding will be more automated, perhaps requiring human attendance only weekly or monthly, and state variable sensors should be much improved. Tracking by the GPS is just being developed for rawinsonde and dropsonde wind finding. Despite its obvious sampling limitations, balloon sounding remains a cost-effective means of sampling the atmosphere, and the deterioration of the global rawinsonde network is alarming.

Commercial air carriers offer another means of sampling the atmosphere, both at cruising level and during ascent and descent. These measurements offer many advantages over balloon soundings, including much better spatial and temporal coverage and potentially low operating costs. One disadvantage compared to balloons is that sounding data are restricted to cruising altitudes and below. We strongly encourage observing system simulation experiments and data denial experiments aimed at determining the importance, or lack thereof, of routine measurements above standard aviation cruising altitudes.

Another means of obtaining atmospheric measurements is by remotely piloted aircraft. Unmanned aircraft have been used by the military for about 50 years; they played a vital role in reconnaissance and as decoys in the Yom Kippur war and in Desert Storm. Technological advances in low-Reynolds-number aerodynamics, propeller design, carbon fiber epoxy construction, and power plants now make it possible to build unmanned aircraft that can cruise at 18 km altitude for two or three days, carrying many hundreds of kilograms in payload, including GPS-based dropwindsondes. These aircraft may soon enable plentiful direct measurements up to the lower stratosphere, at relatively low cost. They are particularly well suited to obtaining soundings over the ocean and over sparsely populated land. They would be instrumental in observation or numerical forecast systems that made use of adaptive observations. A major hurdle to be cleared is the problem of coordinating unmanned aircraft operations with the air traffic control system.

A number of technological developments promise much-improved measurements of atmospheric water vapor. Differential absorption lidar (DIAL) operates by transmitting laser pulses in and slightly off a water vapor absorption band, and comparing the intensities of the received return. A major limitation of the technique is eye safety; this requires transmitting low-power and/or broad-beam pulses, necessitating integration of the return over relatively long periods to achieve a reasonably high signal-to-noise ratio. Current estimates place the minimum error of water vapor estimates by this technique at about 1 g/kg, with vertical resolutions of the order of 100 m and a maximum altitude of about 3 km. The technique cannot be used to retrieve water vapor profiles above cloud base. Even so, DIAL offers much improved sampling of lower-tropospheric water vapor.

Some information about atmospheric water vapor content can be obtained from the GPS. A single GPS receiver is capable of measuring the time delay between reception and transmission of the signal from one or more satellite-borne transmitters. This delay is due to electromagnetic effects in the ionosphere, the total atmospheric mass, and water vapor. The ionospheric delay can be corrected by using two different frequencies and comparing the two time delays, whereas the atmospheric mass component can be accounted for if surface pressure is known to an accuracy of greater than about 0.3 millibar (mbar). The remaining delay is proportional to the vertically integrated water content. At any one time, about six GPS satellites are visible from a single location, so that the different elevation angles of the satellites can be used to make some inferences about the vertical distribution of water vapor. The maximum vertical resolution from this technique is limited to about 1 km. Finally, a satellite-borne GPS receiver can be used to estimate vertical profiles of water vapor by observing the occultation of satellite-borne GPS transmitters, provided an independent evaluation of the vertical temperature profile is available. The water vapor content determined by this technique is effectively averaged over about 200 km horizontal distance, and vertical resolution is limited to about 1 km.

Vertical profiles of virtual temperature can be estimated using the radio-acoustic sounding system (RASS). In this technique, a vertically propagating sound pulse is tracked by radar; since the speed of sound is a function of virtual temperature, the latter can be deduced from the measured velocity of the sound pulse. Vertical resolution and maximum altitude are limited principally by the characteristics of the transmitter, and the data tend to be noisy.

Satellite-Based Measurement Systems

Satellite data fill the space-time gaps within in situ systems more uniformly than data from any other observing system, although information is mostly limited to radiance integrals and cloud-top properties.

Satellite remote sensing will improve in several technical areas during the next decade. Passive microwave observations are promising for detecting precipitation, but the remoteness of geostationary positions presents a major obstacle that must be overcome. Active cooling and better infrared detectors can improve precision from one part in 100 to a few parts in 10,000. Pointing accuracy to the nearest pixel will be possible. Arrays of detectors could deliver "snapshots" of regions within a few seconds, and low-light sensors could deliver high-resolution cloud cover observations on a moonlit night. On-demand "skycam" operations could be directed by local forecast officers with immediate digital data delivery through commercial paths.

During the next decade, satellites will carry infrared spectrometers that can resolve the infrared spectrum and double the vertical resolution to the theoretical limit for passive sensing of temperature and moisture.

Knowledge of the global wind field is widely recognized as fundamental to advancing the understanding and prediction of weather and climate. Several active sensing techniques can be used to detect atmospheric winds. One such technique is Doppler lidar, which operates much like Doppler radar in that signals returned to the receiver from distant targets are analyzed spectrally to recover the Doppler shifts imposed by the motion of the target. The short wavelengths (e.g., 9 µm) involved in lidar, however, mean that the targets can be much smaller than for radar and that comparable Doppler shifts and signal bandwidths are much greater. For wind sensors, the targets are cloud particles or naturally occurring aerosols suspended in the atmosphere, which move at approximately the speed of the wind.

Studies have concluded that tropospheric winds can be measured from space with current lidar technology. Successful experimental demonstrations of a 5-joule class, carbon dioxide (CO_2) laser were conducted in the laboratory as part of design studies for the Laser Atmospheric Wind Sounder (LAWS) instrument.

Sea surface scatterometers can be used to reconstruct surface wind fields over the oceans. Scatterometers are absolutely calibrated radars that measure reflected signal strength from distributed targets. For given operational parameters (wavelength, incidence angle, and polarization), backscatter from the sea surface is primarily a function of the capillary wave spectrum, which is a direct measure of surface wind stress. This can be related to the surface wind vector. Thus, backscatter measured from several perspectives (as provided by an instrument in polar orbit and employing multiple fixed antennae) can be used to infer several possible averaged wind vectors over a portion of the ocean surface. Complementary data and continuity constraints may be used to select the most geophysically likely solution.

Space qualification of the technique dates from the 1978 SEASAT (U.S. sea satellite) mission, although no successor was deployed until the launch of ERS-1 (the European Remote Sensing Satellite) in 1991. The ERS-1 scatterometer samples a swath from 200 to 700 km on one side of the satellite ground track with 50 km resolution for three antennae. A large number of studies comparing SEASAT and ERS-1 scatterometer wind data to those from National Oceanic and Atmospheric Administration (NOAA) buoys and objectively analyzed winds now exists. In the range of 2-3 m/s, scatterometer winds agree to within 2 m/s and 20° of other estimates, barring contamination from sea ice or precipitation. Moreover, assimilation of ERS-1 scatterometer wind data can improve operational weather forecasts, particularly of tropical cyclone formation and location. This improvement is strongest in the data-sparse Southern Hemisphere.

The algorithms for deriving winds and for flagging rain- or ice-contaminated data are empirical. Many groups continue to work to optimize these algorithms in order to provide the best possible wind data. In addition, new algorithms should be developed to extract secondary data products, for example, tracking sea ice to complement images expected from synthetic aperture radar.

In addition to the important scientific advances that would be achieved, there is substantial evidence that significant economic benefit to the nation would occur with the use of better wind data in operational weather forecasting. Two notable examples would be a reduction in fuel consumption by airlines achieved through more accurate wind forecasts in the upper troposphere, and improved hurricane track forecasts that would reduce the area of uncertainty. Satellite-based GPS systems can also be used to measure atmospheric water vapor, as described in the preceding section.

It is vitally important to undertake an analysis of the potential costs and benefits of various systems that have been proposed for enhancing atmospheric wind information. Here, more than anywhere else, we must be able to compare the costs and benefits of existing and proposed systems that span many federal agencies without regard for the needs and goals of individual agencies. Once again, we stress the desirability of using models to make a priori estimates of the impact of new observing systems on forecasts, as one step toward devising an optimal combination of observing systems, where "optimal" includes the associated costs.

Computers

Massively Parallel Machines To circumvent the inherent limitations (the speed of light and the minimum size of a unit) of single central processor computers, the concept of multiple processors performing tasks in parallel has been introduced, and several such machines are now on the market. Thus far, their performance has not lived up to expectations, at least as far as meteorological applications have been concerned. The problems are twofold: (1) Learning how to program these machines is an investment of time and effort that most scientists tend to avoid. (2) Current experience with atmospheric general circulation models shows performance that is basically comparable to a Cray YMP system (1.2 gigaflops), although better performance (10 gigaflops) can be achieved on more specialized problems. If numerical weather prediction is to be done with kilometer-scale resolution over synoptic-scale (1,000 km) domains, massively parallel machines are presently the only ones that could in theory deliver the 1,000-gigaflop speed required. Solutions to fundamental problems with this technology remain to be worked out by the computer industry, and better software has to be developed by users to reach the needed speed.

Workstations In recent years, the appearance of high-performance workstations (approaching the speed of a single-processor Cray YMP) have made possible truly *local* numerical weather prediction. High-resolution [grid spacing O (5 km)] forecast models, run over domains large enough (500 km) to avoid contamination by artificial signals from the domain edge, can provide significant enhancement of short-term forecasts (3-12 hours). Possible applications that are being investigated by researchers include enhancement of emergency response systems, more detailed local forecasts for military operations, thunderstorm forecasting, and daily weather forecasts where terrain, coastlines, and/or other physiographic features exert a significant influence on the weather. Given the trend in costs and the ease of access to forecast models, local weather forecast offices could each have such a system by the turn of the century.

CONCLUSION

Basic and applied research in atmospheric science has yielded dramatic improvements in weather forecasts and warnings over the past several decades and is now poised to make even more spectacular advances. The main stumbling block to realizing significant progress in basic research and operational meteorology is the need for better measurements of the atmosphere, oceans, and land surface, and the need to better understand and delineate optimal combinations of measurement systems for specific forecast problems. Our nation has invested heavily in environmental satellites, and this investment has been paid back many times in improved understanding of the atmosphere and better warnings of hazards ranging from hurricanes to severe thunderstorms and tornadoes. However, observing systems of great importance, some of low cost, have been allowed to deteriorate. Examples include research Doppler radar facilities, global rawinsonde coverage, small research aircraft for boundary layer studies, and research surface mesonets. Meanwhile, the measurement of atmospheric water vapor continues to be vastly inadequate for a number of purposes, ranging from quantitative precipitation forecasting to climate prediction. In some cases, we have just begun to realize the potential benefits of certain types of measurements, such as soil moisture and the detailed structure of the tropopause. We must stand back and take a hard look at the costs and benefits of *all* existing and proposed measurement systems, from the perspective of basic scientific progress and societal need, with a blind eye toward the objectives and budgets of individual federal agencies.

If we elect to take a rational and well-thought-out approach toward observations in support of basic research and operational objectives, there is every reason to believe that the potential exists for great advances in understanding and prediction. A proper accounting of land surface physics and irreversible processes in the atmosphere may lead to large increases in the skill of seasonal forecasts. Better measurements of atmospheric water vapor and of cloud microphysical processes, particularly those involving ice, may allow us to solve a number of outstanding problems such as predicting the development and movement of mesoscale convective systems and the response of atmospheric water vapor and cloud cover to climate change. Advanced applications of ensemble and adjoint techniques to numerical weather prediction may reveal, in near real time, those parts of the atmosphere that are particularly susceptible to initial error, allowing us to target such regions for observational scrutiny and thereby greatly reduce numerical forecast errors. Better in situ observations in the atmospheric and oceanic environment of hurricanes may lead to dramatically improved forecasts of the motion and intensity of these great hazards. These are but examples of what we can expect to achieve in the coming decades if we take the right approach now.

PART II

4

Upper-Atmosphere and Near-Earth Space Research Entering the Twenty-First Century[1]

SUMMARY

This Disciplinary Assessment identifies those research essentials with the strongest societal and environmental impacts that derive from the scientific disciplines covered by the National Research Council's (NRC's) Committee on Solar-Terrestrial Research (CSTR) and Committee on Solar and Space Physics (CSSP). These committees are concerned with the areas of solar and heliospheric physics, magnetospheric physics, ionospheric physics, middle- and upper-atmospheric physics, and cosmic-ray physics.

---

[1]Report of the Committee on Solar-Terrestrial Research and the Committee on Solar and Space Physics. Committee on Solar-Terrestrial Research: M.A. Geller (Chair), State University of New York, Stony Brook; G.P. Brasseur, National Center for Atmospheric Research; J.V. Evans, COMSAT Laboratories; P.A. Evenson, Bartol Research Institute, University of Delaware; J.F. Fennell, The Aerospace Corporation; J.T. Gosling, Los Alamos National Laboratory; S.R. Habbal, Harvard-Smithsonian Center for Astrophysics; M. Hagan, National Center for Atmospheric Research; M.K. Hudson, Dartmouth College; G. Hurford; California Institute of Technology; M.C. Kelley, Cornell University; J.U. Kozyra, University of Michigan; N.F. Ness, Bartol Research Institute, University of Delaware; A.D. Richmond, National Center for Atmospheric Research; T.F. Tascione, Sterling Software; and R.K. Ulrich, University of California, Los Angeles. Committee on Solar and Space Physics: J.G. Luhmann (Chair), University of California, Berkeley; S.K. Antiochos, Naval Research Laboratory; T.I. Gombosi, University of Michigan, Ann Arbor; R.A. Greenwald, Applied Physics Laboratory; R.P. Lin; University of California, Berkeley; M.A. Shea, Air Force Phillips Laboratory; H.E. Spence, Boston University; K.T. Strong, Lockheed Palo Alto Research Center; and M.F. Thomsen, Los Alamos National Laboratory.

Major Scientific Goals and Challenges

When societal and environmental impacts are considered, the dominant scientific and technical goals in upper-atmosphere and near-Earth space can be identified as the following:

- to understand the physical, chemical, and dynamical processes that determine the interactions between the stratosphere, climate, and the biosphere;
- to develop the infrastructure that will permit operational forecasting of "space weather";
- to understand the relationships between changes in the middle and upper atmosphere and the Earth's surface and lower-atmospheric climate; and
- to study solar variability and its influence on the middle and upper atmosphere.

Key Components of the Scientific Strategy

The components of the strategy to address the major scientific issues in upper-atmosphere and near-Earth space science are developed on the basis of four national goals:

1. To study atmospheric processes using observations, laboratory research, theory, and modeling.
2. To have the necessary observations, understanding, modeling capability, and transfer to operations to permit skillful forecasts of "space weather."
3. To document middle- and upper-atmospheric change and produce models that consistently simulate these changes along with those of the lower-atmosphere-surface system.
4. To document changes in solar output, determine how these affect lower-atmosphere and surface climate, and compare these with the climate record.

Scientific Requirements for the Coming Decade(s)

Role of the Stratosphere in Climate, Weather Prediction, and Tropospheric Chemistry

The stratosphere plays two roles in the climate system. The first involves the impact of stratospheric trace gases and aerosols, including those of anthropogenic origin, on radiative fluxes through the tropopause. The second role of the stratosphere in the climate system is through the dynamic coupling between the troposphere and the stratosphere. Considerations of the stratospheric role in various aspects of climate and weather include the following:

- modeling and observational studies of how the stratosphere determines various aspects of climate,
- determinations of how the stratosphere should be correctly represented in numerical forecast models,
- analysis to determine if present and anticipated stratospheric data are sufficient for climate and weather forecasting purposes, and
- analysis to determine how stratospheric change will affect tropospheric chemistry.

Space Weather

In order to "nowcast" and ultimately forecast key aspects of the near-Earth space environment with the goal of mitigating the negative effects of space weather on human life and technology, progress must be made on the following fronts:

- achieving a basic understanding of the relevant physical phenomena and processes so that physical models of the near-Earth system can be developed;
- putting in place the infrastructure to convert research models into operational models;
- obtaining the necessary data to assimilate into and test numerical models of space weather;
- improving on existing statistical models that specify "space climate"; and
- producing nowcasting and numerical forecasting capabilities and using them to develop mitigation strategies.

Global Change in the Middle-Upper Atmosphere

It is critical to understand the effects of natural variability and anthropogenic effects on the ozone layer, the influences of the stratosphere on tropospheric climate, and the impact of upper-atmospheric changes on space-based systems and telecommunications. Scientific requirements in this area include

- analysis of historical data from systems operating from the 1940s to the 1960s;
- monitoring sensitive parameters in the middle and upper atmosphere;
- monitoring inputs to the middle and upper atmosphere from space above and the lower atmosphere below;
- understanding atmospheric phenomena that are now poorly understood, such as "sprites"; and
- developing models that correctly treat disparate and interacting processes important for coupling the middle and upper atmosphere with regions above and below.

Long-Term Solar Variability and Global Change

The Sun undergoes a variety of small changes in its radiant and corpuscular energy output. Long-term, well-calibrated measurements of the outputs and understanding solar variations and the atmospheric response to them are the focus of studies in this area. Scientific requirement for this topic include

- measurement of the solar energy output continuously over at least one full solar cycle,
- investigations of the Earth's temperature and middle- and upper-atmosphere chemical responses to changes in the Sun's energy output, and
- studies comparing solar variations to those of similar stars.

Expected Benefits and Contributions to the National Well-Being

A successful program of research on the upper atmosphere and near-Earth space, with implications for the long-term stability of the ozone layer, will provide insight into issues of the biological effects of increased ultraviolet radiation and the effects of changes in the middle and upper atmosphere on spacecraft operating practices and radio communication.

Upper-Atmosphere and Near-Earth Space Research Tasks

Recommended Stratospheric Research

Stratospheric Ozone
- Deploy manned and unmanned aircraft to make high-precision, high-data-rate measurements.
- Use stratospheric satellite measurements from the Earth Observing System (EOS) to make comprehensive upper-air chemistry measurements.
- Develop three-dimensional, stratospheric models to assess the response of stratospheric ozone to various atmospheric emission scenarios.

Volcanic Effects
- Improve characterization and modeling of volcanic aerosols in the stratosphere for studies of stratospheric heterogeneous chemistry and radiation transfer.
- Improve microphysical models for characterizing stratospheric aerosols to improve atmospheric models.
- Improve currently crude treatments of heterogeneous chemistry in atmospheric models as a fundamental requirement for making atmospheric chemistry models more realistic.

Atmospheric Effects of Aircraft

• Develop three-dimensional stratospheric models including heterogeneous chemistry and microphysics to make these models more realistic.

• Improve characterization of stratosphere-troposphere exchange to correctly treat the exchange of chemical constituents near the tropopause.

Stratospheric Role in Climate and Weather Prediction

• Test effects of including a more realistic stratosphere in numerical predictions to determine how this will affect forecast skill.

• Develop better understanding of upper-troposphere and lower-stratosphere water vapor measurements for the improvement of models.

Recommended Space Weather Research

• Develop a basic understanding of the physical phenomena and processes to provide the basic knowledge required for space weather models.

• Develop better statistical space climate models to provide useful forecasts of space climate.

• Develop nowcasting and numerical forecasting capability to provide more skillful space weather forecasts.

• Evaluate mitigation strategies to ensure the best use of state-of-the-art space weather forecasts.

Recommended Research on Global Change in the Middle and Upper Atmosphere

• Analyze historical data to extend the data record for identifying changes in the middle and upper atmosphere.

• Monitor sensitive parameters in the middle and upper atmosphere to identify parameters that show unexpected variability.

• Model inputs to the middle and upper atmosphere to identify the drivers for middle- and upper-atmosphere change.

• Pursue research on poorly understood processes to determine the significance of global change.

• Understand and model chemical and physical interacting processes to permit the development of comprehensive general circulation models.

• Distinguish between natural and anthropogenic effects to determine their relative importance in middle- and upper-atmosphere global change.

• Investigate the consequences of middle- and upper-atmosphere changes on biotic systems, tropospheric chemistry, and climate.

Recommended Research on Solar Variability and Global Change

• Measure the solar energy output continuously over at least a full solar cycle to establish the range of variation of solar radiant and corpuscular energy.

• Establish the Earth's temperature sensitivity to variations in the solar output to separate anthropogenically produced changes in temperature from solar-induced changes.

- Determine the atmospheric effects of changes in solar x-ray and ultraviolet emissions to determine the response of middle- and upper-atmosphere chemistry and ionization to such changes.
- Determine the Sun's interior dynamics to develop a model of the solar dynamo.
- Explore the variability of solar-type stars to develop statistical estimates of the likelihood of solar variations.

INTRODUCTION

This Disciplinary Assessment identifies those research priorities with the strongest societal impacts that derive from the scientific disciplines covered by the NRC's Committee on Solar-Terrestrial Research and Committee on Solar and Space Physics. These committees cover the scientific areas of solar and heliospheric physics, magnetospheric physics, ionospheric physics, middle- and upper-atmospheric physics, and cosmic-ray physics. A brief description of the coupled Sun-Earth system that is the subject of these research areas is given below. The four research priorities identified are then discussed.

The Sun

The most obvious solar output reaching the Earth is the steady 5,700 K black-body photon emission from the Sun's visible photosphere. Other, more subtle solar-terrestrial connections that originate from different regions of the Sun and through other forms of energy emissions also exist. Above the photosphere, the temperature of the solar atmosphere first falls slowly in the chromosphere and then rises rapidly up to 106 K in the solar corona at a few thousand kilometers above the photosphere. The solar corona is heated from below by mechanisms that are still not well understood. In regions where the solar magnetic field cannot constrain it, the hot ionized gas expands outward to form the solar wind and reaches supersonic speeds at a distance of a few solar radii.

Interplanetary Space

Interplanetary space is permeated by the dilute, yet hot and fast-flowing, solar wind plasma (see Figure II.4.1). Owing to the high electrical conductivity of the plasma, remnants of the solar magnetic field are "frozen" into the solar wind flow. Rooted at one end in the Sun, the interplanetary magnetic field (IMF) is twisted into an Archimedean spiral by the combined effects of solar rotation and the outward solar wind flow. Energetic particles produced in solar outbursts and in interplanetary space are guided by the IMF.

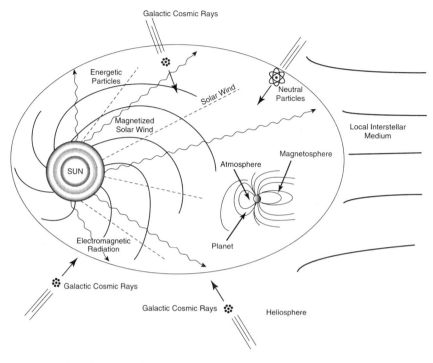

FIGURE II.4.1 Solar connections.

The Magnetosphere

The geomagnetic field presents an obstacle to the solar wind. Its interaction with the solar wind produces a large cavity in the flow, called the magnetosphere (see Figure II.4.1), that surrounds the Earth. This cavity is compressed on the sunward side by the ram pressure of the solar wind, but it is elongated on the night side into a very long magnetic "tail." While shielding the Earth from the incident solar wind, the geomagnetic field also acts as a magnetic bottle that traps and holds plasma that leaks in from the solar wind and escapes from the Earth's ionosphere. These plasmas are heated, accelerated, and transported within the magnetosphere by a variety of processes that are only partially understood.

Of particular concern in this Disciplinary Assessment are dramatic changes to the Earth's magnetosphere occurring as a result of propagating structures in the solar wind. A southward turning of the interplanetary magnetic field increases the transfer of energy from the solar wind into the magnetosphere, resulting in increases in the trapped (Van Allen) radiation, auroras, changes in the surface magnetic field, and heating of the upper atmosphere that creates high-speed winds and composition changes.

The Ionosphere-Upper Atmosphere

Earth's upper atmosphere extends out to several hundreds of kilometers and is partially ionized by extreme ultraviolet radiation from the Sun, creating what is known as the ionosphere. Cosmic-ray and solar energetic particle bombardment and magnetospheric particle precipitation augment solar-produced ionization, especially at auroral latitudes. Daily changes in solar ionizing radiation and heating drive large-scale motions in the upper atmosphere and ionosphere. Electrodynamic coupling between the ionosphere and magnetosphere allows magnetospheric currents and fields to influence ionospheric structure. Collisional, frictional, and chemical changes in the neutral atmosphere also force changes in ionospheric structure and dynamics.

The Middle Atmosphere

The atmosphere is divided into a number of layers associated with obvious changes in temperature structure. This structure (see Figure II.4.2) is determined primarily by the absorption of solar radiation. The atmosphere is mostly transparent to the bulk of solar radiation, which is in visible wavelengths (400 to 700 nm). This leads to most of the solar radiation being absorbed at the Earth's surface. Sensible and latent heat transport from the Earth's surface is responsible for heating the lower atmosphere, and this accounts for the fall-off of temperature with increasing altitude in the troposphere. Ultraviolet (UV) solar radiation (wavelengths less than 242 nm) dissociates molecular oxygen and leads to stratospheric ozone formation. Ozone, in turn, absorbs ultraviolet radiation at slightly longer wavelengths (less than about 300 nm). These processes account for the temperature increasing with height throughout the stratosphere. Above this, the temperature decreases with height in the mesosphere until extreme-ultraviolet (EUV) radiation (wavelengths less than 180 nm) dissociates and ionizes the atmospheric gases, leading again to a temperature increase with altitude in the thermosphere.

Ozone losses occur as a consequence of chemical reactions in which catalytic reactions with hydrogen, nitrogen, and halogen oxides are crucial. Any process that increases the concentration of such reactive species will lead to stratospheric ozone decreases. For instance, industrial chlorofluorocarbon (CFC) emissions have increased the concentrations of reactive chlorine in the stratosphere and led to observable ozone losses that are of societal concern.

Atmospheric waves, on various spatial and temporal scales, are forced mainly in the troposphere. As these waves propagate upward and grow in amplitude, they become very important and comprise a major component of the circulation at higher altitudes. These waves affect the dynamics of the middle atmosphere, which in turn gives rise to transports of many minor atmospheric constituents, including ozone. Since ozone absorbs solar UV radiation but is affected by

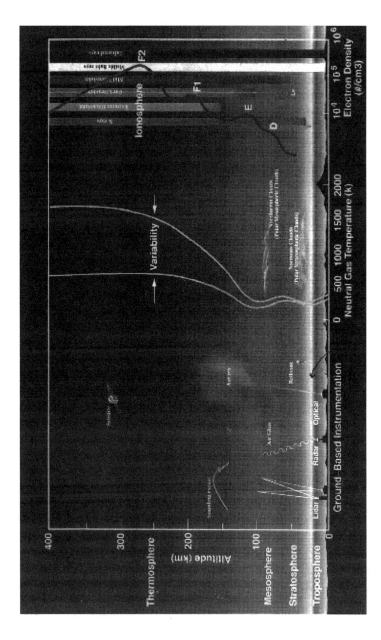

FIGURE II.4.2 Schematic illustration of the atmospheric thermal structure and electron (ion) content. Various processes of interest are depicted, showing the altitude range over which they are observed. These include nacreous and noctilucent clouds, the aurora, and the attenuation of selected solar wavelength ranges that produce the ionospheric layer and excite and dissociate atmospheric species. SOURCE: J.H. Yee and associates, Applied Physics Laboratory, Johns Hopkins University.

chemical and transport processes, middle-atmospheric behavior is determined by rather complex radiation-chemistry-dynamics interactions.

Cosmic Rays

In addition to the neutral gas, plasma, and field environments described above, the Earth is immersed in an extremely tenuous rain of highly energetic charged particles called cosmic rays. Such particles are produced both in our galaxy and in other galaxies. Processes occurring at the Sun and in interplanetary space also occasionally accelerate particles to cosmic-ray energies. The Earth's magnetic field acts as a barrier to this cosmic-ray bombardment, but the shielding effect is imperfect, especially in the magnetically open polar regions and at high altitudes. Owing to their great energy, cosmic rays can be especially dangerous to man and machine throughout space.

Research Priorities

A recent report (NRC, 1995b) of the CSSP-CSTR identified research priorities for overall scientific progress in these areas. Here, we identify imperatives for research, selected from the broader priorities in the earlier report (NRC, 1995b), that specifically address the national goals of

- protecting life and property,
- maintaining environmental quality,
- enhancing fundamental understanding, and
- enhancing economic vitality.

The topics selected with these criteria are, in priority order,

1. stratospheric processes important for climate and the biosphere,
2. space weather,
3. middle- and upper-atmospheric global change, and
4. solar influences.

This prioritization reflects not only national priorities but also other considerations such as timeliness of the research and relevance to other areas of atmospheric research.

Research into the topic "stratospheric processes important for climate and the biosphere" is vital to our understanding of the Earth's atmosphere. Anthropogenically produced substances have been shown to be altering the stratospheric ozone layer; studies in this area are important for maintaining environmental quality. Regulations have been promulgated internationally that ban the future use of chlorofluorocarbons, and decisions concerning future supersonic transport aircraft may hinge on predictions of their effects on stratospheric ozone. Thus, research in this area has important considerations for national economic vitality. Finally, given the fact that stratospheric ozone affects the biosphere's exposure to harmful UV radiation, which can have consequences for human health, and that changes in ozone can affect the Earth's climate, research into this area clearly can have an effect on life and property. Hence, studies of stratospheric processes are relevant to all four of the criteria enumerated above.

"Space weather" encompasses all of the effects associated with the variable release of energy from the Sun in the form of x-rays, energetically charged particles, and streams of plasma with embedded magnetic fields. Through a complex sequence of events, the plasma streams interact with the Earth's magnetosphere giving rise to auroral events, the Van Allen radiation belts, and other geophysical phenomena grouped under the heading "magnetic storms." Such events have caused failures in power transmission grids at high latitudes and the loss of control over communication satellites. Outbursts of x-rays (from flares) or protons pose threats to humans in space. Even the occupants of high-flying aircraft on polar routes are exposed to much higher levels of radiation at certain times.

The space weather imperative is particularly strong because of its timeliness. The reliance on space systems in the civilian sector for meteorological weather forecasting, navigation, and communications is increasing at a rapid rate. This enormous investment of resources is at considerable risk until a coordinated approach to space weather forecasting and improved models of radiation hazards are developed. As a consequence of space research conducted during the past several decades, our understanding of the linkages between the Earth and the Sun has reached a level at which the development of numerical models can now be attempted and efforts undertaken toward using them for prediction. Our inability to relate this topic to "maintaining environmental quality" is the reason this topic has been given second priority, but its importance for manned space flight and for prediction of near-Earth space conditions that affect military communications adds strategic elements that have not been fully explored in this report.

The "middle and upper atmosphere" is the region of our atmosphere extending roughly from 10 km altitude out to several hundred kilometers. This region is subject to long-term changes due to both man-made and solar variability effects. Man-made changes in the ozone layer and in the concentrations of other trace gases are expected to cause major changes in the temperature structure within these regions, which will be most noticeable at high altitudes. These changes will influence the atmospheric circulation, including possible effects on tropospheric climate, and may influence space operations and radio-wave communications. Although much remains to be learned about this region, models are being constructed that include many of the important effects related to dynamics, chemistry, and energetics. This topic was judged to relate less strongly to the stated societal imperatives than those listed previously, so that it was given third priority.

Fourth on our priority list is the topic "solar influences." The Sun, of course, warms our planet and sustains life thereon. Recently discovered small variations in solar irradiance over the sunspot cycle are now suspected to have played a role in past climate fluctuations. Moreover, most space weather events are initiated by changes at the Sun. The understanding of processes at work in our star is perhaps least developed among the topics discussed here. However, with the exception of improved forecasting of some space weather events, it is not clear how studies of the Sun can be translated into short-term, societally relevant impacts, such as enhancing economic vitality and maintaining environmental quality, yet its possibly critical role in driving climate change makes this a subject of considerable concern over the long term.

It is important to note that although we have assigned priorities and selected topics on the basis of our best judgment, it would, in our opinion, be foolhardy to forsake other areas covered in the recent NRC science strategy report (NRC, 1995b). The progress of science has demonstrated repeatedly the wisdom of performing basic research on a broad front, since our ability to judge those areas in which great payoffs are to be found is imperfect, at best. Two examples of this are discussed briefly below to illustrate the point.

The demonstration by Marconi in 1901 that radio waves could traverse the Atlantic Ocean caused Kennelly and Heaviside independently to suggest that a conducting region high in the Earth's atmosphere was responsible for reflecting the waves. A small group in Cambridge, England, under Sir Edward Appleton first measured the height of reflection (of a medium-wavelength British Broadcasting Corporation transmitter's signals) in 1924. Their work was followed by that of another small group at the U.S. Naval Research Laboratory under Breit and Tuve, who developed a pulse-sounding technique for exploring reflection heights as a function of frequency at vertical incidence (i.e., overhead). These results paved the way for widespread use of ionosondes, which proved critical for optimizing high-frequency communications (on which the armed forces depended) during World War II. The pulse-height technique, moreover, is often considered the genesis of pulse radar, which also proved to be of critical importance to the Allies during the war. The origin of this ionospheric work was pure scientific curiosity, with little awareness of its ultimate importance.

Another area of research in which fundamentally important results were obtained despite the absence of high initial priority can be found in both the discovery and the explanation of the Antarctic ozone hole. Although the subject of ozone depletion was already receiving a lot of attention in the middle 1980s and satellite data were being used to look for stratospheric ozone depletions, it was a small group of British researchers, who had been making ground-based observations of total ozone at Halley Bay in Antarctica since the 1950s, that first noticed the precipitous decrease in springtime stratospheric ozone there. They had, moreover, suggested that its cause might be found in industrial chlorofluorocarbons released into the atmosphere. Also, during the 1980s, much research was being performed in stratospheric chemistry, but the great emphasis was on gas-phase chemistry. Fortunately, a few groups had been investigating heterogeneous chemistry (primarily with respect to chemical reactions in tropospheric clouds), and in only a few years it was established that the Antarctic ozone hole was caused by chemical reactions occurring on the surfaces of the aerosols in polar stratospheric clouds. This discovery was made possible by a large, directed research effort involving ground-based and aircraft measurements aimed specifically at finding the reason for the very large ozone decreases observed in the Antarctic. Thus, remarkable progress was built on foundations made by a few small research groups, which shows the value of modestly funded research of no great national priority followed by a very high priority research effort.

With these lessons in mind, this report attempts to highlight a few areas in which we feel that directed research will pay large dividends as we enter the twenty-first century. It is equally important, however, to maintain broadly based research in the other areas of upper-atmosphere and near-Earth space discussed in the earlier NRC report (NRC, 1995b) so that the groundwork can continue to be laid for future knowledge and understanding and for the rare—sometimes important—surprises that can occur.

STRATOSPHERIC PROCESSES IMPORTANT FOR CLIMATE AND THE BIOSPHERE

Natural and anthropogenic variations of the stratosphere can affect the climate and the biosphere in several ways. Firstly, stratospheric ozone (O_3) is the major absorber of UV-B radiation (280 to 320 nm wavelength) in the Earth's atmosphere. Since UV-B is known to damage DNA and thus harm biological systems, any processes, whether natural or anthropogenic, that cause decreases in stratospheric ozone and therefore increases in UV-B are of great concern.

Changes in the stratosphere also affect the climate in complex ways through radiative and dynamical interactions with the troposphere. On the one hand, ozone decreases result in less absorption of solar UV in the atmosphere, thereby allowing more solar radiation to heat the surface. On the other hand, reduced stratospheric ozone leads to a cooler stratosphere that, in turn, radiates less infrared energy downward into the troposphere, resulting in tropospheric cooling. The climate can be changed as a consequence of alterations in these incoming and outgoing radiative fluxes. It is also possible that ozone changes in the stratosphere lead to changes in the stratospheric distributions of wind and temperature and thus affect the dynamical interactions between the troposphere and stratosphere. There is likewise a possibility that reduced concentrations of ozone in the stratosphere could enhance photochemical activity in the troposphere and hence alter the oxidizing capacity of the atmosphere.

Clear evidence now exists that the stratosphere has changed as a result of anthropogenically induced changes in atmospheric composition (WMO, 1995).

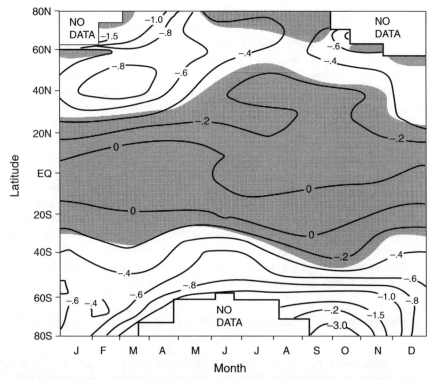

FIGURE II.4.3 Total ozone trends (percent per year) as observed by the Total Ozone Mapping Spectrometer.

Ground-based, aircraft, and satellite observations have clearly shown that the Antarctic ozone hole is caused by increased chlorine and bromine radical concentrations in the lower stratosphere and that these are a result of the anthropogenic emission of huge amounts of halogen-containing compounds in combination with the unique meteorological conditions of the Antarctic lower stratosphere. Similar enhancements in chlorine and bromine radicals are also observed in the Arctic winter lower stratosphere for periods of time, and smaller ozone decreases are seen in this region and at lower Northern Hemispheric latitudes (due to the different meteorological conditions there), but there is no detailed understanding of the reasons for these decreases in ozone at this time. Finally, substantial ozone losses (see Figure II.4.3) have been observed at midlatitudes, particularly in the Northern Hemisphere, exposing large populations to increased levels of UV-B radiation.

Thus, the stratosphere is known to be changing in response to human activities, and stratospheric changes of these types can affect the climate and biosphere (including human health). This situation has led the international community to adopt a number of regulations on anthropogenic trace gas emissions that can affect ozone, with the goal of returning ozone to its natural level in several decades.

Some other outstanding questions in this area include the following:

- How does the increase in concentration of stratospheric aerosol affect climate by changing the radiation balance of the troposphere and altering stratospheric chemistry?
- What is the atmospheric impact of possible future fleets of stratospheric aircraft?

Below, the four specific areas in which research on stratospheric effects is critical (stratospheric ozone, volcanic effects, atmospheric effects of aircraft, and the stratosphere's role in climate and weather prediction) are described, including a brief scientific background, important questions to be answered, and current and future research. Finally, the research imperatives for the twenty-first century that arise from these topics, their contributions to solving societal problems, the programs required, and measures of success are discussed.

Stratospheric Ozone

Atmospheric penetration of UV-B radiation is determined primarily by the amount of absorbing ozone in the stratosphere, even though aerosols, tropospheric clouds, and tropospheric ozone also play an important role. UV-B radiation is damaging to living cellular tissue. Thus, the observed decrease in ozone (Figure II.4.3) implies an increase in the biosphere's exposure to this harmful radiation. For example, Figure II.4.4 shows the estimated daily dose of effective UV radiation for generalized DNA as a function of latitude and season for the average ozone distribution in 1979-1989, along with the trend in daily effective UV dose predicted by ozone trends for this period. Note that although the maximum downward ozone trend at middle northern latitudes occurs in February-March, the maximum absolute increases in UV radiation are seen later in the year due to the annual march in the solar zenith angle. Only with the deployment of the latest generation of research instrumentation has the predicted relationship between ozone and UV-B radiation been verified (see Figure II.4.5). Previous operational networks were not capable of making the necessary measurements.

Ozone in the lower stratosphere acts as a greenhouse gas. As mentioned earlier, it plays two distinct roles—as an absorber of solar radiation and as an emitter of infrared radiation. Because of these roles, the effect of ozone changes on tropospheric climate depends on the altitude at which the ozone changes occur. Models show that the surface climate is most sensitive to ozone changes in the vicinity of the tropopause, where the temperature is the lowest (see Figure II.4.6).

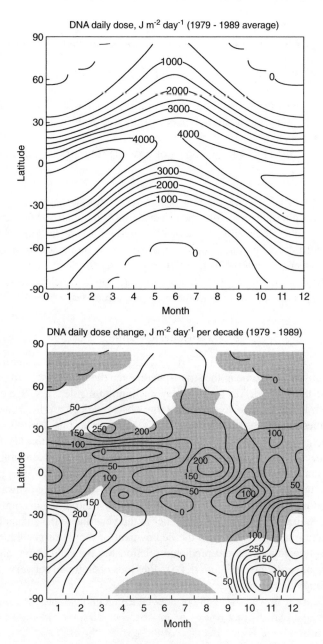

FIGURE II.4.4 Estimated daily dose of effective UV radiation for generalized DNA: seasonal and latitudinal distribution (1979-1989).

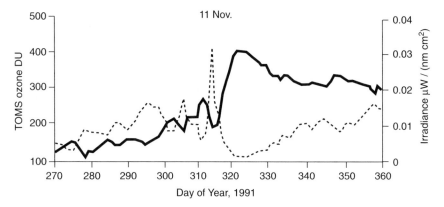

FIGURE II.4.5 Irradiance at 300 nm (dotted line, right axis) and SBUV ozone (heavy line, left axis) during spring season at the South Pole.

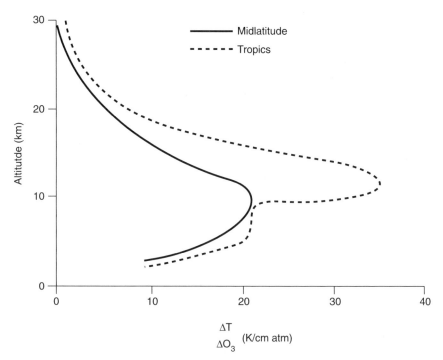

FIGURE II.4.6 Computed change in surface temperature (per unit local O_3 change) as a function of altitude at which O_3 is perturbed. The surface temperature sensitivity curve is shown for midlatitude (solid line) and tropical (dashed line) atmosphere conditions. SOURCE: Wang and Sze, 1980.

These considerations, together with the now well-established fact that the Antarctic ozone hole is caused by anthropogenic emissions of halocarbons, have had profound implications on international policy. The London (1990) and Copenhagen (1992) amendments to the Montreal Protocol and the U.S. Clean Air Act Amendments have accelerated the phaseout of many halocarbons. These actions appear to be having their desired effect since very recent measurements have shown that the rate of increase of the chlorinated fluorocarbons and halons is slowing, while CFC substitutes are beginning to accumulate in the atmosphere (WMO, 1995). These results, documented through a worldwide measurement network, verify that emission controls are beginning to have a beneficial effect on atmospheric halogen levels. Nevertheless, according to models, ozone depletion will likely worsen for at least another decade (WMO, 1995). Decision makers continue to turn to the world's scientific community for advice. How well can we forecast the future of the ozone layer? Responding to this will require research that answers the following critical science questions:

- What processes are causing the observed ozone depletion at midlatitudes in the Northern Hemisphere, which is greater than predicted by current models?
- What will the global ozone losses and surface UV increases be in the next 20 years, during which atmospheric halogen levels are expected to peak?
- What ozone losses and surface UV increases could appear in the Arctic (and the Antarctic) during the same period?
- What is the quantitative relationship among global, lower-stratospheric ozone depletion; radiative forcing of the Earth's atmosphere system; and climate change?
- What are the global impacts of the Antarctic ozone hole (which is expected to continue well into the next century)?
- How well do we understand the role of methyl bromide in ozone depletion?
- When will the ozone layer begin to be rehabilitated?

Currently, there is much research activity in these general areas, and proposals for future programs have been developed. In Europe and the United States, extensive aircraft campaigns to better understand stratospheric processes take place on a regular basis, especially in the Arctic and Antarctic lower stratosphere. These typically involve both high-flying stratospheric aircraft that make in situ measurements and lower-flying aircraft that make remote sensing measurements. Many ground-based measurements are also being made of the stratosphere, both to better understand processes and to establish trends in temperature and composition at a number of locations. A surface UV network has been deployed, but as discussed above, it has proved incapable of making the measurements required for the detection of surface UV trends because of inadequate instrumentation and calibration. Although research instrumentation is capable of making such measurements (Figure II.4.5), little use has been made of this in an operational environment. Photochemical models for stratospheric ozone exist, but they do not correctly predict the observed levels of ozone at upper-stratospheric altitudes or observed ozone decreases in the lower stratosphere. Interactive three-dimensional radiative-dynamics-chemistry models for stratospheric ozone are in their infancy, but much has been learned from simplified versions of these models. Both field measurements and laboratory work are going on to understand anthropogenic chemicals that may pose threats to stratospheric ozone, such as methyl bromide and the CFC substitutes whose atmospheric concentrations are increasing rapidly. Lastly, the Upper Atmosphere Research Satellite (UARS) has provided very valuable observations of chemical compounds never observed globally before, as well as solar flux, energetic particles, and winds, thus enabling a global picture of the workings of the stratosphere that was not available earlier.

Future programs that are needed to make progress in this area include the deployment and utilization of unmanned aircraft, stratospheric satellite observations, stratospheric modeling, and monitoring of surface UV radiation. These programs are described in more detail below.

Aircraft observations have shown their great value in unraveling the physics and chemistry of polar ozone. The ability to deploy these aircraft in specific regions and to make high-precision and high-data-rate measurements of many atmospheric parameters has led to great advances in understanding. Unfortunately, these aircraft are limited to the lower stratosphere, and it is difficult and expensive to deploy them in remote regions. Unmanned aircraft are being developed that will carry significant payloads to higher altitudes, will be less expensive to operate, and will not be subject to the safety limitations imposed on piloted aircraft. It is very important that development of these aircraft continue and that they be used extensively to better understand stratospheric chemistry.

UARS, and the earlier NIMBUS-7 satellites, have shown the impact that satellite observations can have on stratospheric research. There were, however, significant gaps in UARS observations that lacked hydroxy radical (OH) observations and had limited comprehensive chemistry measurements in polar winter. The next opportunity for comprehensive stratospheric satellite measurements will be provided by the EOS, which is scheduled to be deployed in 2002. The atmosphere will have changed a great deal by then, so EOS measurements should show different chemical characteristics. It is crucial that these comprehensive stratospheric measurements be carried out as planned.

Comprehensive three-dimensional models of the stratosphere are just being developed. Very little use has been made of three-dimensional models for assessments of stratospheric ozone. Continued development in this area is needed to make assessments for the future as well as to check that the atmosphere responds to current chemical usage regulations as expected.

Improved UV monitoring instrumentation must be deployed to verify that our understanding of the effects that control UV flux to the surface is correct and to verify calculation methods used for locations where measurements are not being made. This UV flux is crucial for biological investigations.

Volcanic Effects

Discovery of the Antarctic ozone hole in the austral spring spurred intense effort to learn the reason for this unanticipated decrease in total ozone column. Models were developed suggesting that heterogeneous chemical reactions were responsible for converting relatively chemically inactive chlorine species (usually referred to as reservoir species) into reactive chlorine species (radicals) that catalyze ozone loss. These model predictions were later substantiated through observations. The reactions that destroy ozone take place in the lower stratosphere where temperatures are low enough for polar stratospheric clouds (consisting of nitric acid and water) to occur and provide the surfaces on which chemical reactions take place. This knowledge has spurred an increased interest in heterogeneous chemistry.

In the 1980s, there were suggestions that stratospheric ozone decreases occurred following the eruptions of Mt. Agung in 1962 and El Chicon in 1983. It was then proposed that heterogeneous reactions might be occurring on the surfaces of volcanic aerosols that form in the stratosphere after a large eruption such as that of Mt. Pinatubo (in 1991). Comparisons of stratospheric aerosol levels following the eruption of Mt. Pinatubo with those prior to the eruption are available from satellite observations. Later aircraft campaigns did, in fact, confirm the existence of ozone-destroying chemical reactions in the lower stratosphere resulting from heterogeneous reactions occurring on aerosols from Mt. Pinatubo. In addition, it had long been known that after large volcanic eruptions, the enhanced stratospheric aerosol amounts increase both the backscattering of solar radiation and the absorption of solar radiation. Thus, a temperature decrease at the surface and a temperature increase at the height of the volcanic aerosol are expected. Both are, in fact, observed, although great care must be taken to extract the relatively small signal from natural variability.

The current state of our knowledge is that the increased stratospheric aerosol loading resulting from large volcanic eruptions causes perturbations both in the radiation balance of the atmosphere and in stratospheric chemistry. As a result, the following questions must be addressed to further our understanding:

• How has the evolution of aerosol concentrations in the stratosphere over the past decades affected the evolution of stratospheric ozone?
• Are the effects of major volcanic eruptions on tropospheric and stratospheric temperatures, which have been inferred from observations, consistent with modeling predictions?
• Can we observe differences in the distribution of species within chemical

families [e.g., oxides of nitrogen (NO_x)] in the stratosphere due to major volcanic eruptions?
- Can models explain the stratospheric chemical differences that occur in the presence of "background" and volcanic aerosols?
- What are the net chemical-radiative-dynamical effects on the atmosphere of large volcanic eruptions?

Several massive volcanic eruptions have affected the stratosphere in recent decades. In particular, the eruption of Mt. Pinatubo occurred at a time when many observational systems were deployed to observe the stratosphere. Microphysical models have been developed and merged with atmospheric transport models in an attempt to understand the evolution of the volcanically induced stratospheric aerosol. There has been rapid development in understanding the heterogeneous chemical reactions that occur on volcanic aerosols. Some observations have indicated that the chemical partitioning predicted to occur in the presence of stratospheric aerosol particles does, in fact, occur. Some general circulation models have been run to examine the climatic effects expected from volcanic eruptions. Future projects include improved characterization of stratospheric aerosols, improved microphysical models, and improved treatment of heterogeneous chemistry.

There is a paucity of information on the composition, surface characteristics, and so forth, of stratospheric aerosols. Yet, these characteristics are crucial in considerations of heterogeneous chemistry and radiative transfer. In recent years, several microphysical models have been developed for characterizing stratospheric aerosols. More has to be done, and these models should be combined with dynamical and heterogeneous chemistry models for more realistic characterization of the atmosphere and improved predictive capability. Laboratory and field measurements have shown that heterogeneous chemistry is very important in the stratosphere. Yet, the treatment of this chemistry in models remains very crude. More realistic treatment of heterogeneous chemistry in atmospheric models is needed and should be coupled to detailed microphysical models.

Solar Effects

The abundances of ozone and other chemical species in the stratosphere are affected by photodissociation processes and hence by the level of solar ultraviolet radiation penetrating the atmosphere. Since the intensity of solar radiation at short wavelengths varies with solar activity, the concentrations of stratospheric gases (including ozone) are expected to vary with a period of approximately 11 years. Ozone and other chemical compounds are also expected to respond to changes in solar radiation on a time scale of 27 days, corresponding to the apparent rotation period of the Sun. Although these natural perturbations in the chemical composition of the middle atmosphere are generally smaller than some of the recent anthropogenic effects, they are nevertheless significant and are discussed more extensively later.

Quasi-Biennial Oscillation Effects

Atmospheric waves (gravity waves and equatorial planetary waves such as Kelvin and mixed Rossby-gravity waves) interact with the mean flow in the equatorial lower stratosphere to produce an oscillation in the mean zonal winds such that they reverse from eastward to westward with an average period of approximately 28 months. This is known as the quasi-biennial oscillation (QBO). Although the wind oscillation is confined to the lower stratosphere, it appears to modulate global stratospheric circulation, and hence the meridional transport of ozone and polar winter temperatures. This effect modulates the severity of the Antarctic ozone hole, and global ozone concentrations in general, according to the phase of the QBO.

There have also been suggestions that the QBO significantly influences the troposphere, including the number of Atlantic hurricanes. It is uncertain how this influence arises, but a suggestion has been made that the QBO influences upper-tropospheric wind shear in the tropics, which in turn affects the nature of deep convection that occurs during hurricane formation. More research is needed to understand the QBO and its influence on global circulation.

Atmospheric Effects of Aircraft

Much of the impetus for modern stratospheric ozone research has its roots in the supersonic transport research of the 1970s. At the time, the concern was that such aircraft would emit large amounts of water vapor and nitrogen oxides into the stratosphere where they could initiate catalytic ozone destruction. Since then, the airline industry has grown considerably, and this expansion has included great growth in long-distance routes in the Pacific region. At this time, plans are under way for a new generation of supersonic civilian transports. This planning involves assessing the effects of proposed aircraft operations on the stratosphere, with particular emphasis on stratospheric ozone.

The nature of this research puts great emphasis on atmospheric photochemical modeling, since no alternative method for predicting the effects of supersonic transport exists at present. This places particular stress on current modeling abilities. For instance, two-dimensional models have been used extensively to simulate the effects of halogens on stratospheric ozone. This has some justification since the halogen-containing gases are long-lived and should therefore mix extensively within the troposphere before entering the stratosphere. On the other hand, aircraft exhaust products are deposited along the flight path, and since many of the products have short lifetimes, they do not become well mixed. Thus, analysis of the effects of aircraft on the atmosphere entails treating what is inherently a three-dimensional process.

In addition, aircraft exhaust products are deposited in the lower stratosphere, near the tropopause, which is a very difficult region to model correctly because, in this region, the very complex tropospheric chemistry merges with stratospheric chemistry. Further, the exchange of mass between the stratosphere and the troposphere, known as stratosphere-troposphere exchange (STE), is not well understood. In addition, heterogeneous chemical reactions that occur on ambient and aircraft-produced aerosols are a concern. To make modeling efforts relevant to this problem, many reaction rates must be measured in the laboratory under conditions encountered in the stratosphere, and atmospheric measurements must be carried out to test our understanding and the correctness of the models.

The decision on whether or not to build a fleet of high-altitude aircraft, and how to operate them, depends to a large extent on the confidence that the scientific community has in correctly predicting their atmospheric effects. To provide meaningful input in this area, the following critical questions must be answered:

- What will the effect of a large fleet of supersonic aircraft be on atmospheric composition and structure?
- What are the limits imposed by using two-dimensional models to simulate the impact of aircraft on a three-dimensional atmosphere?
- How do we represent stratosphere-troposphere exchange processes properly in chemical transport models?
- What differences would there be in the atmospheric effects of supersonic aircraft under volcanically enhanced stratospheric aerosol conditions and background conditions?
- What will the atmospheric effects of aircraft be in the polar regions?
- What will the effects of aircraft be under different chlorine loading conditions?

During the past several years, extensive programs have been put in place to assess the atmospheric consequences of the present aircraft fleet, as well as to predict the effects of future fleets of subsonic and supersonic aircraft. Experimental facilities have been built to measure the exhaust products of present aircraft and proposed aircraft engines. Laboratory investigations are proceeding to measure the rates of chemical reactions that transform the aircraft exhaust products and influence atmospheric composition. Models are being developed to predict the transformations that take place between the time the aircraft exhaust leaves the tail pipe and the time it mixes with atmospheric gases. Models have been built to predict the changes in atmospheric composition resulting from the deployment of present and future aircraft fleets. At present, these models are mostly two dimensional, but a few three-dimensional models are starting to appear. To make substantive progress, more realistic three-dimensional models are needed, as well as better characterization of STE processes, particularly along flight paths. These models require better treatment of aerosol physics and heterogeneous chemistry. The nature of stratosphere-troposphere exchange is not sufficiently well understood to be incorporated satisfactorily in model treatments at this time. Observational campaigns, together with modeling efforts, will be necessary to gain the required understanding of these processes.

The Role of the Stratosphere in Climate and Weather Prediction

The stratosphere plays at least two roles in the climate system. The first is the impact of stratospheric trace gases and aerosols on the net radiative balance of the surface-troposphere system. Ozone and aerosols reduce the downward radiative flux at the tropopause by absorbing and reflecting solar radiation. On the other hand, ozone, carbon dioxide (CO_2), and several other trace gases increase the downward radiative flux by emitting longwave radiation. The second role of the stratosphere in the climate system is through the dynamic coupling of the troposphere and stratosphere. Stratospheric circulation influences the vertical propagation of tropospheric waves, but the feedback of this process on tropospheric circulation is not well understood.

It is essential that the stratosphere be considered in any effort to understand certain aspects of climate. For instance, CFCs are greenhouse gases, so their increasing concentrations lead to climatic warming. CFCs also lead to stratospheric ozone depletion and are probably responsible for the observed decreases in ozone in the lower stratosphere. Since lower-stratospheric ozone is itself a greenhouse gas, the combined effects of CFCs on radiative transfer and ozone depletion lead to less greenhouse warming when their effects on stratospheric ozone are considered. When comparing trends in lower-stratospheric temperatures with model predictions, it is also important to consider both the CO_2 greenhouse and the ozone depletion effects.

The following are critical questions for characterizing the role of the stratosphere in weather and climate:

- How do processes in the stratosphere affect the prediction of future climate states?
- How can improved representation of the stratosphere be included in numerical forecast models?
- Are the present and anticipated data sources for the stratosphere (e.g., for water vapor and winds) sufficient for climate and weather forecasting purposes?
- Do current models correctly model troposphere-stratosphere interactions?
- How do different models of the troposphere-stratosphere system compare with one another?

Several active areas of research are currently being pursued. Radiation chemistry climate models exist that can be used to address some questions involving troposphere-stratosphere interactions, but most are relatively crude (either one or two dimensional). Some modeling work has also been done on dynamical-radiative interactions, at times with conflicting results. For instance, one general circulation model predicts that doubled CO_2 concentrations will lead to a warmer lower stratosphere in winter, whereas others do not. This question is of great importance in understanding the prospects for the occurrence of an Arctic ozone hole in the future. A few research papers have appeared indicating that better inclusion of the stratosphere in numerical forecasting models leads to greater forecasting skill, but there have been no large-scale operational tests of this type.

To make further progress, the effects of inclusion of a realistic stratosphere in numerical weather prediction models have to be better understood. In addition, models must be tested both against each other and against observations. Better observations of water vapor in the upper troposphere and lower stratosphere are also needed. Both radiative and chemical models require such data.

Key Initiatives

In the following, some key initiatives for the next 15 years are discussed that will enable progress to be made in the research issues identified earlier. These initiatives are organized by scientific area. It should be noted that although they have been listed in specific scientific areas, many of them will also be useful in other areas. In all cases, a strategy that combines observations, laboratory studies, and modeling is needed.

Stratospheric Ozone

- *Deployment and Utilization of Unmanned Aircraft:* Aircraft observations have shown their great value in unraveling the physics and chemistry of polar ozone. The ability to deploy research aircraft in specific regions and to make high-precision and high-data-rate measurements of many atmospheric parameters has led to great advances in understanding. Unfortunately, these manned aircraft have ceiling limits that are low in the stratosphere, and they are difficult and expensive to deploy in remote regions. Unmanned aircraft are being developed that will carry significant payloads to higher altitudes, will be less expensive to operate, and will not be subject to the safety limitations imposed on piloted aircraft. It is very important that unmanned aircraft have a defined role in the study of stratospheric chemistry. Support for other unique platforms such as the National Aeronautics and Space Administration's (NASA's) ER-2 and the National Science Foundation's (NSF's) WB-57 is also important because these research aircraft have the capability to carry large scientific payloads at high altitudes.

- *Stratospheric Satellite Observations:* Satellite observations, such as those from UARS, have made a substantial impact on stratospheric research. However, measurements have been limited by the lack of OH observations and the fact that comprehensive chemistry measurements are available for only one Southern Hemisphere winter. The next opportunity for comprehensive stratospheric satellite measurements will be on the EOS mission.
- *Stratospheric Modeling:* Comprehensive three-dimensional models are just being developed. Very little use has been made of three-dimensional models for assessments of stratospheric ozone. Continued development in this area is needed to make assessments for the future as well as to check that the atmospheric response to current regulations has been as expected.
- *Monitoring Surface UV Radiation:* Improved ultraviolet monitoring instrumentation is required to verify that our understanding of the effects that control UV flux to the surface are correct, as well as to verify the calculation methods that will be used for locations where measurements are not being made. It is this UV flux that is crucial for biological investigations.

Volcanic Effects

- *Improved Characterization of Stratospheric Aerosols:* There is a paucity of information on the composition, surface characteristics, et cetera, of stratospheric aerosols. Yet these characteristics are crucial in considerations of heterogeneous chemistry and radiative transfer.
- *Improved Microphysical Models:* In recent years, several microphysical models have been developed for characterizing stratospheric aerosols. More needs to be done, and these models should be combined with dynamical and heterogeneous models for more realistic characterization of the atmosphere and improved predictive capability.
- *Improved Treatment of Heterogeneous Chemistry*: Heterogeneous stratospheric chemistry has been shown to be very important by laboratory and field measurements as well as in models. Yet, the treatment of heterogeneous chemistry in models remains very crude compared to laboratory understanding. More realistic treatments of heterogeneous chemistry in atmospheric models are needed. Also, these treatments should be coupled to microphysical models.

Atmospheric Effects of Aircraft

- *More Realistic Three-Dimensional Models:* More realistic three-dimensional models of aircraft effects on the atmosphere have to be produced. These models require better treatment of aerosol physics and heterogeneous chemistry, as well as stratosphere-troposphere exchange.
- *Better Information on the Exchange of Material Through the Tropopause:* The decrease of temperature with height and slight stability typical of the tropo-

sphere is reversed in the stratosphere. Although stratospheric temperature increases with height, the resulting stability of the stratosphere does not bar the exchange of materials such as radioactive particles or ozone-depleting chemicals between the two.

- *Better Characterization of Stratosphere-Troposphere Exchange:* The nature of stratosphere-troposphere exchange is not sufficiently well understood to check against model treatments. Observational campaigns together with modeling efforts will be necessary to gain the required understanding of these processes.

Climate and Weather Prediction

- *Weather Prediction:* The effects of inclusion of a realistic stratosphere in numerical weather prediction models must be better understood. Versions of numerical weather prediction models should be developed that include the stratosphere in a realistic fashion. Retrospective and real-time weather prediction testing are required to see how inclusion of the stratosphere affects the forecasting skill of these models.
- *Testing of Models:* General circulation models that include the troposphere and stratosphere have to be better tested against observations. Also, models must be carefully intercompared so that the reasons for their different behavior are understood.
- *Water Vapor:* Better observations of water vapor in the upper troposphere and lower stratosphere are needed. Both radiative and chemical models require such information.

Measures of Success

A successful program will provide a much-improved quantitative understanding of the fundamental chemical, dynamical, and radiative processes that influence the physical and chemical behavior of the middle atmosphere. It will reduce some key uncertainties affecting the behavior of ozone and other chemical constituents in the middle atmosphere, and specifically in the lower stratosphere, where large ozone depletions have been observed. A successful program should also provide the information needed to determine whether it is possible to take measures that would enhance the long-term stability of the ozone layer and the climate of the Earth.

SPACE WEATHER

Earth does not exist in an unchanging and benign vacuum. Rather, it is embedded in the dynamic solar wind that fills interplanetary space with a continuous supersonic flow of plasma from the Sun. The interaction of the solar

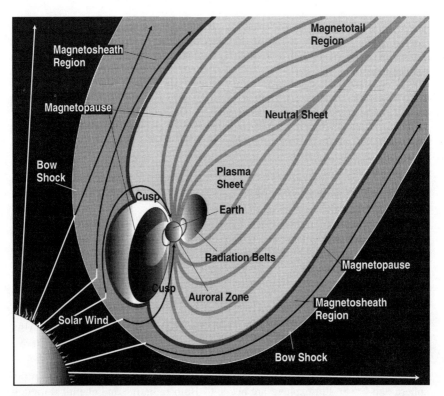

FIGURE II.4.7 Sun-Earth connections.

wind with the Earth's magnetic field (see Figure II.4.7) yields a dynamic structure called the magnetosphere. Spatial and temporal variations in the solar wind outflow, in response to spatial and temporal changes of the Sun's magnetic field, have a profound effect on the magnetosphere. The complex, time-dependent variations of particles and electric and magnetic fields in the solar wind, the magnetosphere, and the ionosphere produced by solar variability are known collectively as "space weather." The domain of space weather is distinct from that traditionally associated with better-known lower-atmospheric weather (tropospheric meteorology). It encompasses those regions noted in the introduction to this Disciplinary Assessment that are most sensitive to transient phenomena originating on the Sun. They are physically linked to one another in various ways; this coupling makes space weather intrinsically an interdisciplinary topic.

As with lower-atmospheric weather, research on space weather has two main thrusts: (1) a basic research focus to explore and understand the physical processes linking the fundamental elements of the solar-terrestrial system, and (2) an applied research focus to develop useful physical models capable of predicting the changes associated with space weather. Relevant to the latter point, various aspects of space weather can have deleterious effects on both ground- and space-based technological assets, as well as on humans operating in space or at high altitudes. As our reliance on space systems continues to grow, so too does our vulnerability to space weather. The first crude, unmanned satellite was launched about 40 years ago. At present, more than 200 sophisticated satellites are operating at geosynchronous orbit alone, and there is a routine manned presence in low Earth orbit. In the decades ahead, we anticipate many hundreds of technologically advanced satellites operating in space and a nearly continuous manned presence.

The effects of space weather on satellite performance and human health are well documented. Some of these effects are listed in Box II.4.1 for two disturbed periods in 1989. Both naturally occurring space radiation and electrical discharges on spacecraft induced by the space environment can damage and/or destroy critical spacecraft components. Ionization by a single energetic heavy ion can randomly change the state of electronic logic circuits and thus place a spacecraft in jeopardy. Changes in ionospheric structure can adversely affect critical civilian and federal navigation and communications systems. Astronauts on board the shuttle or space stations and even people on board aircraft in polar routes can be at risk in a radiation environment that changes in response to solar-terrestrial interactions. Space weather variations also have negative effects at the surface of the Earth. Reliance on large-scale power grids has led to a broad vulnerability to transient electrical currents induced by time-varying magnetic fields in near-Earth space. These vulnerabilities will likely increase as technology advances, as our presence in space intensifies, and as we become dependent on ever more sophisticated systems for communications, navigation, and other critical functions.

Our understanding of space weather, our ability to specify its present state, and our ability to predict changes in this state are at a primitive level, perhaps analogous to that of tropospheric meteorology in the early 1950s. The purpose of the space weather research program is to convert our present fragmentary understanding into a coherent body of knowledge, so that reliable numerical models of the space environment and the changes associated with space weather can be developed. Just as the extension of meteorological capabilities to stratospheric altitudes has been essential for the full exploitation of commercial aviation, so will the application of the meteorological paradigm to space weather be essential for the full exploitation of space technology. It is critical to recognize that both long-term (years) and short-term (minutes) variations are important in this endeavor. These aspects are described below in the context of solar variability and the corresponding changes that occur in the solar wind and in the Earth's magnetosphere, ionosphere, and upper atmosphere.

> **Box II.4.1**
> **Some Consequences of Space Weather Disturbances in 1989**
>
> *March 13-14, 1989*
>
> - Massive power outage darkened most of Quebec Province for up to nine hours. At the same time, power losses occurred on power distribution lines in central and southern Sweden.
> - GOES-7 lost imagery and had a communications outage.
> - Seven commercial satellites required 177 manual operator interventions to maintain operational attitude orientation. This is more than are normally required during a year of regular operations.
> - Numerous LORAN navigation problems. Difficulty in using high-frequency radio communications to alert users to the problem.
> - California Highway Patrol messages were overpowering local transmissions in Minnesota.
> - Large voltage swings in undersea cables.
>
> *September-October, 1989*
>
> - Data from radiation sensors aboard the Concorde indicated that passengers and crew received a radiation dose the equivalent of a chest X-ray.
> - Shuttle ATLANTIS astronauts reported eye "flashes" produced by energetic protons penetrating the optic nerves.
> - Computations indicate that an unshielded astronaut on the Moon would have received "lethal" radiation.

Scientific Background

Space Climate

To understand the dynamical behavior of the coupled solar-terrestrial system and its potentially deleterious effects, it is imperative to understand first its gross time-averaged conditions and the extreme, long-time-base departures from the mean ("space climate"). In the introduction to this Disciplinary Assessment, some aspects of the climatology of the four coupled regions (Sun, interplanetary space, magnetosphere, ionosphere-upper atmosphere) that comprise the solar-terrestrial system were outlined. Space climate models are especially important at present, given that our ability to provide accurate and specific space weather forecasts is rather limited at the present time. Indeed, knowledge of the space climate rather than space weather is used principally by engineers who design and build systems to withstand the equivalent of a hundred-year flood or storm. This manufacturing philosophy may lead to inefficiency and costly overdesign. Ultimately, production of a reliable space weather forecast capability may give designers the confidence to build "smart" systems that take advantage of advanced knowledge of inclement conditions. In the meantime, improving the validity of space climate models is an important first step toward the goal of understanding and mitigating the effects of the space environment.

The Space Weather System

The Sun is a variable star. Driven by dynamics in the solar interior, the solar magnetic field is continually evolving. This evolution produces the well-known ~11-year cycle of solar activity. The solar magnetic field causes the Sun's outer atmosphere, the solar corona, to be highly structured. Even when the Sun is relatively quiet, the solar wind near Earth is highly variable since the solar rotation (with a period near 27 days) produces a progression of different coronal regions facing the Earth. Solar wind flow speeds near Earth vary from approximately 300 to 850 km/s, densities range from approximately 1 to 50 cm^{-3}, and magnetic field strengths vary from approximately 1 to 30 nT; average values are approximately 400 km/s, 8 cm^{-3}, 10 eV (electron volts), and 5 nT. Large deviations of the magnetic field from the standard Archimedean spiral direction are common. These temporal variations in solar wind are usually organized into alternating streams of high- and low-speed flows, with the density and field strength generally being strongest on the leading edges of the high-speed streams as a result of compression that occurs in interplanetary space. When the magnetic field within a compression region on the leading edge of a high-speed stream is directed southward, the solar wind is particularly effective in stimulating geomagnetic activity.

The most dramatic forms of solar activity are solar flares and coronal mass ejections (CMEs) (see Figure II.4.8). Flares are distinguished by enhanced electromagnetic radiation over a broad range of frequencies on time scales ranging from seconds to hours. Often particles are accelerated to high energies during the flare process. CMEs are events in which large amounts of solar material are suddenly injected into the solar wind. They originate in closed magnetic field regions in the solar corona that have not previously participated in the solar wind expansion. Although they are distinct phenomena, flares and CMEs both seem to result from the release of stored energy from unstable magnetic configurations in the solar atmosphere. In particular, flares are usually observed on closed field lines statically bound in the solar atmosphere, whereas CMEs are characterized by mass motions on field lines being opened to interplanetary space. CMEs exhibit a wide range of outward speeds. The faster CMEs usually produce major shock wave disturbances in the solar wind. The strong interplanetary magnetic fields produced by such disturbances are particularly effective in stimulating geomagnetic activity when they contain fields with southward components on their arrival at Earth.

Intense and long-lasting energetic particle events, usually called solar energetic particle (SEP) events, are often observed in interplanetary space in associa-

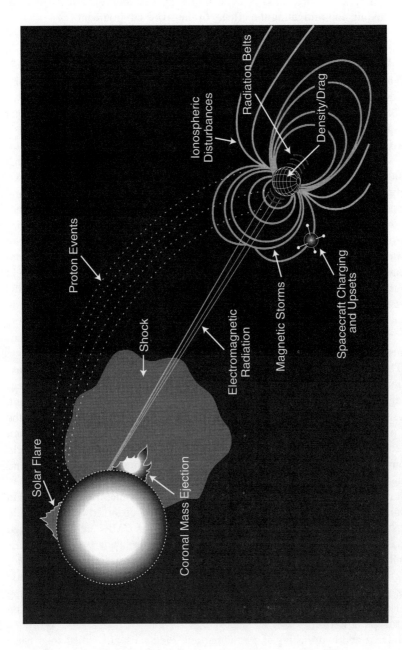

FIGURE II.4.8 Solar flares and coronal mass ejections.

tion with shock disturbances driven by fast CMEs. The temporal profiles of these particle enhancements differ from event to event, depending on the position of the Earth relative to the propagation direction of the interplanetary disturbances. Typically, however, major energetic particle events in interplanetary space begin shortly after fast CMEs lift off from the Sun and continue until well after shocks driven by the CMEs pass the Earth several days later. Although some of the energetic particles observed in major SEP events often are accelerated near flare sites at the Sun, most of the energetic particles in major events appear to be the result of a shock acceleration process that occurs in the outer solar corona and in interplanetary space.

The various manifestations of solar variability in interplanetary space produce both magnetospheric and ionospheric responses, as illustrated in Figure II.4.8. The magnetospheric regions of particular relevance to space weather are shown in Figure II.4.9 and include the magnetopause, the outer boundary of the magnetospheric cavity; the tail plasma sheet, a region of warm plasma extending across the midplane of the geomagnetic tail on the night side of Earth; the near-Earth plasma sheet, the sunward extension of the tail plasma sheet into the region surrounding the inner magnetosphere, out to the dayside magnetopause; the plasmasphere, a region of dense, cold plasma relatively near the Earth populated primarily by particles that escape from the ionosphere; the radiation belts, consisting of magnetically trapped ions and electrons of very high energies ($\sim 10^6$ eV); and the ring current, a region of quasi-trapped, high-temperature plasma that carries a current large enough to be detectable at the surface of the Earth. Each of these regions is connected along geomagnetic field lines to low altitudes. For example, the ovals where auroral emission occurs are low-altitude projection(s) of the plasma sheet at high geomagnetic latitudes. Field lines in regions near the geomagnetic poles are interconnected to the interplanetary magnetic field and are said to be magnetically "open."

When the interplanetary magnetic field at Earth contains a southward component, energy transfer from the solar wind to the magnetosphere increases and the magnetosphere becomes stressed. When the magnetosphere relaxes from this stressed state, strong plasma heating and particle acceleration occur in the near-Earth plasma sheet, enhanced particle precipitation into the upper atmosphere occurs at high latitudes with a concurrent brightening and motion of auroral forms, and electric current is diverted from the magnetotail down to the nightside ionosphere. This global release of energy is known as a magnetospheric substorm. Because the magnetic field embedded in the solar wind often contains a southward-directed component even when the solar wind is not disturbed, magnetospheric substorms occur at a typical rate of one to a few per day.

As noted above, particularly strong geomagnetic responses are triggered by the strong southward-directed fields often contained within compression regions on the leading edges of high-speed streams and by interplanetary shock disturbances driven by fast coronal mass ejections. Geomagnetic activity stimulated in

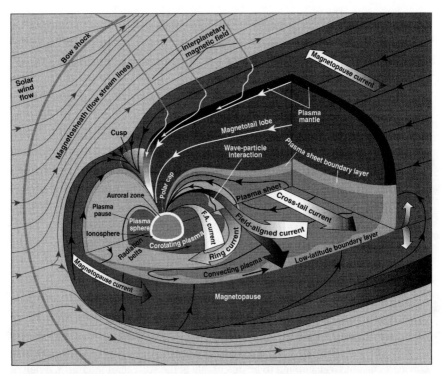

FIGURE II.4.9 Magnetospheric regions of particular relevance to space weather. NOTE: F-A = field-aligned.

this way often persists for several days at a time; such intervals are known as geomagnetic storms. The electric and magnetic field perturbations associated with large geomagnetic storms can extend to the Earth's surface, driving strong ground currents. New, transient radiation belts may also be formed in the largest such events.

Because solar and interplanetary events vary in frequency and intensity with the solar activity cycle, so too does geomagnetic activity. The most severe geomagnetic storms usually are associated with interplanetary disturbances driven by fast CMEs and thus are most common near the maximum of solar activity. Geomagnetic storms associated with high-speed stream compression regions are generally less severe, but tend to recur at the 27-day rotation period of the Sun, particularly on the declining phase of the solar activity cycle. Moreover, for unknown reasons, recurrent storms are much more effective in accelerating electrons to million-electron-volt energies in the outer reaches of the radiation belts. Hence, it is during the approach to solar activity minimum that the fluxes of these electrons with million-electron-volt energies are particularly elevated within the magnetosphere, and these flux enhancements tend to recur at 27-day intervals.

The ionosphere-upper atmosphere system also responds to changes in the external space environment at all local times and latitudes. The upper reaches of Earth's atmosphere lie at the low-altitude extension of the magnetosphere and include both neutral and charged constituents. The charged component is called the ionosphere and is collocated with the upper neutral atmosphere (Figure II.4.2). The ionosphere responds to and affects both the magnetosphere and the neutral atmosphere and thus plays a crucial role in coupling the two.

The physical properties of the ionosphere and upper neutral atmosphere are affected dynamically by changes in both solar radiation and magnetospheric electrodynamics. Solar ultraviolet radiation ionizes the neutral atmosphere, creating the ionospheric structure noted in Figure II.4.2. The ionosphere extends from altitudes of about 90 to 500 km, with local peaks in electron density near 100 and 250 km. Electrical currents, electric fields, and particle precipitation are all imposed onto the ionosphere from their magnetospheric source regions in the polar magnetic cusp and boundary layers, the geomagnetic tail, and the inner magnetosphere. Time variations in magnetospheric convection, especially during magnetospheric substorms and storms, couple electrodynamically with ionospheric motions via magnetically field-aligned currents in the auroral zone. Horizontal ionospheric currents act to couple auroral latitudes to lower latitudes. During large geomagnetic disturbances, activity at relatively high magnetic latitudes can thus affect the nature of the near-equatorial ionosphere. These effects include enhanced or decreased ionization, enhanced or reduced winds, composition changes, heating, gravity wave generation, plasma irregularities and instabilities, and enhanced atmospheric density. These may affect communications, electric power distribution, navigation, space system operations, satellite drag, geomagnetic surveys, and radiation dose.

The ionosphere-upper atmosphere also responds to rapid changes in solar ionizing radiation and energetic particle precipitation that accompany transient solar events. These events, including solar flares and CMEs, produce significant changes in electron density at lower altitudes (80 to 90 km), which in turn inhibit or block high-frequency radio communication at all daytime latitudes.

From a practical standpoint, much of what is relevant to the human condition in space weather involves the Earth's ionosphere. This blanket of ionized matter, or plasma, around the Earth is more dense than that around any other planet, and one must approach the surface of the Sun to find a comparable plasma environment. In our ever-increasing dependence on communications involving satellites that lie far above the ionosphere, this medium must be traversed by electromagnetic waves, which carry our messages and even information about where we are located. In addition to buffeting from above by the solar atmosphere, the ionosphere is subjected to enormous winds borne aloft by tidal and atmospheric gravity waves generated in the dense atmosphere of the Earth. Just as waves breaking on the Earth's ocean shores create turbulence and severe rip currents, these upward-propagating disturbances create ionospheric space weather that matches the severity of solar-induced effects.

Space Weather Effects

To understand and assess the effects of the space environment on systems, it is necessary to know both the environment and the way in which technological systems interact with it. The following material provides several examples of the types of interactions that occur; the different classes of effects are outlined below. Such a list is not exhaustive; indeed, it continues to grow as new and more sophisticated technologies are put to use in human endeavors.

The principal space weather hazard to humans in space and at high altitudes is exposure to ionizing radiation. The exposure gained in high-altitude aircraft is lower in a spacecraft because of the shielding effect of the atmosphere above the aircraft. The primary regions of concern for aircraft are the high magnetic latitudes, where energetic cosmic rays and solar particle events are not shielded by Earth's magnetic field. Manned space flight programs are also very concerned about the exposure of astronauts to radiation. For missions that leave low Earth orbit, the ability to traverse quickly known concentrations of radiation, such as the Earth's radiation belts, and to predict the occurrence of energetic solar particle events is of extreme importance.

Varying levels of solar UV during the solar cycle change the altitudinal profile of the ionosphere, thereby changing ground-to-ground transmission paths for radio communications that use the ionosphere as a reflection medium, and limit the maximum usable frequency (MUF) for such systems. In addition, magnetospheric storms create enhanced and localized ionization levels in the ionosphere, while the variations in the electric fields in the atmosphere-ionosphere electric circuit lead to instabilities that structure the ionospheric ionization. This structured ionization can cause time-variable disruptions in ground-to-satellite and ground-to-ground transmission paths.

High-frequency (HF) radio communications continue to be used by the Department of Defense, shortwave broadcasting authorities, mariners, and others. The ionosphere, which supports such communications, is subject to disturbances during which there can be degraded capability and even a total communications blackout. Sudden ionospheric disturbances caused by intense solar flares are short-lived (minutes to an hour) interruptions caused by increased absorption in the lowermost (D-) region of the Earth's ionosphere. Longer-lived disruptions occur as a result of heating of the upper atmosphere at auroral latitudes. Composition changes carried by enhanced winds depress levels of ionospheric density (in the F-layer) at midlatitudes, resulting in poorer HF communications. This effect can last up to several days in severe magnetic storms.

Both modern and traditional navigation technologies can be affected by space weather. For example, variations in the strength and location of ionospheric currents and the currents that couple the ionosphere and magnetosphere cause significant errors in navigation by magnetic compass systems. The variability of the ionospheric electron density, discussed above, causes phase shifts and time delays in global positioning system (GPS) signals, which can lead to ephemeris and position errors for the user and decreased reliability and accuracy of GPS products.

Electric power systems can be affected by currents induced in the Earth by enhanced ionospheric currents during magnetospheric storms and substorms. These effects are large enough to damage parts of power networks (e.g., transformers) and disrupt power distribution systems.

Space weather can similarly affect the function of modern telecommunication systems. For example, magnetospheric storms and substorms drive ionospheric currents that can in turn induce significant voltages in long transmission lines (e.g., transoceanic cables) and potentially result in loss or limitation of function. To protect against this possibility, electrical design limits must be set very high with cost impacts on the systems.

Satellites experience several different types of space environment effects. Some are climatological—for example, the degradation of electronics, solar cells, and materials that results from long-term radiation exposure or the erosion of materials via oxygen bombardment when traversing the oxygen-rich upper atmosphere. Similarly, polymerization and embrittlement of some materials by UV exposure or the single-event effects induced in electronics by galactic cosmic rays (see Figure II.4.10A) can be considered climatological effects.

Other satellite effects occur during transient space weather events. For example, satellite charging (both surface and deep dielectric charging) occurs when a satellite is rapidly immersed in a hot plasma or the energetic electron radiation is significantly enhanced above average levels for extended periods. If the charging level exceeds the dielectric strength of a component, an electrostatic discharge can occur and result in operational anomalies (see Figure II.4.10B) or even system failures (see Figure II.4.10C). Enhanced solar cell radiation damage can be caused by energetic solar particles. Accelerated decay of satellite orbits is caused by increased atmospheric density at satellite altitudes as a result of atmospheric heating via sporadic solar x-ray and UV input or dumping of magnetospheric energy into the ionosphere-atmosphere system (see Figure II.4.10D).

Geomagnetic surveys from aircraft and on the Earth's surface are an important tool used by commercial companies in their searches for natural resources. The variation in strength and position of the ionospheric and magnetosphere-ionosphere coupling currents can create significant errors in such surveys. For example, they can create strong signatures in the survey data that are related not to subsurface features but to transient ionospheric currents operating at the time of the survey.

Ionospheric disturbances driven from below produce very severe effects in

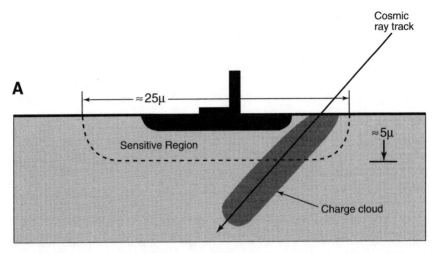

FIGURE II.4.10A Illustration of radiation effect; electron-hole generation near a sensitive region as a result of local ionization produced by a traversing cosmic ray or energetic particle.

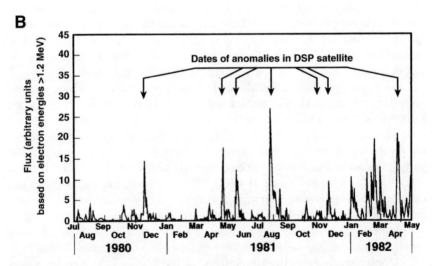

FIGURE II.4.10B Plot of the energetic electron flux in geosynchronous orbit, showing the correlation between flux enhancements and satellite anomalies. SOURCE: V.A. Joselyn and E.C. Whipple, Vampola, private communication, 1990.

two latitudinal bands that straddle the equator. These disturbances, which are called Appleton anomalies after their discoverer, were first photographed during the Apollo mission. Created by a fountain effect due to the geometry of the Earth's magnetic field, the anomaly zone often becomes extremely disturbed

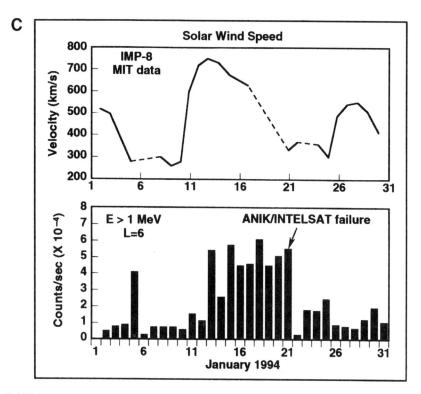

FIGURE II.4.10C Average solar wind speeds.

after sunset when severe convective storms disrupt the entire low-altitude zone. Since most of the world's people live in this region, such severe space weather must be understood and predicted for commercial and military purposes.

Critical Science Questions

The capabilities of our current space weather and climate prediction services are quite rudimentary compared to societal, commercial, and governmental needs. Because our basic understanding of fundamental physical processes is not well developed, integrated physical models do not currently exist at operational facilities, and many of the data required to drive these models are not available. Thus, the space weather science plan must address some basic questions that must be answered before adequate space weather support can be delivered.

As described above, major solar events have a profound effect on space weather in the vicinity of the Earth. In addition, there are effects driven by the quasi-stationary solar wind. Because the disturbances evolve as they propagate

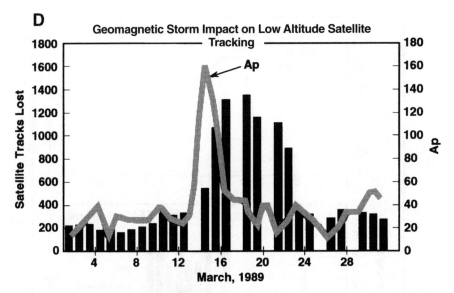

FIGURE II.4.10D Geomagnetic storm impact on low-altitude satellite tracking (Ap is the geomagnetic index). SOURCE: Chen et al., 1993. Reprinted with permission of the American Geophysical Union.

from the Sun to the Earth and the evolved structure affects near-Earth space, it is necessary to understand the physical processes that control this evolution.

- What are the fundamental causes of CMEs and flares, and what controls their properties (size, energy release, etc.)?
- Can we predict when CMEs and flares will occur, and are there specific precursor events that can facilitate prediction of the occurrence and effects of a given CME or flare?
- What determines the solar activity cycle and its amplitude?
- How can we predict the orientation and magnitude of the magnetic field and the plasma flow speed associated with transient solar events and their arrival time at Earth?
- How is the solar wind accelerated in coronal holes? Can we predict the size of the coronal hole and the speed of the flow?
- What factors control the characteristics of solar energetic particle events, and can we predict them on the basis of solar observations and numerical models?
- What is the most crucial factor in determining the overall effect of a CME on the space weather system: speed, mass, energy content or magnetic field?

In addition to understanding the causes and properties of space weather, a crucial element in progress toward quantitative and accurate space environment predictions is a comprehensive physical understanding of the global magnetospheric response to external variations and internal reconfigurations. The spatial and temporal evolution of the magnetospheric space environment is driven by and responds to variations in the solar wind and interplanetary magnetic field as described above. Consequently, the element that is crucial for progress toward quantitative space environment predictions of sufficient accuracy and specificity is a comprehensive physical understanding of the internal magnetospheric response to external variations. The dynamism of the magnetospheric space environment is governed by physical transfer mechanisms operative largely within the magnetospheric boundary layers, which separate interplanetary and magnetospheric regimes. Presently, the major physical processes at these interfaces are fairly well known; however, a comprehensive and quantitative measure of the relative role of each transfer process is not well specified, as a function of either space or time.

Magnetospheric space weather depends not only on the physics of the driver (solar wind) and of the mass, momentum, and energy filter (boundary layer processes), but also on the complex internal magnetospheric response to the external, filtered stimuli. These internal reconfigurations may be tightly coupled or may have a nonlinear response to the driver input. Regardless of the form of the response, most space weather effects are strongly connected with the most dynamic elements of geomagnetic storms and substorms. As such, there are specific scientific questions relevant to storms and substorms that require more complete answers than presently known. Each of these questions requires a quantification and sophistication beyond those presently available in order to improve space environment predictions. Questions include the following:

- What in detail are the coupled processes by which mass, momentum, and energy are transferred from the interplanetary medium to the magnetospheric system?
- What physical processes or boundary conditions differentiate geomagnetic storms from substorms?
- What determines the location of, and what process triggers, substorm onset?
- What are the mechanisms responsible for kilo-electron-volt particle energization (both in the ring current and in the auroral zone) as well as particle loss?
- What processes are responsible for million-electron-volt outer-zone electron modulation?
- What processes are responsible for the formation of new inner-zone radiation belts on the scale of particle drift times?

When discussing the electrodynamic response of the magnetosphere to external forces, it is imperative to recognize the importance of the coupled ionosphere and, to a lesser extent, the neutral atmosphere (inasmuch as it is coupled to the ionosphere by collisions between ions, neutral atoms, and molecules). First, we need improved global models of the bulk macroscopic properties of the atmosphere and ionosphere. We must also make progress in understanding their dynamic responses to the space weather effects discussed above. Unlike the magnetospheric environment, the ionospheric-atmospheric environment is responsive to solar variations both directly (e.g., photoionization) and indirectly (e.g., auroral joule heating). Both aspects are important and have several associated outstanding scientific questions, the former dependent primarily on photons and the latter on charged particles and fields.

Predicting ionospheric space weather and its effects both in space and on the ground requires, in part, accurate descriptions of electric fields and currents present in the ionosphere (at both auroral and subauroral latitudes). As noted above, the magnetosphere and ionosphere are tightly coupled electrodynamically and must be treated as a system. The critical issues of the magnitude and location of time-dependent ionospheric currents and electric fields thereby rely on the physics operative both locally and globally. At present, our understanding of the individual components is maturing; however, much work is needed to achieve the synthesis required to move to the next stage of physical understanding. Specific outstanding questions follow:

- Can we develop a predictive understanding of the connection between magnetosphere-ionosphere coupling processes and the production of ionospheric irregularities and scintillation (particularly for ionospheric disturbances associated with auroral bombardment ionization)?
- What physical processes determine the electrodynamic structure of the ionosphere during geomagnetic storms, magnetospheric substorms, and quiescent convection? How are high-latitude, ionospheric variations transferred to low latitudes? Are the processes predictable? Do data exist to evaluate them?
- How does the flux of solar ionizing radiation vary in time? Are direct measurements or proxy data available to provide accurate answers?
- What is the role of the ionosphere-upper atmosphere in magnetosphere-ionosphere coupling?
- What are the factors that control the day-to-day variability of severe low-altitude ionospheric disturbances?
- What is the role of atmospheric gravity waves in seeding severe ionospheric weather in the equatorial and anomaly zones, and can these waves be predicted?
- Progress is needed in numerical space; existing space weather models must be implemented so that deficiencies can be identified and rectified.

Another agent of ionospheric-atmospheric space weather effects, whose role is poorly quantified, is the very energetic charged particles that interact with ionosphere and upper atmosphere. Some ionospheric disturbances are thought to be initiated by the deposition of relativistic electrons and solar protons to low altitudes that enhance ionization and cause plasma instabilities. These instabilities lead to complicated ionospheric structure that can affect communications. Some of these energetic particles may even contribute to modifying mesospheric ozone indirectly through chemical and transport influences.

History and Current Research Activities

Over the past nearly 35 years of basic research, the solar-terrestrial and space physics communities have developed a broad empirical and theoretical understanding of solar-terrestrial relationships and the space environment through a balanced program of spaceflight experimentation, data analysis, and theory. This advance has been motivated largely by the intrinsic scientific merit of these studies. In the past decade or so, increased emphasis has been placed on applying this basic knowledge to societal concerns about the space environment both in the private sector and in several national agencies [e.g., Department of Commerce (DOC), National Oceanic and Atmospheric Administration (NOAA), Department of Defense (DOD), U.S. Air Force, Department of the Interior (DOI), U.S. Geological Survey, NASA, NSF, Department of Energy (DOE)]. As a result, the first numerical models are now being developed to specify, nowcast, and forecast the space environment. In response to this need, a National Space Weather Program (NSWP) is being formed through the coordinated efforts of many government agencies.

To see how the NSWP might evolve, it is instructive to compare first the field of space physics (specifically, space weather) with the development of atmospheric physics (in particular, dynamic meteorology). Since their inception, numerical weather prediction models have shown a steady improvement in both their accuracy and their specificity of tropospheric weather over the last 35 years. One standard figure of merit is the so-called S1 score (a measure of the 36-hour prediction of the geopotential height at 500 mbar), which when converted to a percentage accuracy has improved from approximately 28 percent predictive in 1956 to approximately 94 percent predictive in the early 1990s. This achievement was accomplished by a rigorous effort wherein each element supported and motivated the others: basic research, model development, model testing, application, data gathering, and assimilation. Throughout this effort, a strong customer base was established and continued to grow as forecast and specification capabilities improved.

The interagency NSWP is now at the same crossroads that confronted the meteorology community in the early 1950s. However, in at least one respect, speedier success might be anticipated because computational resources today are orders-of-magnitude more powerful and sophisticated than they were in the 1950s.

In addition, a broad customer base that recognizes the importance of space weather to its operations or products has been established over the past 30 years.

For significant improvement to be made in numerical space weather predictions, we must make progress on the same type of issues that confronted dynamic meteorologists 40 years ago. It is important to implement existing space weather models quickly so that deficiencies can be identified rapidly and remedied. Concurrently, a vigorous research program should continue to explore the basic physics of the comprehensive space environment, and new advances should be included in improved operational numerical models through the NSWP. Another necessary element for progress is the identification of critical input parameters and data needed for models and the development of experimental programs to provide these data. To summarize, the following efforts must be implemented that are supportive of a well-balanced space weather initiative (many of which are already ongoing):

• Establish new experimental spacecraft missions to provide the inputs needed to improve prediction capabilities, for example:

1. We cannot currently observe Earth-directed CMEs.
2. We cannot currently measure coronal magnetic fields.
3 We cannot currently measure near-Sun solar wind properties.

• Embark on comprehensive observing programs to elucidate the flow of mass, momentum, and energy from Sun to Earth.
• Establish a window on the polar ionosphere using remote sensing capabilities.
• Exploit existing data bases; update and improve existing empirical and statistical models to serve as climatological models.
• Develop improved physical models of components of the comprehensive system through focused research campaigns.
• Promote experimental, theoretical, and analytical research programs that are supportive of space weather needs.

Key Initiatives

A space weather program should achieve the following: (1) increase humankind's understanding of space weather processes and problems to a level high enough to implement numerical space weather prediction codes; (2) continually improve the capability to specify, nowcast, and forecast key aspects of the space environment; and (3) through a combination of items 1 and 2, mitigate the negative effects of space weather on human life and technology. To achieve these goals, progress must be made in five areas:

1. Develop and disseminate a better basic understanding of the relevant physical phenomena and processes.

2. Generate statistical models, based on comprehensive measurements, that specify the average space environment properties and the range of anticipated values (space climate).

3. Produce nowcasting capabilities that permit the instantaneous state of the magnetosphere to be described on the basis of specific near-real-time observations.

4. Build numerical forecasting capabilities that provide accurate predictions of space weather properties with enough advance warning to allow mitigating actions.

5. Evaluate mitigation strategies based on a synthesis of scientific understanding, engineering considerations, and operational guidelines.

Several specific initiatives must be pursued to accomplish these tasks:

- We must establish and foster communications between scientists and those affected by space weather, including industry, government agencies, and the public. A start in this direction is the NSWP currently being developed as a cooperative program of NSF, DOC, DOD, NASA, and other agencies (OFCM, 1995, 1997).
- We must identify and make the key measurements needed both for progress in the fundamental understanding of space weather processes and for use in monitoring and forecasting space weather conditions. A number of existing and planned satellite programs can contribute to this initiative (e.g., the International Solar-Terrestrial Program satellites and geosynchronous satellites operated by DOD and NOAA). Additionally, key measurements are required to image coronal mass ejections as they leave the Sun, and a secure, reliable source of real-time, continuous, upstream solar wind data must be established and maintained.
- Support must continue for the development of numerical models of the solar-terrestrial relationship chain, as well as their integration into full-scale predictive tools. NSF's Geospace Environment Modeling (GEM); Coupling, Energetics, Dynamics of Atmospheric Regions (CEDAR); and SUNRISE programs are directly targeted toward this effort (NSF, 1986, 1988, 1990).
- Models must be tested, evaluated, and verified continuously against space weather measurements.
- User-specific and user-friendly space environment products must be developed, including environmental specification models; educational tools for both the public and engineering users; and expert system design tools.
- Physical models must be translated into operational nowcasting and forecasting codes as soon as possible.

Qualitative measures of the success of this initiative fall into two basic categories, scientific and societal:

1. *Scientific:* The general measure of scientific success is assessment of the degree to which our basic understanding of the solar-terrestrial connection and the space environment has improved. Specifically, have we developed more accurate numerical models, predictions, and specifications of

- near-Earth interplanetary conditions,
- storm onset timing and magnitude,
- substorm onset time and location,
- magnetospheric particle flux profiles, and
- spacecraft orbit alterations due to upper-atmospheric changes?

2. *Societal:* The general measure of societal success is assessment of the degree to which increased basic understanding of the solar-terrestrial connection and the space environment is used to modify and improve applications that are of benefit to society at large, specifically:

- Have science-based applications or products been developed that are used to provide accurate determinations of the space environment?
- Are potential users optimally reaping the benefits of increased knowledge in this field (e.g., designs of environment-tolerant systems, optimizing of resource management, reducing asset risk)?

Over the past ten years, new technologies such as personal computers, lasers, and the telecommunications revolution have profoundly changed the day-to-day nature of all our lives in ways we could foresee only dimly at first. Presently, our lives are being reshaped by the emerging technologies of computer networking, cellular communications, and the GPS, to name a few. Increasingly, these new technologies will contain key space-based elements. Thus, success in achieving a better understanding of both space climate and space weather will have broad and profound effects. The mechanisms by which such benefits will accrue may not always be obvious, but given the routine reliance on space-based systems, they will be profound. Failure to address space weather issues will lead to equally profound but deleterious effects.

Direct economic benefits will accrue from the use of improved tools for spacecraft design, which generate more cost- and performance-effective space-based assets. Economic benefits will result from improved reliability of satellite, communications, navigation, and power systems. Management of space-based assets will be enhanced through improved environmental and orbital predictions, both for critical short-term operations and for long-term reliability. The radiation safety of Earth-orbiting astronauts and high-altitude aircraft crew and passengers will be enhanced. Furthermore, in the longer term, an increased understanding of the space weather environment will be needed as we approach the establishment of lunar colonies and the manned exploration of Mars.

Finally, the program outlined here will result in an increased and broader intellectual understanding of our environment, an environment whose boundaries have expanded and continue to expand upward from the ground, through the lower atmosphere early in this century, to the upper atmosphere, and into the near-Earth space environment. This program will provide a driving motivation to unify the diverse fields of solar physics, space physics, magnetospheric physics, and atmospheric physics using the tangible test of prediction as a metric to judge the success of our understanding.

MIDDLE-UPPER ATMOSPHERE GLOBAL CHANGE

The lower, middle, and upper atmospheres form a single, highly coupled physical system. In attempting to understand global changes in the atmosphere resulting from natural and anthropogenic influences, one should consider the changes in all of these regions, because if we cannot explain the changes occurring in all regions of the atmosphere, our understanding is incomplete. The middle and upper atmosphere undergo climatological changes that are often larger than those in the troposphere. Some of these changes are the result of natural variability in the UV and EUV radiation from the Sun; others are thought to be induced by anthropogenic effects. It is critical to understand the nature of these changes because of the importance of the ozone layer for terrestrial life, because of subtle influences of the stratosphere on tropospheric climate, and because of the impact of the upper atmosphere on space-based technological systems and radio telecommunications.

Scientific Background

Climatological change has already occurred in the middle and upper atmosphere. Some direct and indirect measurements of mesospheric temperatures have suggested a cooling on the order of 2-4 K per decade (Figure II.4.11) over the past 10 or 20 years, considerably greater than predicted by current models that consider the enhanced radiative cooling due to increased atmospheric carbon dioxide. The stratosphere has also cooled measurably owing to increased carbon dioxide and decreased ozone. Noctilucent clouds were virtually unknown before 1885 but are commonly observed today, and their frequency of occurrence is apparently increasing from decade to decade (Figure II.4.12). The cause appears to be partly the decreased mesopause temperature, but more importantly, the increased mesospheric water vapor concentration that is expected from observed increases in the atmospheric methane burden. Observations suggest that the density of exospheric hydrogen may also have increased substantially over the

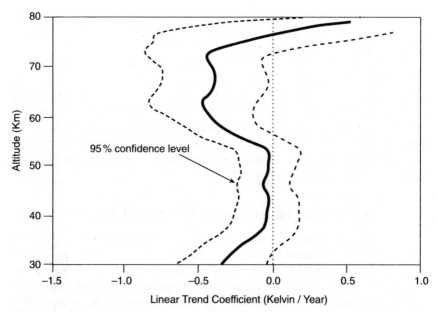

FIGURE II.4.11 Linear trend of the lidar temperature during summer months (data from 1979 to 1990).

past 20 years, even more than would be predicted by models based on the effects of increased methane. A quasi-biennial oscillation in the atmospheric semidiurnal tide has only existed since after about 1905.

One of the most dramatic changes resulting from human activities has been the growth of the springtime Antarctic ozone hole. This and other aspects of stratospheric ozone have been discussed earlier. Stratospheric radiative cooling by greenhouse gases can be expected to exacerbate the Antarctic ozone hole, and possibly lead to an Arctic ozone hole, by increasing the occurrence of polar stratospheric clouds that help catalyze ozone destruction.

The sensitivity of the middle and upper atmosphere to global change results in part from the relatively large changes that may occur in forcings from above and below. Far-ultraviolet radiation from the Sun, which is absorbed in the middle and upper atmosphere, is much more variable than visible solar radiation that reaches the ground. The upper atmosphere responds strongly to cyclic changes in solar ultraviolet and x-radiation. Weaker changes in upper-stratospheric temperatures and ozone concentrations are also related to the solar cycle. The upper atmosphere is influenced by auroral energy inputs that undergo large long-term, as well as short-term, variations. Infrequently, large fluxes of highly energetic particles penetrate into the mesosphere and stratosphere at mid- to high latitudes, producing changes in atmospheric photochemistry and ionization. The

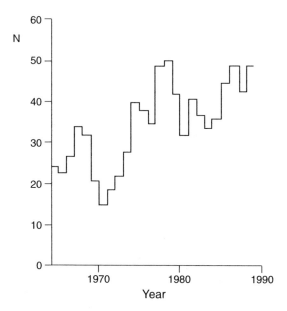

FIGURE II.4.12 Number of nights per year N on which noctilucent clouds were reported from northwest Europe for 1964-1988. SOURCE: Gadsden, 1990. Reprinted with permission of Elsevier Science Ltd., Oxford, England.

global electric circuit and the electric potential difference between the ionosphere and the ground can be directly affected by changes in global lightning activity, which may be very sensitive to surface temperatures. The concentrations of a number of anthropogenic gases that are important for physical and chemical processes in the middle and upper atmosphere have been increasing rapidly in recent years, especially methane, halocarbons, carbon dioxide, and nitrous oxide.

Besides being subjected to relatively large variability in external forcing, the middle- and upper-atmospheric regions respond very sensitively to a number of these forcing influences. The sensitivity of stratospheric ozone concentration to the presence of chlorine compounds, to stratospheric temperature changes, and to the presence of aerosols is one example. Another example is the greater cooling of the middle and upper atmosphere than heating of the troposphere, due to increased concentrations of carbon dioxide. Figure II.4.13 shows calculated changes in upper-atmospheric temperature and density for a doubling of CO_2 (solid lines), as well as for a halving of methane (CH_4) (dashed lines), the latter being possibly representative of a glacial maximum. The occurrence of clouds in the polar mesosphere is highly sensitive both to the temperature and to the increasing amount of water vapor associated with methane.

There are major gaps in our understanding of the interacting physical and

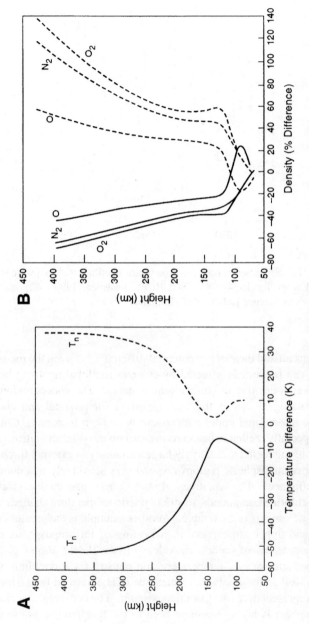

FIGURE II.4.13 Calculated changes from the base case to (A) atmospheric temperature T_n and (B) densities of atomic oxygen (O), molecular oxygen (O_2), and molecular nitrogen (N_2), when atmospheric carbon dioxide and methane are doubled (solid lines) or halved (dashed lines). SOURCE: Roble and Dickinson, 1989. Reprinted with permission of the American Geophysical Union.

chemical processes occurring in the middle and upper atmosphere; as a result, there are grave deficiencies in our ability to predict the nature, magnitude, and consequences of changes that may result from altered external forcings. For example, depending on the nature of the physical parameterizations used in three-dimensional models of the middle atmosphere, one can find either heating or cooling of the winter polar stratopause and summer polar mesopause in conjunction with doubled atmospheric CO_2. Unexplained electrical phenomena, such as optical flashes in the stratosphere and mesosphere ("jets" and "sprites") and strong electric fields in the ionosphere above the areas of electrical storms show how incomplete our knowledge is of electrical processes in the middle and upper atmosphere and of the way these may couple to lower-atmospheric electrical phenomena. Electrical discharges, the presumed source of optical flashes, may be an important and previously unappreciated means of moving charges between the troposphere and middle atmosphere, and may impact middle-atmospheric chemistry.

Large uncertainties exist in other areas of middle- and upper-atmospheric science. The microphysics involved in the formation of polar mesospheric clouds, such as the roles of meteoric dust and cluster ions, is not well known. The structures of turbulence and of mesoscale motions, and their effective roles in heat balance and in the transport of minor species in the mesosphere and lower thermosphere, remain poorly understood. The generation of gravity waves by orographic, convective, and baroclinic sources has not been quantified on a global basis, even though such waves are now known to play a critical role in middle-atmospheric circulation and the production of turbulence as the waves grow and break. The causes of variability in atmospheric tides and planetary waves are not fully understood, even though these global-scale waves can be the dominant form of dynamical variation in the mesosphere and thermosphere. Thermal balance in the upper atmosphere is strongly affected by nonlocal thermodynamic-equilibrium radiative processes that have been difficult to quantify in models.

Critical Science Questions

What physical processes determine the state of the middle and upper atmosphere? How are atmospheric regions coupled to those above and below? How do middle-atmosphere electrodynamics, heterogeneous chemistry, polar mesospheric cloud chemistry, wave-mean flow interactions, and turbulence affect the state of the mesosphere and its response to inputs from space and the lower atmosphere? How can these effects be incorporated into global models of the coupled system to predict short- and long-term variability?

What changes have already occurred in the state of the middle and upper atmosphere? What are the current and expected future trends? Is it possible to separate the responses of the middle atmosphere to natural, as opposed to anthropogenic, forcing by looking at characteristic time scales for the variability (i.e., the 11-year solar activity cycle)?

How are climatological changes in the state of the middle and upper atmosphere related to variability in the forcing of this region from above and below due to solar variability; the diffusion of trace gases; changing patterns of tides, gravity waves, and planetary waves propagating upward from the lower atmosphere; and electrodynamic coupling to higher altitudes?

How do short- and long-term changes in the state of the middle atmosphere impact lower altitudes (i.e., weather prediction, penetration of harmful UV radiation to the Earth's surface) and affect the near-Earth space environment and, through it, U.S. space assets (i.e., satellite lifetimes, space station reboost activities, aerospace plane operations) as well as other relevant technologies?

Key Initiatives

A combination of theoretical and observational initiatives are needed to address the critical questions posed above.

• *Analyze Historical Data:* Many long-term records exist that relate to the state of the middle and upper atmosphere, and analyses of these records have already determined certain climatological changes, such as the solar-cycle variations of the upper atmosphere and the growth of the ozone hole. The existence and magnitude of other possible changes (e.g., changes in the mean temperature and wind structure of the middle and upper atmosphere) are less certain. Careful analysis of historical data that are related either directly or indirectly to these possible changes is required to explore and determine their magnitudes. This knowledge is needed to test simulation models that might be used to predict future states. The analysis of historical data is often fraught with problems of changing data quality, calibrations, and measurement locations, as well as problems of data gaps, uncertain data locations, and data that are not in machine-readable form. In addition, some data types give only indirect information about the state of the middle atmosphere. For example, surface pressure records have been analyzed for tides from which inferences about the QBO could be drawn. The interpretation of these indirect data requires a good understanding of the interacting processes in the middle and upper atmosphere and the ability to accurately model these processes. Some examples of long-term data bases that might be useful in this analysis are

1. ionosonde observations since the 1940s of the charged particles in the ionosphere;

2. topside sounder observations from Alouette 1 and 2, and ISIS 1 and 2, that give historical information on the ionosphere above ~500 km altitude;

3. a variety of observations of the charged and neutral components in the thermosphere and above since about 1962 including satellite drag;

4 incoherent scatter data since the mid-1960s on a regular basis; and

5. magnetic variation data since the nineteenth century that contain information about the variability of atmospheric tides.

Figure II.4.14 gives a history of the number of ionosonde stations since they came into regular use in the 1940s. Figure II.4.15 is a time line from 1965 through 1989 (more than two solar cycles), indicating the satellites in low Earth orbit and the altitude range addressed by their instruments.

• *Monitor Sensitive Parameters of the Middle and Upper Atmosphere:* Clearly, long-term monitoring of middle- and upper-atmospheric parameters that may be sensitive to change, using well-calibrated techniques and ensuring the continuity of observations, will greatly simplify future studies of long-term trends. Because of the complex nature of spatial and temporal variations, it will be important to make long-term observations at many geographical locations. This can be accomplished most effectively by a combination of spaceborne observations, with the advantage of global coverage, and ground-based sensors needed for maintaining well-calibrated measurements over a long period of time. Medium-frequency (MF), mesosphere-stratosphere-troposphere (MST), incoherent scatter (IS), and meteor radars are particularly useful in observing the middle atmosphere and examining its coupling to higher-altitude regions. Because of the importance of the polar region, measurements in the Arctic are key components for studies in the polar middle and upper atmosphere. Lidars provide important information on the distribution of temperature and trace species. Improvements in sensor technology have enabled instruments to probe previously inaccessible spectral regions to obtain information on important chemically and radiatively active species in the middle atmosphere both from space and from the ground, opening up this region of the atmosphere to detailed study. Some of the parameters of greatest interest are temperature; winds; aerosols; concentrations of important constituents such as ozone, water vapor, hydrogen, nitric oxide, halogens, and the hydroxyl radical; heights and densities of ionospheric layers; and ionospheric electrical potential. One of the requirements for research is thus to establish and/or maintain long-term measurement programs for middle- and upper-atmospheric parameters that may be sensitive to change. We need to closely monitor the occurrence and latitudinal extent of polar mesospheric clouds as a marker of global change.

• *Monitor Inputs to the Middle and Upper Atmosphere:* To understand the causes of any climatological changes in the middle and upper atmosphere, we must know how the forcing has changed. Thus, it is critical to establish and/or maintain long-term programs to make stable, accurate measurements of parameters that influence the middle and upper atmosphere, including solar ultraviolet

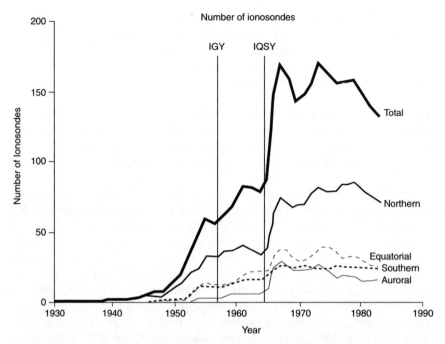

FIGURE II.4.14 Number of ionosondes operational since 1930. NOTE: AUR = number at auroral latitudes; EQU = number at equatorial latitudes; IGY = International Geophysical Year 1957-1958. SOURCE: Bilitza, 1991. Reprinted with permission of the Society of Geomagnetism and Earth, Planetary, and Space Sciences.

and x-ray fluxes, cosmic rays, auroral particles and fields, tropospheric trace gases, global thunderstorm activity, and injections of volcanic material. Some, but not all, of these inputs will be monitored in the Mission to Planet Earth satellite program.

- *Understand Uncertain Processes:* Knowing the trends of atmospheric parameters and of changing inputs will not be enough. A number of processes are occurring within these atmospheric regions that we cannot yet predict with any reasonable degree of confidence. We do not understand whether and how they may be significant to mechanisms of global change. We must aggressively pursue research on these poorly understood processes to determine the roles they may play. Areas in which understanding is particularly deficient include middle-atmospheric electrodynamics, heterogeneous chemistry, polar mesospheric clouds, wave-mean-flow interactions, and turbulence.

- *Understand and Model Interacting Processes:* Even though the nature of many atmospheric processes is reasonably well understood (e.g., basic dynamics, chemistry, radiation, and ionospheric electrodynamics), the manner in which

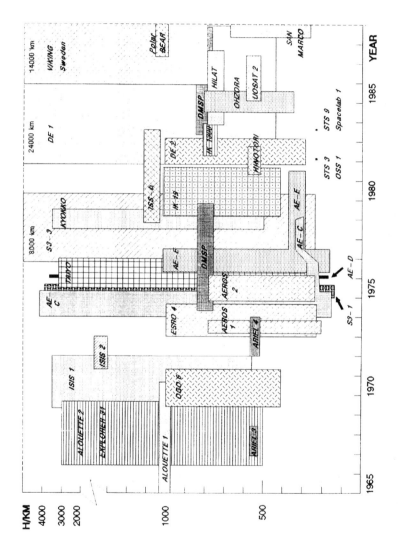

FIGURE II.4.15 Altitude-time chart of satellites measuring in situ in the ionosphere and thermosphere. NOTE: Alouette 1,2, ISIS 1,2, IK19, and ISS-b carried topside sounders measuring the electron density down to the F peak; ISIS topside ionograms were recorded up to 1989. SOURCE: Bilitza, 1991. Reprinted with permission of the Society of Geomagnetism and Earth, Planetary, and Space Sciences.

these processes interact in their response to changing forcing influences is extremely complex and not well understood. Thus, there must be a broad-based program of research with the goal of understanding the middle and upper atmosphere as an integrated physicochemical system, including the interactions with regions above and below. A critical part of this research program must be the development of general circulation models of the middle and upper atmosphere that incorporate all of the important physical and chemical processes, with the aim of continually reducing the need for ad hoc parameterizations so that valid predictive capabilities can be attained.

- *Distinguish Between Natural and Anthropogenic Effects:* One of the most important goals of research on global change in the middle and upper atmosphere is to determine the relative importance of natural and anthropogenic sources of the change, since it is only the latter that may be altered through policy decisions. The clearest distinction between natural and anthropogenic effects would come from detailed knowledge of the variability in all of the various forcing elements, together with an accurate, comprehensive modeling capability. However, even before this ideal situation is achieved, progress can be made by carefully analyzing the temporal and spatial characteristics of the different forcing functions and comparing these with the temporal and spatial characteristics of the atmospheric response. For example, solar radiation influences have a strong 11-year cyclical component that often helps identify these influences in the atmospheric response. Similarly, observed changes in the middle and upper atmosphere that are found to be most marked in recent decades may be associated with the rapid increase in certain anthropogenic gases. However, distinguishing the source of different effects merely by comparison of temporal trends will never be completely convincing, so that the development of detailed knowledge and comprehensive simulation models remain essential.

- *Understand the Consequences of Middle- and Upper-Atmosphere Global Change:* The uncertainties associated with global change in the middle and upper atmosphere concern not only the nature and magnitude of the changes that might be expected, but also the possible consequences of these changes on biological systems, on tropospheric chemistry and climate, and on space-based technological systems. Research is required to determine the relative importance of these consequences.

Contributions to the Solution of Societal Problems

Changes in the state of the middle and upper atmosphere may have a variety of impacts. Reductions in ozone concentrations can increase the intensity of solar ultraviolet radiation that reaches the troposphere and the ground. In addition to having biological effects, the increased UV radiation will alter tropospheric chemistry, including the atmospheric oxidizing capacity and hence the lifetimes of species such as methane. Altered middle-atmospheric structure can affect the propagation conditions of global-scale planetary and tidal waves, which in turn are likely to influence atmospheric circulation. The lifetimes, and thus atmospheric concentrations, of some long-lived greenhouse gases such as nitrous oxide and CFCs may be affected by changes in stratospheric circulation and solar UV intensity, the latter due both to solar irradiance variations and to altered absorption by ozone. Upper-atmospheric cooling will lead to reduced drag on spacecraft and space debris, increasing the lifetimes of both at a given altitude and thus affecting space operations and planning. The altitudes of the ionospheric layers might decrease, causing changes in high-frequency radiowave propagation conditions. Increased hydrogen densities in the exosphere may further affect ionospheric densities, as well as possibly increasing satellite drag and affecting the rate of loss of protons from Earth's radiation belts. Understanding the sources, nature, and magnitudes of these impacts will be necessary to plan mitigation strategies.

The ability to predict anthropogenic sources of possibly harmful influences on the state of the Earth's atmosphere at a stage early enough to allow intervention is a powerful tool for protecting the environment and preserving the quality of life. The middle atmosphere is a particularly sensitive indicator of perturbations to trace gases originating in the lower atmosphere that diffuse upward and disturb the sensitive balance in this region.

Measures of Success

An aggressive and successful research program will enable us to do the following:

- Identify changes in the state of the middle and upper atmosphere that have already occurred through careful analysis of well-calibrated historically available and targeted observations of atmospheric parameters and external inputs to the region.

- Achieve increased accuracy in predictive physical models by improving our knowledge, and thus accuracy, in representing physical phenomena that are poorly understood at present. Targeted areas include the effects of sprites and jets on the global electrical circuit and middle-atmospheric chemistry, the formation of aerosols in the stratosphere and mesosphere, the rates of important heterogeneous chemical reactions, and the role of wave-mean-flow interactions in the atmospheric circulation.

- Establish the nature of the relationship between changes in the middle and upper atmosphere and changes in the troposphere, the role of the global electrical circuit in climate, and the response of the middle and upper atmosphere to changed inputs.

SOLAR INFLUENCES

The Sun modifies the Earth's environment in ways that are both obvious and subtle. That the Sun's radiant energy is essential to life on Earth is obvious. The major component of this energy is steady and is taken to be a baseline constant for the average environmental conditions. However, the Sun undergoes a variety of small changes in its output, and the identification and measurement of the causes and effects of these variations are the focus for the study of solar influences. We note that the topics of shorter-term variations such as coronal mass ejections and solar flares are included in the discussion of space weather. The study of solar influences does not have a long history because solar variations are small and their potential impacts on the low Earth atmosphere are easily masked by the larger intrinsic variations of the weather system. Within the past few decades it has become possible to monitor solar variability through space-based observations, and most recently, large-scale observations of the Earth's upper and middle atmosphere have provided evidence of a terrestrial atmospheric response to the solar output. In building on this new data base, several activities that are poised for significant progress:

- Measure the solar energy output with space-based monitors continuously over at least a full solar cycle. Adequate accuracy in the measurement can be achieved only by simultaneous operation of two instruments in space. The changes in the Sun's irradiance are so small that they can be detected only as a variation of a single instrument. The transference of the absolute scale from an aging space-based instrument to its replacement requires that both be operating in space simultaneously. At present, no adequate time series of the full solar cycle is available because previous monitors of solar output ceased operation before their replacements could be placed in orbit so no calibration intercomparison was possible.
- Investigate the Earth's temperature sensitivity to variations in the solar energy output. Knowledge of this dependence is essential to separate global warming effects due to greenhouse gases from temperature changes caused by solar output variations.
- Determine the response of the Earth's middle- and upper-atmosphere chemistry and state of ionization to variations in the Sun's UV and x-ray emissions. Production of ozone and other gases, as well as the Earth's global electric circuit, depends in part on the Sun's hard x-radiation, which is produced in the complex outer solar atmosphere.
- Measure the Sun's interior dynamics, and develop a model of the solar dynamo that both agrees with the Sun's observed internal dynamical state and reproduces the pattern of solar magnetic activity. Helioseismology now provides knowledge of the Sun's interior dynamics, that is inconsistent with the assumptions required by previous models of the solar dynamo.

- Study possible long-term changes in solar behavior through the observation of solar-type stars. The Sun has undergone periods of decreased activity typified by the Maunder Minimum. Observations of solar-type stars can provide a statistical estimate of the future likelihood of such behavior through the large number of realizations available at any given time.

Solar Energy Output over a Solar Cycle

The total solar irradiance is the energy flux crossing a surface at the average distance between the Sun and the Earth. Since this parameter is the average rate at which radiative energy is provided to the atmosphere, it is fundamental to the Earth's environment. This energy flow was long assumed invariant; indeed, it was called the "solar constant" until recently. However, the possibility that the solar constant was actually variable remained a logical one that, in principle, could have a profound effect on the Earth's climate. Attempts to measure variations of the total solar irradiance from the ground were made by Abbott during the early part of this century but were thwarted by the variabilities of the Earth's atmospheric transparency.

Space-based monitoring of the Sun's total energy output, which began in 1978, established that the total solar irradiance undergoes changes on time scales of days to years. As the Sun rotates, both bright features (faculae) and dark features (sunspots) are carried across the visible face of the solar surface. These produce measurable changes of up to 0.5 percent in the solar irradiance that can be reproduced approximately based on the positions and strengths of the visible features. The changes associated with the solar rotation include at least two parts: deficits in energy output due to sunspots, typically of 0.3 percent, and enhancements in energy output due to faculae, typically by 0.08 percent. There may be other components that are widely distributed over the solar surface and do not show rotational modulation. When averaged over longer temporal periods that are a fraction of a solar cycle, the total irradiance shows a trend wherein it is greatest by 0.1 percent at the height of the solar cycle. Evidently, the widely distributed surface brightness enhancement (e.g., plages and faculae) is able to overcome the obvious darkening of the sunspots.

Figure II.4.16 shows a summary of the total solar irradiance measurements. The data in this figure are compiled from all available instruments. Only the sequence from the Earth Radiation Budget Experiment (ERBE) sensor on the NIMBUS-7 environmental research satellite is continuous, and it was not designed to provide calibrated long-term stability in its total solar irradiance measurement. The Active Cavity Radiometer Irradiance Monitor (ACRIM I and ACRIM II) instruments were designed for high precision and stability, but do not provide a continuous data set because UARS was launched after the reentry of the Solar Maximum Mission (SMM) spacecraft due to the pointing problems of the SMM spacecraft between late 1980 and spring 1984.

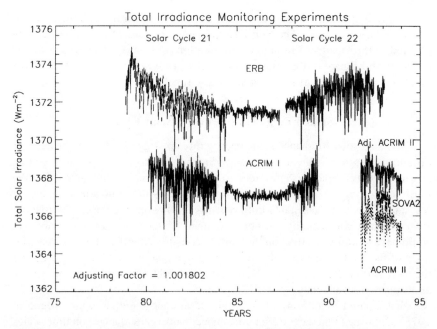

FIGURE II.4.16 Summary of current irradiance observations. ERB instrument on NIMBUS 7 is the only one present continuously. Solid line for ACRIM II has been adjusted by the factor indicated; the need for this adjustment emphasizes the importance of having overlapping space-based observations.

Instruments that monitor the total solar irradiance have great precision and stability but are difficult to calibrate on an absolute scale. In addition, these instruments often undergo an initial period of variation because of processes such as spacecraft outgassing and detector degradation due to solar radiation. Consequently, irradiance values observed with different instruments cannot be combined directly to maintain a long-term high-precision data base. In Figure II.4.16, only the upper curve from the ERBE radiometer on NIMBUS-7 spans the full period of a solar cycle. The other more precise and better-calibrated instruments have had to use the ERBE results to carry the time sequences across gaps in 1980, 1984, and 1989-1991. Parts of the ACRIM I curve and ACRIM II results shown in this figure have been multiplied by normalization factors based on ERBE results. The overall range of variation for the total solar irradiance from minimum to maximum solar activity has thus far not been determined with complete reliability.

To firmly establish the range of variation of the total solar irradiance, it is imperative that space-based monitors be deployed with adequate regularity to ensure overlap in their periods of observation. Cross-comparison is the only way to determine the longer-term variations reliably. Satisfying this imperative may require the development of small and easily deployed spacecraft that carry a basic solar irradiance monitor. The space station may also provide a platform for such monitors, although contamination of the space environment during shuttle visits could introduce accelerated periods of detector degradation that would reduce the effectiveness of this approach.

Separating Solar and Anthropogenic Effects

The question of global warming has been a major public issue during the past decade as measurements of greenhouse gas concentrations have shown a systematic increase capable of eliciting an important temperature response. As described in the previous section, the total solar irradiance also varies. The role of these variations in influencing the global temperature has not received as much attention as greenhouse gases for at least two reasons:

1. the climate system displays such a large natural variability that it can easily mask the effects of solar forcing, and
2. the amplitude of variation of the total solar irradiance is not known on the appropriate time scales.

Over longer periods, where increasing greenhouse gas concentrations should have a greater effect, there are no measurements of total solar irradiance. Over shorter periods, the rate of change in the climate forcing function during either the rising or the falling phase of the solar cycle is comparable to the rate of change in the forcing function due to changes in the greenhouse gas concentration.

The relative contributions of solar and anthropogenic effects to climate forcing are illustrated in Figure II.4.17, which shows estimates of combined effects of solar and anthropogenic variations. The quantities plotted here have all been estimated from very crude models. A similar figure based on sound data and theory would represent success in understanding the roles of solar and anthropogenic effects in the global climate. In evaluating the possible role of the Sun as a climate forcing function from historical records, we are limited by the lack of direct measurements of key quantities. The most extensive record is in the form of sunspot numbers. Figure II.4.18 shows a reconstruction of this parameter from the early 1600s to the present. It is noteworthy that even though the dominant variation is on an 11-year time scale, long-term trends are also evident. Possible correlation of solar activity with sea surface temperature anomalies on decadal time scales has been indicated, as shown in Figure II.4.19.

To compare the theoretical effects of solar and anthropogenic climate forcing to observed effects, we need to translate climate forcing into a temperature change. This parameter is referred to as the climate sensitivity coefficient, and

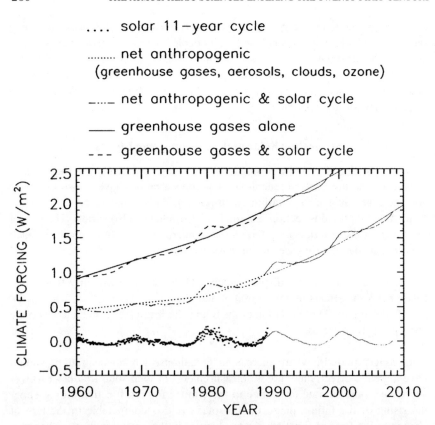

FIGURE II.4.17 Estimated climate forcings during the three recent decades of the twentieth century owing to measured changes in greenhouse gases (solid line); net anthropogenic forcing from greenhouse gases, aerosols, clouds, and ozone changes (dotted line); and solar irradiance variations associated with the 11-year solar activity cycle alone (lowest line). Combined greenhouse plus solar (dashed line) and net anthropogenic plus solar (dash-dot line) forcings are also shown. In each case, the thin lines are projections. The solar forcing is from the empirical model of Foukal and Lean (1990), which accounts for irradiance changes during the 11-year cycle caused by dark sunspots and bright faculae, but does not include additional variability sources acting on longer time scales. Zero point of solar forcing is the 1978-1989 mean. SOURCE: Hansen and Lacis, 1990; Hansen et al., 1993b. Reprinted with permission of Macmillan Magazines Limited.

current general circulation models (GCMs) give a number in the range of 1 K/(W m^2). Although a single number like this is easiest to quote, it is unlikely to represent the true situation where the time scale of change and the wavelength of the changing radiation are certain to modify the temperature response. Similar uncertainties affect the modeling of climate response to greenhouse gas concen-

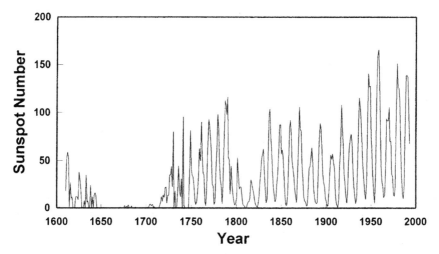

FIGURE II.4.18 Reconstruction of historical record of sunspot number including times of near absence of sunspots during the seventeenth century. SOURCE: Hoyt et al., 1994. Reprinted with permission of Springer-Verlag New York.

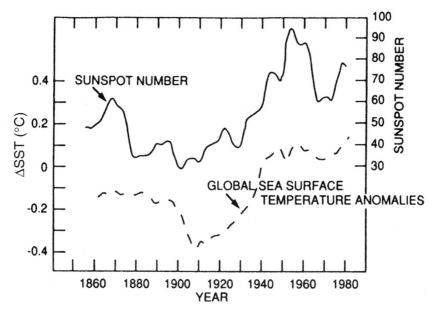

FIGURE II.4.19 The 11-year running mean of the sunspot number and global average sea surface temperature anomalies. In a one-dimensional model of the thermal structure of the ocean, consisting of a 100 m mixed layer coupled to a deep ocean and including a thermohaline circulation, a change of 0.6 percent in total solar irradiance is needed to reproduce the observed variation of 0.4°C in sea surface temperature anomalies. SOURCE: Reid, 1991. Reprinted with permission of the American Geophysical Union.

trations since each gas component interacts with the radiation flow through the Earth's atmosphere in a unique manner.

Our current knowledge of the sensitivity of climate to both solar and anthropogenic effects is limited by the difficulty in isolating the effect of a small-amplitude total irradiance variation from the intrinsic short-term variability of the weather system. Any atmospheric response to forcing function variations will become evident only after the weather (or natural climate) variability has been averaged out. Moreover a large amount of the energy in the climate system is stored in the Earth's oceans, and this will tend to smooth out the effects of small changes. Many of the records of long temporal duration refer to a limited geographical region and thus are especially vulnerable to local effects. The fact that the effects, if present, have to be extracted from a highly variable background has made it difficult to detect a signal identifiable with solar influences. However, longer-term effects may, in fact, be present in the system at a significant level.

Another critical problem in understanding the effects of solar variability comes from the greater variability in the Sun's ultraviolet (UV) output than of visible radiation. This problem is magnified because in those parts of the Earth's atmosphere where UV radiation is absorbed, its effects are dominant. The solar UV output is strongly dependent on the phase of the solar cycle and is often strongly modulated by solar rotation. These separate natural time scales for solar variations can be used as a tool for the identification of terrestrial responses although neither time scale is ideally suited for study. The shorter periods are well suited to solar observations but easily masked by terrestrial variability. The longer periods have less reliable records for solar output and terrestrial response, but the amplitude of response should be largest. In addition, solar UV variability is not well described by a single parameter since it is affected by the details of active region size, strength, and position on the solar disk, as well as by the strength of the widely distributed magnetic network. Thus, a multivariate analysis is required in principle even though the existing single-variate analyses have produced only tentative correlations.

Progress toward the goal of separately measuring climate sensitivity and its response to both solar and anthropogenic forcing variations can be made by fully utilizing the opportunity provided by UARS data, which measure simultaneously the solar inputs and the atmospheric response. This UARS opportunity is of limited duration and has sampled only the declining phase of the solar cycle. The largest and strongest sunspots typically appear during the rising phase of the sunspot cycle, and the effect of sunspots on UV and EUV fluxes has not been monitored and studied in detail yet. An extended UARS mission or future space-based observations could provide the missing observations during the rising phase of the next sunspot cycle.

Progress in the study of longer-term variability requires a separate approach. Current space-based measurements provide high-quality data that address shorter time-scale variations but cannot help with the study of longer time-scale problems. Historical reconstructions of the atmospheric and ocean thermodynamic state, atmospheric composition, and details of the solar output are required to disentangle the interrelationships on longer time scales. These reconstructions will build on the base provided by the verification from UARS and future spacecraft observations. Although such reconstructions based on proxy models are critical to the analysis of historical data, they will require verification by future direct observations and cannot be used as replacements for such observations.

Solar Influences on the Earth's Upper and Middle Atmosphere

The concentration of chemical constituents in the atmosphere is affected by photodissociation and photoionization processes, and hence by the solar flux that penetrates the Earth's atmosphere. Since the shortest wavelengths, which are most affected by solar variability, are absorbed in the highest layers of the atmosphere, the chemical response of the atmosphere is expected to be greatest at high altitudes (i.e., in the thermosphere and mesosphere). However, species such as ozone in the stratosphere, which is produced from the photolysis of molecular oxygen, are also sensitive to changes in extraterrestrial solar flux.

Solar activity has a direct impact on the Earth's ionosphere. Substantial increases in solar EUV radiation and x-rays, associated with enhanced solar activity, lead to substantial increases in the concentrations of ions and electrons in the D-, E-, and F-regions of the ionosphere. In the stratosphere, the largest ionization source results from the penetration of galactic cosmic rays, which is modulated by the solar cycle; hence, the stratospheric ionization rate is reduced during periods of high solar activity.

The abundance of neutral species is also affected by solar activity in the middle and upper atmosphere. For example, the concentration of nitric oxide, a constituent produced by ionic and photolytic processes in the thermosphere, is significantly enhanced in this region of the atmosphere during high solar activity. In the mesosphere, significant variations associated with solar variability affect the concentration of water vapor, a molecule that is photolyzed by shortwave ultraviolet radiation. Finally, in the case of ozone, the response is significant and results from the combination of several processes. In the stratosphere and the thermosphere, the ozone concentration increases with solar activity as a result of enhanced photolysis of molecular oxygen. In the mesosphere, the ozone response is dominated by the enhanced ozone loss caused by hydroxyl (OH) and hydroperoxyl (HO_2) radicals produced by a more vigorous photolysis of water vapor during high solar activity. The resulting change in the vertically integrated ozone concentration (column abundance) over a solar cycle is not greater than 1 or 2 percent.

As middle-atmosphere heating results primarily from the absorption of solar ultraviolet radiation by ozone, the temperatures of the stratosphere and mesosphere are also affected by solar activity. Amplitudes of the temperature variation on the 11-year time scale have been inferred from ground-based lidar and from satellite observations. Temperature amplitudes derived from atmospheric models are not in agreement with values deduced from observations. A major scientific question that remains unsolved is the potential dynamical response of the atmosphere to the 11-year solar cycle. Although substantial changes in dynamical patterns within a period of 11 years have been reported in the stratosphere and even in the troposphere on the basis of statistical analyses, no mechanism has yet been identified to explain these variations, which cannot be reproduced by atmospheric models.

Although evidence for the response of chemical compounds such as ozone to the 11-year solar cycle is provided by long-term observations, definitive quantitative response has not yet been established experimentally, because of insufficient precision in the data and the limited lifetime of the instrumentation. The observational evidence is much better established for the ozone and temperature response on the 27-day time scale, where analyses of satellite observations and model calculations are in fairly good agreement.

The direct measurements of the solar UV and x-ray irradiance provided by current satellites allow study of the current atmospheric composition, but comparable observations are not available for other time periods and may not be available in the near future. Consequently, it is important to be able to model and reproduce these irradiances based on observations of other solar parameters. These quantities are mapped on a regular basis so their positions on the solar disk can be used in the models. Other integrated measurements (e.g., the 10.7 cm flux), which indicate the strength of various integrated UV, EUV, and x-ray fluxes, are needed for middle- and upper-atmospheric chemistry studies but cannot provide definitive measurements with the necessary precision.

The need for detailed knowledge of the distribution and strength of the activity comes from the fact that solar ultraviolet radiation is produced in regions on the solar surface where magnetic fields are concentrated. Often these have sunspot groups at their center, but sometimes the area of higher-than-average magnetic field is a remnant of previous sunspots. An individual sunspot typically is identifiable for a period of up to 30 to 60 days. However, there are regions of enhanced activity that can persist for one to two years. It is common for the solar surface to be covered very unevenly by sunspots so that one hemisphere will emit a high level of ultraviolet radiation while the other hemisphere is very quiet. This configuration produces a strong rotational modulation to the solar UV flux.

The data bases of adequate direct measurements for both the total solar irradiance and the solar UV irradiance are limited to the last 10 to 20 years. Prior to this, the state of the solar output had to be deduced from proxy information. The most readily available proxy is the one shown in Figure II.4.18—the sunspot number. This index is based on the visible distribution of sunspot area and position and does not take into account the more widely distributed magnetic field, which is typically associated with sunspot groups. Other regions of enhanced fields are sometimes found unassociated with sunspots and indeed can be at higher latitudes than sunspots. They are also present during times of sunspot minimum. To fully evaluate the state of the Sun, more than one proxy parameter is required. Current models of the solar UV irradiance include components due to the quiet Sun, sunspots, active regions, and a fourth widely distributed component that comes from linear boundary regions between large convection cells.

Physical Basis of the Solar Activity Cycle

Fundamental to the question of solar influences on the Earth's environment is the occurrence on the Sun of an 11-year cycle of magnetically driven activity in the form of sunspots, solar atmosphere temperature changes, and unstable eruptions. Because of terrestrial responses to solar activity, the functioning of the solar cycle can impact society. The occurrence of periods of low solar activity, such as those shown in Figure II.4.18, indicates that the solar cycle must involve some complex nonlinear processes that affect solar irradiance with or without magnetic activity. If the periods of low activity coincide with periods of low total solar irradiance, then entry of the Sun into a new quiet period could produce global cooling. Indeed, the previous low period of the Maunder Minimum coincided with a time of unusually low temperatures in Europe, sometimes referred to as the Little Ice Age. Figure II.4.20 illustrates this relationship. Without a fundamental understanding of the solar cycle, neither the probability of such future behavior nor the occurrence of possible precursors can be recognized. Should there be a change in the apparent behavior of the solar cycle at some time in the future, we would want to know if this signaled the onset of a Maunder Minimum period or some less significant statistical variation.

The most prominent indicators of solar activity are sunspots. Within these spots the temperature is much lower than that of the surrounding atmosphere, and the emergent flux of visible radiation is substantially reduced. Evidently the convective motions that bring energy from the Sun's interior to the surface elsewhere are suppressed in the spots, and the reduced spot temperature results from the absence of an efficient process to replace the radiation emitted into space. Sunspots typically occur in pairs of opposing polarity. Each spot pair in the Sun's Northern Hemisphere has one east-west orientation and those in the Sun's Southern Hemisphere have the opposite orientation. This configuration is naturally interpreted in terms of a source toroidal magnetic field, with each sunspot pair being an arch that breaks the solar surface. The direction of the field within each torus is opposite in the Sun's Northern and Southern Hemispheres. In addition, there is a weak background solar polar field. The directions of the two toroidal fields and the weak polar field all reverse every 11 years. In addition, the Sun rotates differentially, with an inertial period of 24 days at its equator and 35 days in the polar regions.

Although the above sequence of observed changes through the solar cycle has

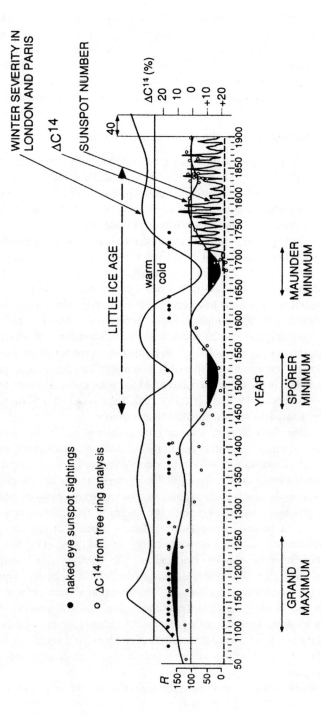

FIGURE II.4.20 Relationship between severity of winter in Paris and London (top curve) and long-term solar activity variations (bottom curve). Shaded portions of curve denote times of Spörer and Maunder Minima in sunspot activity. Dark circles indicate sunspot observations by the naked eye. Details of solar activity variation since 1700 are indicated in the bottom curve by sunspot number data. Winter severity index has been shifted 40 years to the right to allow for cosmic-ray-produced ^{14}C assimilation into tree rings. SOURCE: Daddy, 1976. Reprinted with permission of the American Association for the Advancement of Science.

been known for many years, there is as yet no successful theory of the process. The principal components of the solar activity cycle must come from the interaction of convection and rotation. The pattern of differential rotation should be derived from fundamental hydrodynamic theory but usually is just postulated as part of the input to a model. More importantly, the pattern of internal rotation has only recently become known, at least in a preliminary fashion. This was a critical free parameter, and it was expected that the Sun rotated more rapidly below the surface than at the surface. Such a pattern made it possible to reproduce the solar cycle, but that pattern is now known to be incorrect. Speculation about the driving region for the solar cycle has shifted from the zone just below the solar surface to the interface between the convection zone and the radiative deep interior at a distance of about 70 percent of the way from the Sun's center toward the solar surface. These theoretical ideas are at a very primitive stage of development and have not even been able to reproduce such essential aspects of solar activity as the 11-year period, the direction of sunspot migration, or sunspot size.

Reproduction of the most basic features of the solar cycle must be the first objective of any modeling effort, but this is only a step toward a more urgent goal: understanding the mechanisms or indicators of changes in the Sun's overall level of activity as measured by the strength of the solar cycle. Two historical changes in the cycle strength—the absence of activity during the Maunder Minimum (and similar earlier minima) and the growth of the cycle strength during the past century—are even further from being understood than the cycle itself but have potentially substantial climatic implications. Because these changes have a very long time scale, high-quality data from the most recent space-based era provide little guidance. Hints in the historical record from the late Maunder Minimum period indicate that there were changes in the Sun's rotation pattern or radius associated with the low level of output. Perhaps most intriguing in guessing the nature of the activity cycle during this low period was the chaotic nature of the first few cycles as the Sun recovered. The 11-year period did not manifest itself until nearly 50 years after the first moderate number of spots were seen. There also seems to be a gradual shortening of the cycle length to 9.5-10 years instead of the nominal 11.

The new tool of helioseismology represents the best hope of making fundamental progress in understanding the processes that govern the solar cycle. With this tool, it has become possible for the first time to measure velocities below the solar surface. Both the largest-scale motions involving flows over the entire convective envelope and the smaller-scale flows associated with active regions are accessible with this tool. By making such measurements over a full solar cycle, it should be possible to obtain clues as to the origin and nature of the solar dynamo. Two major experiments in helioseismology have recently begun with the deployment of GONG (Global Oscillation Network Group) instruments at six sites around the surface of the Earth and with the launch of three helioseismology experiments on the SOHO (Solar and Heliospheric Observatory) spacecraft. The

GONG instruments already are beginning to return data of high quality and maps of internal rotation.

In addition to providing a tool for understanding solar interior dynamics, helioseismology may be able to assist with long-range projections of space weather. Active regions and magnetism ultimately depend on the dynamics of the deep solar interior. The oscillation frequencies and their shifts are dependent on the interior velocity field and on the interior structure, including possible strong magnetic field effects. Thus, helioseismology data have the recently demonstrated capability to detect magnetized regions before they appear at the solar surface. On at least one occasion, precursor changes in the solar acoustic spectrum were measured prior to the arrival of a sunspot group on the solar surface. Similar changes are not seen in control regions where sunspots did not appear. This sequence is shown in Figure II.4.21. This type of observation may provide a means of long-range solar activity forecasting. Additionally, there may be relationships between the Sun's acoustic spectrum and the coronal magnetic configuration. This area is completely unexplored at present. Both the GONG and the SOHO experiments have planned durations of two years with possible extensions to longer periods.

Long-Term Changes in Solar Behavior: Solar-Type Stars

The study of solar variability through observations of the Sun is limited in two ways: there is no easy way to extend the time base beyond the current era, and there is no way to change parameters such as the rotation rate that govern solar dynamics. The study of Sunlike stars can ease these problems by sampling a range of states not exhibited by the Sun because of the natural range in stellar rotation rates. We pay two prices for these benefits: the properties of the stars are not fully and accurately known, and there is no way to obtain spatially resolved information about the distribution of activity over the stellar surface (although Doppler imaging can provide some information of this type). Estimates of stellar age are most difficult to obtain and represent the greatest uncertainty in this technique, with rotation rate being adopted as the best available indicator. Stellar observations consist of two parts: (1) a regular measurement of the strength of the ionized calcium emission (at the H and K wavelengths) and (2) a regular measurement of broadband stellar brightness. The longitudinal asymmetry in the distribution of active regions is found for stars as well as for the Sun, so that it is routinely possible to measure the rotation rate for stars from the pattern repeat rate in brightening of the ionized calcium features. More stars have been followed by using calcium emission features than broadband photometry. The set with extensive enough data in both ionized calcium and broadband contains just 10 stars. By adding the Sun to this set, there are 11 stars.

An important question that can be addressed with this stellar sample is whether the amplitudes of the solar chromospheric and total irradiance variation

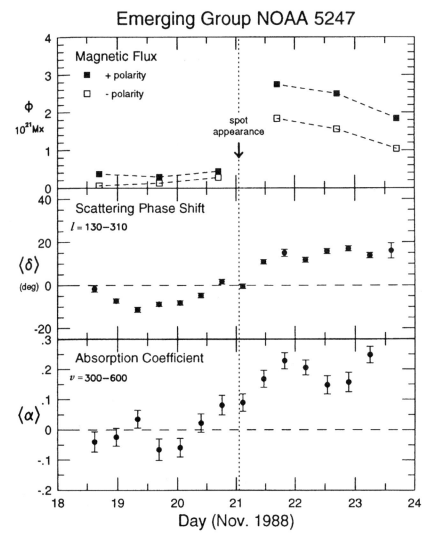

FIGURE II.4.21 Time evolution of magnetic flux and p-mode scattering for the emerging sunspot group NOAA 5247. Top panel shows magnetic flux as measured from KPNO magnetograms versus time. Middle and bottom panels show l- and v-averaged values of scattering phase shifts δ and absorption coefficient α, respectively. Time of appearance of the spot is indicated by vertical dashed line. Negative phase shifts in middle panel indicate a signature of p-mode scattering prior to emergence of the sunspot group. Positive phase shifts observed in the emerged spot are consistent with previous measurements of phase shifts in other (mature) sunspots.

are typical for stars of its type. Figure II.4.22 illustrates the relationship between the amplitude of variation in these two quantities. This figure shows that

- the Sun is near the low range of variability in its ionized calcium index, and
- the correlation between brightness variation and variation in the ionized calcium index is somewhat atypical for the Sun in the sense that the broadband variation is less than the ionized calcium index variation for the Sun. Interpreted literally, this implies that the Sun could in fact have had a much higher level of change in its energy output than has been observed recently.

Data of the type plotted in Figure II.4.22 can be obtained only through long-term studies. At present, the number of stars for which an adequate set of observations has been obtained is very small and does not permit any statistically significant conclusions. One difficulty in the use of stellar data is the estimation of stellar age. A larger sample of stars would permit better determination of the age through statistical use of the rotation rate, spectroscopic characteristics, and stellar motion indicators.

Key Initiatives

Solar influences on the Earth's environment are subtle and require careful measurement to be detected reliably. Nonetheless, as its ultimate energy source, the solar input is fundamental to the Earth's atmosphere and climate system, and understanding solar influences on the Earth's environment requires that we do the following, which has been discussed earlier in more detail.

- Measure the solar energy output with space-based monitors continuously over at least a full solar cycle.
- Investigate the sensitivity of the Earth's temperature to variations in the solar energy output.
- Determine the response of the Earth's middle- and upper-atmosphere chemistry and state of ionization to variations in the Sun's UV and x-ray emissions.
- Measure the Sun's interior dynamics, and develop a model of the solar dynamo that both agrees with the Sun's observed internal dynamical state and reproduces the pattern of solar magnetic activity.
- Study possible long-term changes in solar behavior through the observation of solar-type stars.

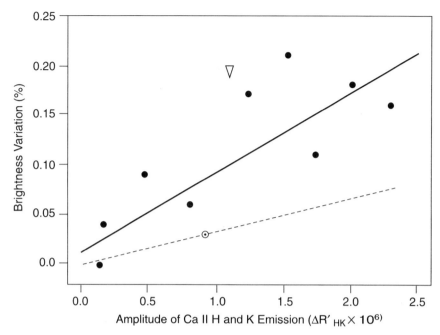

FIGURE II.4.22 Possible relationship between the total brightness variability of the Sun and solar-type stars as measured by $\Delta R'_{HK}$, which depends on chromospheric emission of Ca II H and K spectroscopic lines. Chromospheric emission is sensitive to magnetic activity, whereas the total brightness variation includes sunspot blocking, chromospheric brightening, and other less well understood effects. Quantities shown are root-mean-square variations; peak-to-peak variations are roughly three times larger. Position of the Sun as indicated by ⊙ is taken from SSM measurements and the solar $\Delta R'_{HK}$. Dashed line defines the solar brightness chromospheric activity change ratio based on yearly averaged data of SSM and NSO from 1980 to 1988. Inverted triangle (∇) is longer-term upper bound of the solar total irradiance variation from 1967 to 1984 taken from rocket and balloon measurements; corresponding value of $\Delta R'_{HK}$ is estimated from combined solar measurements of MWO (1967-1978) and NSO (1976-1984). Solid line is linear regression using all data except the upper limit for the Sun. Most noteworthy in this figure is the fact that the variability of solar total output seems to be less than that for other solar-type stars with similar chromospheric activity. SOURCE: Soon et al., 1994. Reprinted with permission of Springer-Verlag New York.

Contributions to the Solution of Societal Problems

The program of research described above will lead to greater understanding of the nature of solar variability, and will improve our ability to predict future states of the Sun. Understanding of the way in which solar variability affects the Earth and its climate will be enhanced, and we will have more confidence in our ability to distinguish anthropogenic effects from effects caused by solar influences.

PART II

5

Climate and Climate Change Research Entering the Twenty-First Century[1]

SUMMARY

Climate is variable on time scales of seasons to centuries and over longer time intervals. Both climate variability and climate change can have significant societal impact. Climate influences agricultural yields, water availability and quality, transportation systems, ecosystems, and human health. Climate variability and change are a product of external factors such as the Sun, complex interactions within the Earth system, and anthropogenic effects. The mission of climate research is to understand the physical, chemical, and ecological bases of climate in order to characterize and predict the nature of climate variability from seasonal and interannual to decadal and longer time scales, and to assess the role of human activities in affecting climate and of climate in influencing human activities and environmental resources.

A central goal of climate research is prediction. The objectives are to understand the mechanisms of natural climate variability on time scales of seasons to centuries and to assess their predictability, to predict the future response of the

---

[1] Report of the Climate Research Committee: E.J. Barron (Chair), Pennsylvania State University; D. Battisti, University of Washington; R.E. Davis, Scripps Institution of Oceanography; R.E. Dickinson, University of Arizona; T.R. Karl, National Climatic Data Center; J.T. Kiehl, National Center for Atmospheric Research; D.G. Martinson, Lamont-Doherty Earth Observatory of Columbia University; C.L. Parkinson, NASA Goddard Space Flight Center; S.W. Running, University of Montana; E.S. Sarachik, University of Washington; S. Sorooshian, University of Arizona; K.E. Taylor, Lawrence Livermore National Laboratory; P.J. Webster, University of Colorado.

climate system to human activities, and to develop improved capabilities for applying and evaluating these predictions.

The climate research of the past few decades drives the requirements for future research by focusing our attention on the remaining uncertainties and on the importance of climatic research for society:

- Climate variability, such as El Niño, can be characterized by significant economic and human dislocations. Modeling studies over the past two decades suggest that aspects of this climate variability may be predictable. In cases where El Niño/Southern Oscillation (ENSO) events were predicted in advance, immediate practical benefits were realized through human response and adaptation.

- Analyses of historical records have revealed a number of interesting cases of longer-period fluctuations for North America and other parts of the world, while model studies have demonstrated that ocean-atmosphere and land-biosphere-atmosphere interactions are plausible mechanisms to explain decade-to-century variability. Historical and paleoclimatic data, as well as coupled models, indicate the potential for significant climate variability on long time scales. Such changes can be expected to occur in the future, irrespective of human impacts on climate. Current observational capabilities and practice are inadequate to characterize many of the changes in global and regional climate. An enhancement of current observational capability and improved knowledge of the coupled Earth system will therefore likely increase our understanding of climate variability on all time scales and lead to a greater realization of practical benefits.

- The effort to predict the climate response to increases in greenhouse gases has both demonstrated the importance of this problem to society and focused attention on many of the most important limitations of current climate models. Increased concentrations of greenhouse gases and changes in land use and land cover are directly and indirectly tied to human activities. Current model projections based on increases in greenhouse gases and aerosols and on land cover change indicate the potential for large, and rapid, climate change relative to the historical and paleoclimatic records, with concomitantly large influences on human activities and ecosystems. Although remarkable progress in developing these climate models has occurred over the past two decades, current climate models are characterized by a great number of uncertainties. Improved predictive capability is likely to have a positive impact on economic vitality and national security because of its potential to minimize risk and maximize benefit associated with the impacts of any climate change.

A comprehensive analysis of the remaining scientific questions and uncertainties and of the societal drivers for climate research leads us to four major imperatives for the twenty-first century. Each imperative is associated with a series of basic requirements:

1. We must work to enhance current observational capabilities and to build a permanent climate observing system.

• Where feasible, adopt consistent data collection and management rules to ensure the utility of operational and research system measurements for climate research.
• Develop and adopt interagency plans to ensure the protection of critical long-term observations, to limit gaps in continuity due to small budget changes in single agencies, and to recognize the value of these observations in a balanced, integrated research program.
• Provide strong U.S. support and participation in the development of a global climate observing system (GCOS).
• Ensure full and open international exchange of data and information.
• Maintain major research observation systems, such as the Tropical Ocean Global Atmosphere (TOGA) Tropical Atmosphere Ocean (TAO) array, that have demonstrated clear predictive value.
• Focus on key opportunities for reducing major uncertainties in climate models, including improved observations of water vapor.
• Ensure full interagency commitment to both the in situ and the satellite observations necessary to address the major uncertainties in our understanding of the climate system, including a commitment to long-term Earth observations of critical variables such as the major climatic forcing factors.

2. We must extend the instrumented climate record through the development of integrated historical and proxy data sets.

• Widely sample the alpine glaciers and ice caps before this important repository of information on natural variability is lost.
• Continue efforts to collect and analyze data from around the world from tree rings, lake sediments, corals, and ice cores, and actively pursue high-resolution records from ocean sediments.
• Focus research efforts on the development and validation of proxy indicators.

3. We must continue and expand diagnostic efforts and process study research to elucidate key climate variability and change processes.

• Enhance cross-disciplinary communication and collaboration.
• Develop clearly articulated linkages between strategies for observation, analysis, model development, and application of predictions to evaluating consequences of climate change.
• Implement focused research initiatives on processes and in regions that are identified as important in understanding variability in the climate system.
• Implement and analyze new observations necessary for understanding the

processes that couple the components of the Earth system and improve our understanding of climate variability on decade-to-century time scales.

• Develop focused process studies with the objective of addressing key uncertainties associated with boundary layer processes and vertical convection; improved linkages coupling the atmosphere, oceans, and land surface; and more explicit representation of land surface processes, including vegetation and soil characteristics.

• Support the development and implementation of a comprehensive research program to study and advance seasonal-to-interannual prediction. Such a program is currently the objectives of GOALS (Global Ocean-Atmosphere-Land System) of the World Climate Research Programme (WCRP).

• Support the development and implementation of a comprehensive research program to study the mechanisms of decadal-to-century variability and its implications for longer time-scale predictability. Currently, the planning for this element is incorporated in the Dec-Cen (study of climate variability on decadal-to-century time scales) and anthropogenic climate change components of the WCRP.

4. We must construct and evaluate models that are increasingly comprehensive, incorporating all major components of the climate system.

• Improve opportunities and enhance efforts at model observation and model-model comparisons that pay particular attention to simulating observed changes associated with solar irradiance, aerosol loadings, and greenhouse gas concentrations.

• Develop mechanisms that promote formal interaction between physical scientists and social scientists, by working on common problems to improve the applications and assessments of climate change impacts.

• Enhance the computational infrastructure and focused efforts to develop climate system models that include explicit representation of the atmosphere, ocean, biosphere, and cryosphere.

• Focus on key opportunities for reducing major uncertainties in climate models, including greater understanding of climate-water vapor feedbacks and improved representation of atmospheric chemistry and indirect chemistry-climate interactions.

• Focus effort on improving the credibility and usefulness of climate model predictions at spatial scales relevant to analysis of the responses of ecosystems, socioeconomic systems, and human health to climate change predictions.

• Develop and construct high-resolution, regional climate models along with empirical methods for producing estimates of climate change characteristics of immediate relevance to humans.

These four imperatives offer a general framework, while the specific objectives and requirements for each characterize more specific opportunities to promote significant advancement in climate and climate change research. To some, the list of requirements outlined above may appear overly ambitious and without priority. However, a comprehensive climate research program that serves societal needs is clearly within our grasp. In many cases, the programs required to achieve these objectives are in place. In other cases, changes in requirements can be implemented with minimum budgetary impact. In still other cases, objectives can be fulfilled by increased collaboration and closer interagency planning and linkages. However, even some of the more logical, minimal-impact issues appear to be problematic. For example, in terms of the requirement for continuity and quality as part of the climate observing system, current policies verge on becoming a national and international embarrassment. Addressing these issues must be a priority. Finally, with careful planning to achieve greater efficiencies, the full spectrum of climate objectives should be realizable. Although each of the listed requirements has substantial merit, we recognize that improvements and augmentations of the U.S. climate research programs must still be paced, based on budgetary and other considerations. Consequently, the requirements described above are placed in a prioritized framework in the remainder of this Disciplinary Assessment. This prioritized framework is based on a relatively simple perspective. Improvements that have minimal budgetary impact but substantial merit should be implemented without hesitation. Requirements with significant programmatic or budgetary implications should have identifiable levels of priority or clear trade-offs with current efforts.

INTRODUCTION

Three general categories of climate variability and change have been adopted by the World Climate Research Programme: seasonal-to-interannual climate variability, decadal-to-centennial climate variability, and changes in global climate induced by the aggregate of human activities that change both the concentrations of greenhouse gases and aerosols in the atmosphere and the pattern of vegetative land cover. Humans, as individuals and societies, and ecosystems are affected by and respond to each of these three categories of variability and change.

Useful predictive skill for seasonal-to-interannual climate variability has been demonstrated. Moreover, early indications of human influence on global climate warming are emerging from the background of natural climate variability. The possibility that human activities have the potential to modify natural climate variability and long-term climate trends on a global scale is a research issue of high priority. Results of such research will have very high utility for informing the public and decision makers of appropriate response strategies.

Climate is defined as the long-term statistics that describe the coupled atmosphere-ocean-land weather system, averaged over an appropriate time period. For example, the averaged daily mean, minimum, and maximum temperatures recorded for a given month at a specified place are some important manifestations of climate. Likewise, the daily average hours of sunlight, cloud cover, rainfall, ground water saturation, snowpack, and runoff observed for a given month at a specified locality are other important climate characteristics.

Climate variability refers to fluctuations in climate statistics with reference to a very long time average. Thus, the average summer temperature over a region may differ from year to year (interannual variability) or may manifest a fluctuation that spans a number of years (decadal variability). Natural climate variability has been observed on a range of time scales from months to seasons to centuries and more.

A climate trend refers to a long-term secular change in average climate statistics or a change in their statistical variation about the average. A climate trend may be forced by a cause external to the climate system, such as a change in the solar radiative output, or by human-induced changes in the atmospheric composition of trace gases and aerosols or the structure of vegetative land cover. A climate trend may also be forced by an internal change in the climate system, which could result, for example, from a change in ocean circulation patterns.

A climate quantity is predictable when a significant fraction of its variations can be consistently explained by a physical theory or mathematical model. Meaningful predictive skill is usually based on correlation between the predicted time series and the verifying time series of the quantity. Since climate statistics are strongly correlated with boundary quantities (e.g., sea surface temperatures), the boundary quantities may be considered climate quantities.

Seasonal-to-interannual variability, such as the phases of ENSO, is associated with widely distributed weather anomalies and sometimes severe conditions. These anomalies may persist for many months and can result in significant economic and human dislocations from Australia through tropical and semitropical South America to parts of Africa. Historical records and paleoclimatic data sources indicate the occurrence of significant climate variability on time scales of decades to centuries. Climate variability on these time scales has produced marked shifts in human well-being recorded in history over the past several centuries and can be expected to result in significant economic and human dislocations in the future. Current climate model projections based on anthropogenic increases in greenhouse gases and land cover changes indicate the potential for large, and rapid, climate change relative to the historical and paleoclimatic records, with concomitantly large influences on human activities and ecosystems.

Climate change can lead to significant changes in energy use, air pollution, crop yields, water quality and availability, the frequency and intensity of severe weather events, and the occurrence and spread of infectious diseases. Improved knowledge of the climate system offers the potential to enhance our predictive capability, which could support societal efforts to adjust to, forestall, or even eliminate some of the negative impacts of projected climate change. An enhanced capability to predict future climate will have a positive impact on economic vitality and national security.

Progress in understanding the physical, chemical, and ecological bases of climate during the past few decades is clearly a result of a wide variety of research efforts. A clear set of scientific objectives and requirements can be formulated for the coming years. Nonetheless, significant progress in achieving the mission of characterizing and predicting seasonal to century time-scale variability in climate, including the role of human activities in forcing this variability, is likely to take a decade or more. Some aspects of the problem will continue to be intractable for considerably longer periods.

The remainder of this Disciplinary Assessment articulates a mission and identifies the principal issues and related scientific questions that challenge the climate research community entering the twenty-first century. Seven scientific and programmatic objectives intended to guide this community over the next decade are presented.

MISSION STATEMENT

Human endeavors have come to depend on familiar global and regional environments. In fact, much of the fabric of our society is tied directly to climate through agriculture, water resources, and energy utilization. We have long recognized that climate is variable on time scales of seasons to centuries, and even longer intervals, and that this variability can have significant societal impact. El Niño events, the 1930s drought in the United States, the Sahel droughts, and variations in the monsoons over the most populous areas of the globe provide examples of the importance of natural climate variability for human activities and well-being. The nature of global and regional climates is also subject to change because of human activities, most notably in response to the observed changes in atmospheric composition (e.g., greenhouse gases and aerosols) and land use, characteristic of the last century. The potential impact of these changes is great and spans such diverse issues as agricultural yield, water resource availability, transportation systems, water quality, energy production and utilization, frequency and magnitude of extreme weather events, natural ecosystem viability, and even the nature of infectious diseases and their spread by agents that are influenced by climate.

The magnitude and timing of human-induced climate change remain active research topics. Large gaps in our knowledge of interannual and decade-to-century natural variability hinder our ability to provide credible predictive skill or to distinguish the role of human activities from natural variability. Narrowing these uncertainties and applying our understanding define the mission of climate and climate change research and education for the twenty-first century.

The mission of climate research is to understand the physical, chemical, and ecological bases of climate in order to characterize and predict the nature of climate variability from seasonal and interannual to decadal and longer time scales, and to assess the role of human activities in affecting climate and of climate in influencing human activities and environmental resources.

The scientific uncertainties, coupled with the potential significance of climate variability and climate change, indicate the importance of developing a scientific strategy for monitoring changes to the climate system, addressing key scientific uncertainties, enhancing our understanding of the impact of human activities, assessing societal vulnerability to climate change, and minimizing risk and maximizing benefits to society. Our primary goal is to enhance our capacity to predict climate variability and climate change, which implies understanding the impact of human activities in influencing climate.

PERSPECTIVES FOR THE TWENTY-FIRST CENTURY

To determine the imperatives for research in the coming decades, one must note the results of the past few decades of research, including both the explicit advances in knowledge and the increased potential to address the remaining critical uncertainties, and must recognize the importance of climatic research for society.

Insights of the Twentieth Century

A broad interest in climate variability and climate change was awakened in the early 1970s and during the 1980s due to a large number of weather-related disasters in widely scattered parts of the world and to accumulating evidence that human activities are altering the concentrations of radiatively important trace gases in the atmosphere. This awakening resulted in a large dedicated effort, through both the WCRP and national efforts, such as the U.S. National Climate Program and the U.S. Global Change Research Program (USGCRP), to enhance and analyze observations, conduct process studies, and improve climate models. The principal goal has been to develop credible methods to predict climate variability and change. The insights gained from these efforts are diverse and numerous. The three sections that follow illustrate the state of the science.

Seasonal-to-Interannual Variability and the El Niño/Southern Oscillation

ENSO is a major global-scale signal of seasonal-to-interannual climate variability. ENSO consists of both warm and cold phases, with the warm El Niño phase attracting most public attention. The El Niño phenomenon is an anomalous warming of surface ocean waters in the central to eastern equatorial Pacific Ocean accompanied by large-scale anomalies in rainfall (Figure II.5.1). El Niño occurs irregularly with a typical time period of three to six years. It has been known throughout the twentieth century, mostly through its detrimental effects

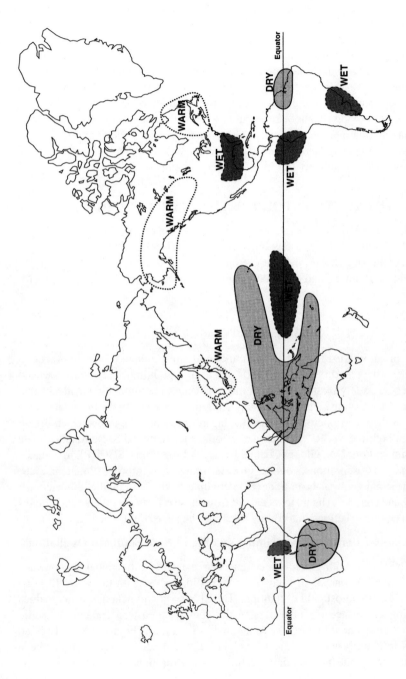

FIGURE II.5.1 Schematic of large-scale climate anomalies associated with the warm phase of the Southern Oscillation during Northern Hemisphere winter. Based on Ropelewski and Halpert (1986, 1987) and Halpert and Ropelewski (1992). SOURCE: NRC, 1994a.

on the fisheries, agriculture, and water resources of countries bordering the tropical Pacific, but only in the past 20 years has major progress been made in understanding the mechanisms that create ENSO and observing its occurrence and wide-ranging impacts.

The 1982-1983 warming, the largest of the twentieth century, was neither predicted in advance nor recognized until nearly at its peak. The enormous worldwide damage directly attributable to this warming (floods in Peru, collapse of the Peruvian anchoveta fishery, devastating drought, and forest fires in Australia and Borneo) gave impetus to an emphasis on observing the tropical Pacific in real time and on predicting the phases and intensity of ENSO.

As a result, the international TOGA program of the WCRP was developed. The accomplishments of TOGA, including major contributions by U.S. scientists, are many (NRC, 1996c):

1. The TOGA observing system, consisting of 65 TAO moorings, expendable bathythermographs (XBTs), drifting buoys, tide gauges, upper-air integrated sounding systems, and volunteer observing ships (Figure II.5.2)—all telemetering to the global telecommunication system (GTS) in real time—allows an unprecedented look at the state of the atmosphere, sea surface and subsurface tropical Pacific in real time (McPhaden et al., 1998).

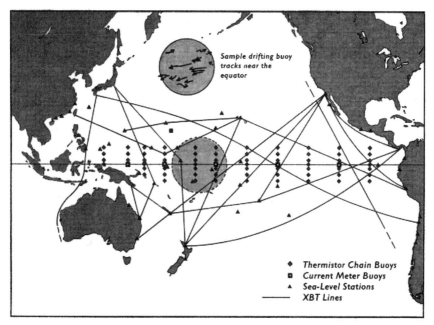

FIGURE II.5.2 The TOGA observing system (TAO). SOURCE: NRC, 1996c.

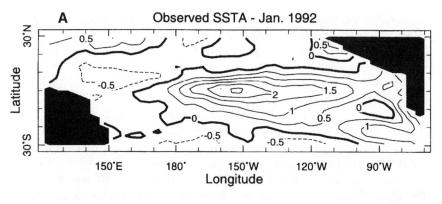
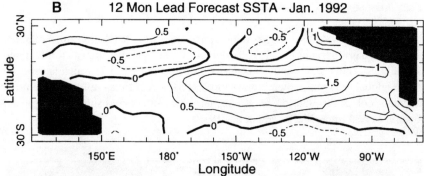

FIGURE II.5.3 (A) Observed sea surface temperature anomalies (SSTA) in tropical Pacific and (B) prediction made 12 months in advance by Cane and Zebiak (1987). Reprinted with permission of the Royal Meteorological Society.

2. A set of theories about ENSO has been developed and the mechanisms that may be responsible for its irregularity have been identified (Battisti and Sarachik, 1995; Neelin et al., 1998).

3. Connections between warming in the equatorial Pacific and climate phenomena in other parts of the world have been demonstrated, and the dynamical mechanisms responsible for these connections are beginning to be understood (Lau and Nath, 1994; Trenberth et al., 1998).

4. Coupled atmosphere-ocean models have been developed that are capable of simulating the major features of ENSO in the tropical Pacific (Zebiak and Cane, 1987; Delecluse et al., 1998).

5. Significant skill beyond persistence has been demonstrated in predicting sea surface temperature anomalies (SSTA) in the eastern to central tropical Pacific as much as a year in advance (Figure II.5.3) (Latif et al., 1994, 1998).

6. Prediction systems, consisting of coupled atmosphere-ocean models, data

assimilation and initialization techniques, and validation methods, are under development (Kleeman et al., 1995; Rosati et al., 1997).

7. Regular and systematic prediction of aspects of ENSO have been implemented (Ji et al., 1996).

8. Short-range climate variability predictions are beginning to be applied for the social and economic benefit of countries influenced by ENSO (Moura, 1994).

The development of predictive skill a year or more in advance is a monumental achievement that has vast implications both for science and for the applications of these predictions for the benefit of humankind. Consequently, we are in a position to consolidate our experience into ongoing prediction efforts and to expand our horizons and more fully explore global seasonal-to-interannual variability, its predictability, and the applications of this predictability.

Decade-to-Century Variability

Widely scattered instrumental records that extend back more than a century and more extensive observations of the past few decades provide a reasonably strong sense of interannual variability. The study of tree rings, ice cores, corals, and lake sediments demonstrates that climate variability on decade-to-century time scales has also occurred over the past millennia; such variability will certainly continue into the future. Our documentation and understanding of these longer-period variations is much weaker than for interannual variability. However, over the past two decades, research on natural climate variability on time scales of decades to centuries has grown because the complexities and uncertainties associated with both detecting and projecting the nature of future climate change have been recognized. Efforts to document and understand natural climate variability on time scales of decades to centuries provide important insights for climate research in the twenty-first century:

1. Analyses of historical records illustrate a number of interesting cases of longer period fluctuations for North America including (a) significant increases in temperature and decreases in precipitation in the 1930s; (b) decreases in tropical storm intensity for the East Coast in the 1960s and 1970s; (c) increases in interannual variability, mean winter temperatures, and total precipitation in 1975-1985; and (d) changes in lake levels over the last several decades (Figure II.5.4).

2. Ocean time series, although limited in availability, demonstrate significant decadal and longer time-scale variability, such as an abrupt change in the ocean surface state during 1976-1977 in the North Pacific, fluctuations in the sea ice limit of the Northern Hemisphere, and the Great Salinity Anomaly in the North Atlantic.

3. Careful study of historical records has also identified false jumps or

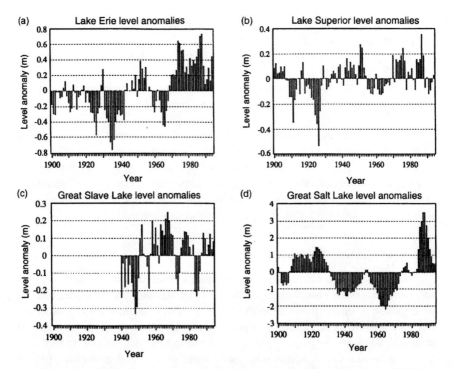

FIGURE II 5.4 Lake-level anomalies for selected North American lakes. SOURCE: Nicholls et al., 1995.

discontinuities in climate time series due to changes in observation practice, such as the relocation of observation sites or changes in instruments (Figure II.5.5).

4. Concerted efforts have resulted in long records of natural variability (e.g., 200,000-year records from ice cores and a greater than 1,000-year record from tree rings). The analysis of ice cores demonstrates century and multicentury variability and also provides remarkable evidence for abrupt (as short as 1 to 10 years) climate changes of regional to global significance (Figure II.5.6). A tree ring study for the midlatitude Asian continent suggests that the last half of the twentieth century is the warmest period of the past millennium for this region. Advances in the study of climate proxies indicate substantial potential to assess climate variability prior to the historical record.

5. Model studies have demonstrated that ocean-atmosphere interaction is one plausible mechanism for decade-to-century variability. The idea that the asymmetry in coupling between ocean and atmosphere, which is associated with heat and moisture fluxes, would create modes of variability has now been seen in simple model experiments. There are strong feedbacks associated with surface heat flux and temperature changes. They are, however, complicated (e.g., the

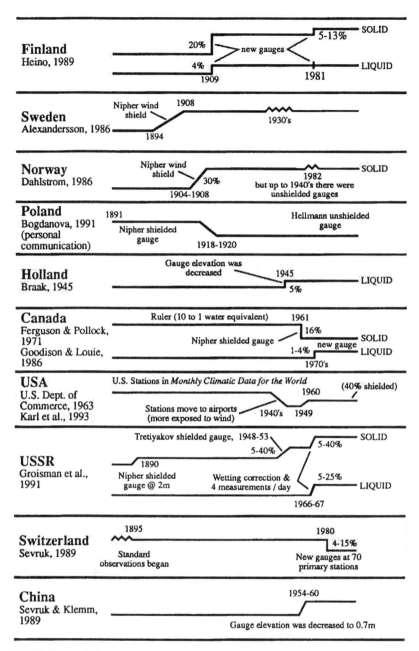

FIGURE II.5.5 Selected set of time-dependent precipitation measurement biases that have affected various countries. SOURCE: From Karl et al., 1993.

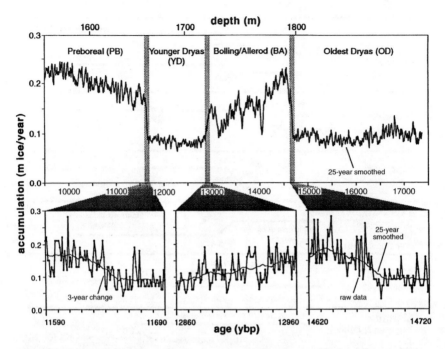

FIGURE II.5.6 Snow accumulation in central Greenland, showing abrupt changes. SOURCE: Alley et al., 1993. Reprinted with permission of Macmillan Magazines Limited.

freshwater flux due to precipitation is a significant influence on the ocean circulation, whereas changes in ocean salinity have little direct impact on the atmosphere).

6. Long simulations of coupled ocean-atmosphere and ocean general circulation models show evidence of centennial variability associated with the thermohaline circulation, as well as evidence for shorter (multidecadal) time-scale variations. These studies, coupled with both observations and other model experiments, identify the North Atlantic and its associated deepwater formation as a focal point of decade-to-century variability.

7. Climate model studies suggest that the feedbacks between land surface characteristics and the atmosphere may also be significant factors in decadal-scale variability (e.g., prolonged Sahel drought).

8. Direct observations of aerosol properties and radiative forcing from the Mt. Pinatubo volcanic eruption have offered a unique opportunity to increase our understanding of the climate system (Figure II.5.7). The global radiative forcing from Mt. Pinatubo of -4 W m^2 is equal but opposite to that due to the doubling of carbon dioxide. Climate model studies have been successful in predicting the correct magnitude of the response to this forcing, thus lending credibility to the

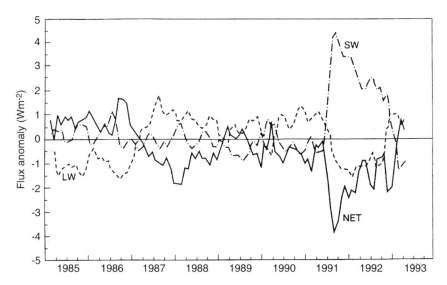

FIGURE II.5.7 Time series of smoothed wide field of view Earth Radiation Budget Experiment long-wave (LW), short-wave (SW), and net (LW-SW) irradiance anomalies between 40°N and 40°S relative to the five-year (1985-1989) monthly mean. The deviation starting in mid-1991 is mainly due to the Mt. Pinatubo eruption—the net anomaly in August (about -4 W m^{-2}) is almost three times higher than the standard deviation computed between 1985 and 1989. SOURCE: After Minnis et al., 1993; updated by Minnis, 1994.

models' predictive capability. In Figure II.5.8, the eruption is indicated by the vertical dashed line. The stratospheric temperatures [Figure II.5.8(a)] are from satellite observations and show the 30 mbar zonal mean temperature at 10°S; they were supplied by M. Gelman, National Oceanic and Atmospheric Administration; model results represent the 10-70 mbar layer at 8 to 16°S. The zero is the mean for 1978 to 1992. The tropospheric temperatures [Figure II.5.8(b)] are from satellite observations, and model results are essentially global. The zero is given by the mean for the 12 months preceding the eruption. The surface temperatures [Figure II.5.8(c)] are derived from meteorological stations; observations and model results are essentially global. The zero is given by the mean for the 12 months preceding the eruption. Note that the model results use a simple prediction of the way the optical thickness of the initial volcanic cloud varied with time, rather than detailed observations of the evolution of the cloud.

9. Observations and models suggest substantial low-frequency variability in the period and amplitude of ENSO, which implies a decadal modulation of the predictability of aspects of ENSO. In particular, the first half of the 1990s exhibited a large-scale warming in the eastern Pacific (within ±30° of the equa-

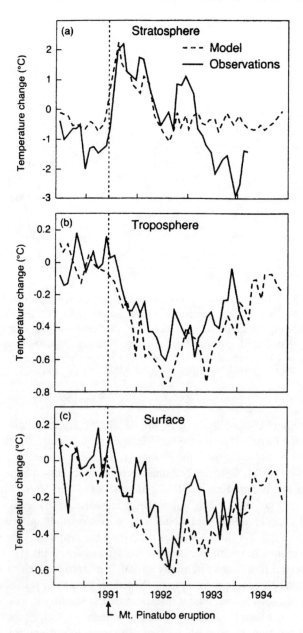

FIGURE II.5.8 Observed and modeled (from the GISS general circulation model) monthly mean temperature changes over the period of the Mt. Pinatubo eruption. The eruption is indicated by the vertical dashed line. Updated from Hansen et al., (1993a). SOURCE: M. Gelman, National Oceanic and Atmospheric Administration.

tor) that coincided with a degradation of the skill of prediction of sea surface temperatures in the tropics. The nature of these decadal modulations is unknown.

10. Observations and models have clarified part of the complexity of natural variability. Isolation of the mechanisms is challenging because of the uncertainty in characterizing the magnitude of many forcing factors and the variety of potential causes. Uncertainties include variability within individual components of the Earth system, variability associated with the coupling of system components with different response times, and forced variability (e.g., solar variation and volcanic eruptions).

An understanding of natural variability is essential to the wise use of resources, human health, agricultural productivity, and economic security. Research in the twentieth century has elucidated much of the complexity of natural variability and has begun to document its scope. These results demonstrate the importance of additional research to reduce the uncertainties associated with detecting and projecting future climate change.

Assessing the Human Role in Climate

Greenhouse gases absorb and reemit the infrared radiation emitted by atmospheric gases, clouds, and the Earth's surface. Atmospheric concentrations of greenhouse gases, including carbon dioxide (CO_2), methane (CH_4), nitrous oxide (N_2O), and halocarbons, have increased significantly above preindustrial levels. [Water vapor is also an important greenhouse gas, which is discussed below (Raval and Ramanathan, 1989; Chahine, 1992; Stephens, 1990).] This increase is clearly due to anthropogenic activities. Because of their infrared absorption, increased concentrations of these gases promote global warming. The debate is not over whether these gases promote global warming but, rather, over the more difficult problem of the timing, magnitude, and regional patterns of the climate change, including regional changes of climate extremes such as tropical and extratropical storms, severe local storms, hail, floods, droughts, and heat waves. The prediction of future climate change is problematic because of a number of significant uncertainties (IPCC, 1996) including (1) the natural variability of climate (NRC, 1995c); (2) the difficulty in predicting future greenhouse gas and aerosol concentrations (NRC, 1993, 1996a); (3) the potential for unpredicted (e.g., volcanic eruptions) or unrecognized factors (e.g., unknown human influences and unrecognized climate feedbacks); and (4) a lack of understanding of the total, coupled climate system. Because of these uncertainties, future greenhouse warming is usually described in terms of a range of increase in globally averaged temperature for specific scenarios of greenhouse gas emissions (e.g., 1.5 to 4.5°C temperature increase for a doubling of CO_2 concentration). Although the past few decades of research have yielded substantial progress in projecting future human-induced climate change, there is much to be learned about the regional or local characteristics of global climate change (Giorgi and Mearns, 1991).

Efforts to understand the forcing and response of the climate system and the role of human activities in effecting climate change provide a number of important insights:

1. Intercomparison of the magnitude of cloud feedback in a number of global climate models (Figure II.5.9) indicates nearly a fourfold range of uncertainty, with some models predicting strong positive cloud feedback and others a weak negative feedback to the climate system. This range of uncertainty in cloud feedback contributes to the range of uncertainty in climate model prediction. Calibrated five-year Earth Radiation Budget Experiment (ERBE) observations have documented that clouds have a net global radiative cooling effect on the Earth-atmosphere system. Regional cloud forcing data have contributed significantly to diagnosing deficiencies in global climate model treatment of cloud radiative interactions. This regional cloud forcing has a significant impact on the magnitude and direction of ocean heat transport (Gleckler et al., 1995; Hack, 1998). The net effect of changing atmospheric concentrations of greenhouse gases on these other factors is an outstanding scientific issue.

2. Water vapor behavior and feedback analysis have been advanced on theoretical, observational, modeling, and methodological grounds, spurred by the understanding that water vapor in today's atmosphere is its most radiatively important greenhouse gas. Observational uncertainty in relative humidity, particularly in upper levels of the atmosphere, contributes substantially to the uncertainty in predicted climate changes at the surface and as a function of altitude.

3. Improved representation of the land surface in climate models—from early parameterizations of specified albedo, emissivity, and simple "bucket" hydrology to fully coupled biosphere-atmosphere transfer schemes with multiple soil layers—has served to demonstrate that changes in the land surface with land use change, and through vegetation-climate feedbacks, can be significant because of their impact on energy, moisture, and greenhouse gas fluxes.

4. Experimentation with ocean general circulation models and coupled ocean-atmosphere models demonstrates the potential for "surprises" in future global change. For example, model studies have demonstrated the possibility of more than one stable mode for the ocean circulation associated with changes in the hydrologic balance in the North Atlantic. The importance of incorporating explicit ocean heat transport and oceanic processes, such as resolution of the thermohaline circulation and accurate representation of surface moisture and energy fluxes, is clearly evident from these studies.

5. Incorporation of the radiative effects of specific trace gases and their distribution in the atmosphere improves climate simulations substantially. Further, research during the last decade has demonstrated the importance of interactions between atmospheric chemistry and climate and the need for improved

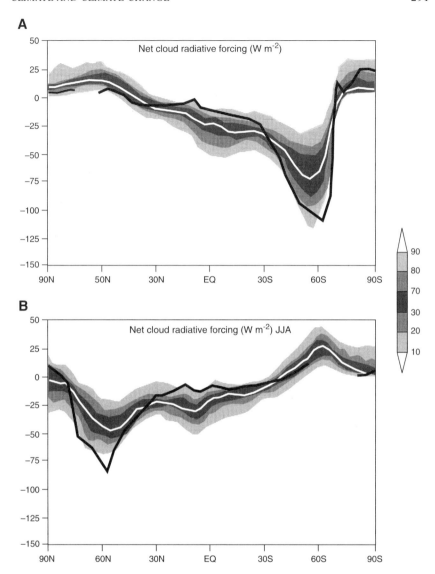

FIGURE II.5.9 Zonally averaged net cloud-radiative forcing (W m^{-2}) as observed (black line) and as simulated by Atmospheric Model Intercomparison Project (AMIP) models for December-February (top) and June-August (bottom). Mean of model results is given by white line; the 10, 20, 30, 70, 80, and 90 percentiles are given by shading surrounding the model mean. Observational estimates are from ERBE data for 1985-1988. Climate simulations completed by participants in AMIP were carried out with sea surface temperatures prescribed for 1979 through 1988. Note that most models overestimate the cooling effect of clouds in the tropics and underestimate the cooling effect of clouds in the summer midlatitudes. SOURCE: Harrison et al., 1990. Reprinted with permission of the American Geophysical Union.

representations of atmospheric chemical interactions within climate models. In this regard, models have shown the importance of temperature-dependent reaction rates in determining ozone concentrations, which in turn affect temperatures. Chemical interactions are also important in determining the distribution and concentration of sulfate aerosols and the hydroxyl radical (OH).

6. Various ice- and snow-related feedbacks, based largely on the high albedos of ice and snow and their insulating qualities, make the sea ice, ice sheet, and snow cover particularly important to the climate system. Incorporation of simple parameterizations of sea ice and simple relationships between snow cover and albedo have yielded marked effects on the polar simulations of general circulation models (GCMs), confirming the expected importance of sea ice and snow cover to the model results. Such results further identify the need for more complete parameterizations of sea ice, including ice dynamics as well as thermodynamics, plus more complete snow cover parameterizations and incorporation of dynamic ice caps.

7. The focus on transient climate model simulations, extending from 100 years in the past to 100 years into the future, including time-varying forcing, provides much stronger tests of model capability. For example, comparison of atmospheric GCMs with the time evolution of temperature identify important limitations of these models. As another example, ocean tracer data in conjunction with the transient integration of climate system models can be used to evaluate the mechanisms by which carbon, mass, and heat enter the ocean interior. Transient integrations also provide more reliable estimates of the effect of increasing greenhouse gases on the climate system.

8. Anthropogenic aerosols can affect climate directly by scattering sunlight and indirectly by serving as condensation nuclei for cloud drops, thereby potentially altering cloud radiative properties and lifetimes. Increases in aerosols tend to cool climate at least locally and contribute to historically observed changes in climate; they may be a factor in explaining some of the differences between observations and model predictions of the warming due to increases in carbon dioxide (Figures II.5.10-II.5.12). However, characterization of the distribution and nature of aerosols in the atmosphere is currently inadequate.

9. The evidence from improved weather forecasting suggests that the use of finer-scale models and empirical techniques that allow models to incorporate regional features, such as improved topography, vegetation, and soil characteristics, has the potential to improve climate model predictions (Giorgi and Avissar, 1997).

10. In addition to the improved recognition and documentation of the scope of natural variability, comparison of paleoclimatic data and model experiments suggests a range of climate sensitivity in geologic history to a variety of climatic forcing factors, including carbon dioxide, which is very similar to the range given by the Intergovernmental Panel on Climate Change (IPCC) assessment.

11. The linkage of climate models with models designed to assess the im-

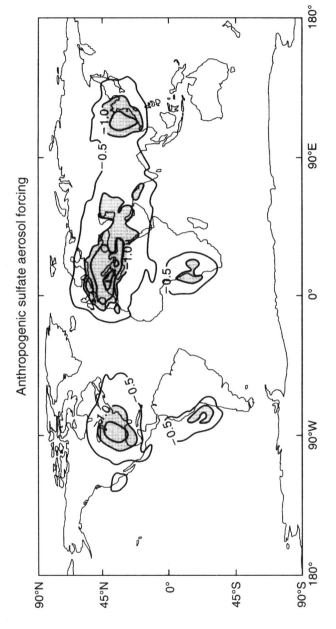

FIGURE II.5.10 Annual mean direct radiative forcing (W m^{-2}) resulting from anthropogenic sulfate aerosols. SOURCE: Shine et al., 1995.

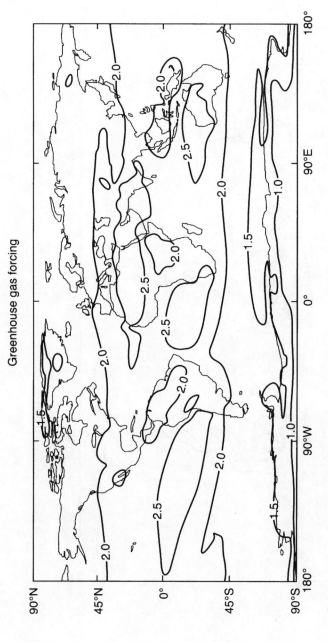

FIGURE II.5.11 Annual mean instantaneous greenhouse forcing (W m^{-2}) from CO_2, CH_4, N_2O, CFC-11, and CFC-12 from preindustrial to present. NOTE: CFC = chlorofluorocarbon; CH_4 = methane; CO_2 = carbon dioxide; N_2O = nitrous oxide. SOURCE: Shine et al., 1995.

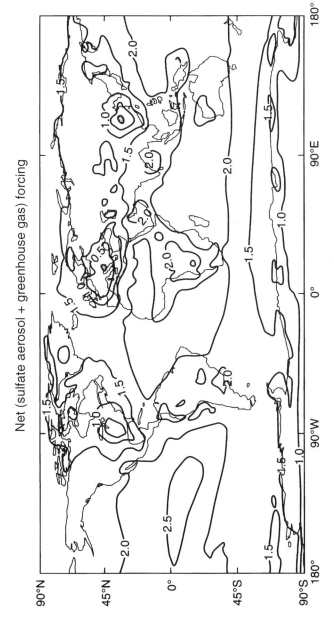

FIGURE II.5.12 Annual mean instantaneous radiative forcing (W m^{-2}) since pre-industrial times due to changes in both greenhouse gases and sulfate aerosols. SOURCE: Shine et al., 1995.

pact of global change on agriculture, water resources, ecosystems, health, and the economy, and quantification of the positive and negative effects of climate change, have been substantially improved and show promise in the development of integrated assessments.

Progress over the past decade in projecting future human-induced climate change is clear. However, many critical climate questions remain. The answers to these questions cannot be achieved without (1) a long-term commitment of resources; (2) a comprehensive, integrated program of observations, process studies, and modeling; and (3) a dedicated effort at improving prediction.

The Scientific Questions

The scientific uncertainties, coupled with the potential significance of climate variability and climate change, demand a careful, focused scientific strategy. The foundation of this strategy must be based on addressing key scientific questions:

- What is the nature of global and regional climate variability on seasonal-to-decadal and longer time scales? What are the spatial and temporal characteristics of this variability? What are the extremes in variability? What are the climate phenomena that acutely affect societies on a regional scale, and what are the probability distributions of these phenomena?
- To what extent are these variations predictable? For which parameters, locations, and times of the year are prediction skills highest? What data and model characteristics enhance predictive capabilities?
- What data are needed to evaluate these predictions?
- What is the climate history of the Earth and what caused it?
- What are the human-induced and natural forcing changes in the global climate system? How well do they explain the observed climate record?
- What is the response of the climate system, including its variability and extremes, to projected changes in greenhouse gases, water in all its phases, aerosols, and other human forcing? For example, how does the addition of radiatively important gases affect the intensity and frequency of ENSO cycle fluctuations? How does this manifest itself throughout the world?
- To what extent can climate change be simulated at a scale appropriate to assess its impact on human activities?
- What are the expected impacts of climate change on the rest of the global system, especially elements of immediate relevance to humans (e.g., growing seasons, agricultural yields, spread of diseases)? What information is needed to maximize the benefit to society of future predictions of climate variability and climate change?

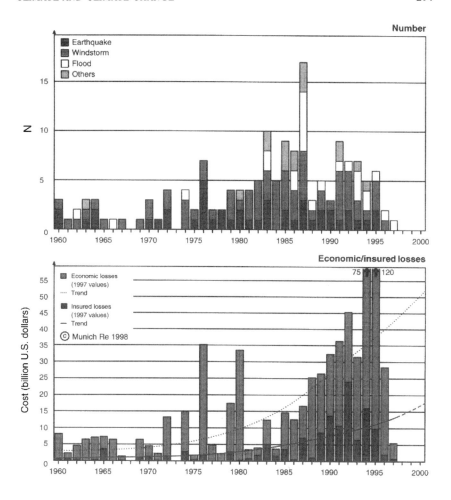

FIGURE II.5.13 Number and cost of great natural catastrophes. Economic losses have been adjusted for inflation. SOURCE: From G. Berz, 1998.

Key Drivers for Research in the Twenty-First Century

The observations and the insights derived from the twentieth century can be used to describe the impetus for climate and climate change research entering the twenty-first century. The impetus centers around increased recognition of (1) the evidence for climate variability and the potential for change, (2) the economic and societal impact of this variability and change (Figure II.5.13), (3) the opportunities for progress in enhancing our predictive capabilities, and (4) the potential for immediate practical benefits as a result of increased predictive skill. The primary drivers are the following:

- Seasonal to interannual variability (e.g., the phases of the ENSO cycle) is associated with widely distributed weather anomalies and sometimes severe conditions, which are characterized by significant economic and human dislocations.

Tropical Pacific sea surface temperature (SST) changes on interannual time scales associated with ENSO events produce both local, or nearby, and remote effects. Local effects are robust and determined mostly by the association of rainfall with warm water. When the SST increases in the eastern and central Pacific, the area of heavy rainfall expands eastward to lie over the warmest water, leaving the western end of the tropical Pacific drier and the central and eastern end wetter. The local effects of a warm phase of ENSO include excess rainfall and storminess over central Pacific islands, increased rainfall over the normally semiarid plains of coastal Peru and Ecuador, drought in Northeast Brazil, excess rain in southern Brazil and Uruguay, drought over the Indian peninsula during the summer monsoon, and drought in Australia and Indonesia. The remote effects are well documented, but the mechanisms are less well understood. Remote effects of a warm phase of ENSO frequently include anomalous warmth in Newfoundland and in the northwestern sector of North America, dry conditions in the southeastern part of the African subcontinent (including South Africa and Zimbabwe), and a weaker Asian monsoon. The economic effect of the 1982-1983 warming has been estimated to be of the order of $13 billion (1983 dollars).

- Modeling efforts over the past two decades to couple the atmosphere, ocean, and land system indicate that aspects of the coupled system, especially SSTs and rainfall, may be predictable.

Starting with the first forecast of Cane and Zebiak (1987) of the 1986-1987 warm phase of ENSO one year in advance, prediction systems have been demonstrating increases in forecast skill over the existing record. A typical prediction system involves a coupled atmosphere-ocean model, data for initialization provided by the TOGA observing system and other sources, a data assimilation and initialization method (including quality control of data over the period of record), and an evaluation procedure to compare forecasts with actual results. Recent advances in all of these areas have indicated that forecast skill can be useful as much as a year in advance but that the skill varies decadaly in ways that are not presently understood. Also, the limits of predictability are not understood. Current skill levels for process-based physical models are comparable to those achievable by the best statistical regression-based models, so it should be anticipated that further increases in skill will result from an increased process-level understanding of the coupled ocean-atmosphere system and incorporation of this understanding into physical models.

- In cases where ENSO events were predicted in advance, immediate practical benefits were realized through human response and adaptation.

Peru, Brazil, and Australia routinely make use of predictions of SST in the tropical Pacific (e.g., Peru regularly passes laws regulating agricultural policy). As a result of the forecast of warm rainy conditions in 1986-1987, rice was favored over cotton, and the agricultural output, normally adversely impacted by warm phases of ENSO, remained normal. Northeast Brazil, which is usually semiarid and marginal for agriculture, has learned to use these forecasts for agricultural planning and now, despite severe droughts, is able to maintain its agricultural output near normal and therefore avoid the traditional plague of human migration, endemic to Brazilian history. Of course, such successes are also limited by the skill of predictions. Care must be exercised in extrapolating from a few successes. Long time series are required to establish the level of skill of a prediction system, especially since ENSO is only a part of the explanation of interannual variability. For example, precipitation in Northeast Brazil is highly correlated to tropical Atlantic variables such as sea surface temperature. ENSO appears to explain only 20 to 25 percent of the variation in rainfall in northeastern Brazil.

- Historical and paleoclimate data sources, and coupled atmosphere-ocean model experiments, indicate the potential for significant climate variability over long periods of time. Irrespective of human impacts, future climate can be expected to vary significantly, with the expectation that human and economic dislocations will also be significant.

Paleoclimatic records from a variety of sources reveal a rich history of climate variability on time scales of decades to millennia. Coupled model experiments contribute to our understanding of the mechanisms that are responsible for these longer time-scale variations. Given the causes, from forced changes (e.g., volcanic eruptions) to internal variability associated with coupling the different components of the Earth system, there is every reason to believe that such variability will continue into the future. Paleoclimatic records indicate that the magnitude of regional and global variability can exceed observed interannual variations that are known to result in significant human and economic dislocations.

- Improved knowledge of the coupled Earth system will increase understanding of natural variability on all time scales and lead to a greater realization of the practical benefits of enhanced predictive capability.

Isolation of the mechanisms that produce decade-to-century variability is challenging. One major mechanism is the nonlinear coupling of system components that have different time constants or whose coupling is not symmetric (i.e., the coupling of temperature and salinity between the atmosphere and ocean). The coupled modeling of the components of the Earth system is in its infancy. Improved knowledge of this coupling is, therefore, likely to enhance our understanding of at least one of the major mechanisms associated with decade-to-century variability. In this area, increased predictive capability is possible, whereas some of the forced elements of natural variability (e.g., volcanic eruptions) may remain unpredictable.

- Increased concentrations of carbon dioxide, methane, nitrogen oxides, chlorofluorocarbons (CFCs) and aerosols, and changes in land use and land cover, are directly and indirectly tied to human activities.

The observed concentration of atmospheric carbon dioxide is 30 percent higher than preindustrial levels as measured directly and in ice cores. The major anthropogenic sources (fossil fuel consumption and deforestation) are significantly larger than anthropogenic sinks. Carbon isotope studies demonstrate that this increase is due to fossil carbon and biomass reduction. Methane concentrations are more than 100 percent higher than preindustrial levels. Anthropogenic sources, such as agriculture, energy production, and energy use, and knowledge of potential sinks are consistent with these measured increases. Nitrous oxide concentrations are about 10 percent above preindustrial levels. Again significant anthropogenic sources have been identified (nylon production and agriculture). Preindustrial concentrations of halocarbons are zero because there are no natural sources. Emissions of sulfur dioxide (SO_2) have increased dramatically over the past 50 years; anthropogenic emissions began to exceed the global natural source of SO_2 around 1940. This large increase in atmospheric SO_2 has led to a substantial increase in sulfate aerosols.

- Current model projections based on the increases in greenhouse gases and aerosols, as well as land cover changes, indicate the potential for large and rapid climate change relative to the historical and paleoclimatic records, with concomitantly large influences on human activities and ecosystems.

The radiative effects of greenhouse gases are well known. Because of their infrared absorption, increased concentrations of these gases should act to warm the air. Current climate models provide the most comprehensive projections of the magnitude and timing of climate change associated with increases in greenhouse gases and aerosols. The range of model experiments and their assessment by the IPCC suggest that global mean surface temperature will increase by about 0.9 to 3.5°C by the end of the twenty-first century. The best estimates for a climate in equilibrium with a doubling of CO_2 is 2.5°C with a range of 1.5-4.5°C. These projected warmings are large compared to the historical and most recent paleoclimatic records and would produce the warmest global climate of the past 200,000 years. For comparison, the temperature change associated with the last ice age is of a magnitude similar to some greenhouse climate projections (the last ice age had a globally averaged temperature approximately 3-5°C cooler than present day), but the changes from glacial to interglacial periods occurred over thousands of years rather than a century. Use of these results to examine potential human impacts indicates some significant changes in crop yields, the availability of energy and water resources, natural ecosystems, and other factors such as the potential distribution of infectious disease vectors.

- Remarkable progress in developing climate models has occurred over the past two decades, but current climate models are characterized by a large number of remaining uncertainties.

The differences between climate models of a decade ago and current versions are considerable, especially in spatial resolution, treatment of oceans, hydrologic cycle, land surface processes and vegetation-atmosphere interactions, and clouds. However, most coupled atmosphere-ocean models still exhibit significant drift, indicating continuing problems with the component models and their coupling. Sensitivity studies have indicated that the response of climate models to a given forcing can vary significantly due to differences in the way these processes are parameterized. In many cases, model improvements have actually revealed uncertainty because we have come to recognize the importance of a specific interaction (e.g., vegetation-climate), yet our understanding of the processes involved is insufficient to define the magnitude of the effects. To improve the predictive capability of these models, refinements in key model processes are required.

• Improved knowledge of the fully coupled climate system can lead to enhanced predictive capability and the possibility of minimizing risk and maximizing benefit associated with the impacts of projected climate change. An enhanced ability to predict future climate is likely to have a positive impact on economic vitality and national security.

Advances in ENSO prediction, improvements in weather forecasts associated with incorporation of specific system components (e.g., soil moisture), and improvements in climate models over the past two decades strongly suggest that enhanced predictive capability will occur as a result of increased knowledge of the coupled Earth system. Just as interannual-to-seasonal forecasts have considerable economic value, increased predictive capability on longer time scales is likely to have a positive impact on economic vitality and economic security.

• The development of international, national, and regional policies will be enhanced by an increased ability to separate natural variability from human-induced climate change.

Uncertainties associated with model predictions of future climate change and an inability to separate clearly the human-induced climate signal from natural variability hinder developing and sustaining optimal policies. Three elements limit our ability to address these uncertainties: (1) the lack of a comprehensive climate observing system, (2) inadequate knowledge of the scope and character of natural variability, and (3) limitations of current climate models.

• Current observational capabilities and practices are inadequate to provide the long-term, continuous, quality observations required to characterize changes in global and regional climate.

Much has been learned from existing observation systems for operational weather forecasting. However, in many cases the operation of these observation systems does not fulfill the climate mission. There are several reasons: the basic observational infrastructure has deteriorated (NRC, 1992, 1994d); standard procedures for collecting side-by-side overlapping measurements are rarely applied when measurement techniques change significantly; inclusion of information about the nature of the observations, station relocations, algorithms, and quality control are inadequate; and observations of some variables have been discontinued. Special attention must be given to the long-term homogeneity of existing climate records. Few cases can be identified in which there is truly a long-term consistent data set for any major climate or hydrologic variable. In addition, in several cases the observation of specific variables should be enhanced in order to address ocean-atmosphere coupling, atmospheric water vapor-climate feedback relationships, and the role of clouds in climate change.

OBJECTIVES AND REQUIREMENTS FOR CLIMATE RESEARCH

The results of twentieth century research yield a set of important remaining scientific questions. The economic and societal importance of climate provides the impetus for developing a comprehensive program for climate research. The scientific questions lead to a set of research objectives. In most cases, these research objectives address multiple scientific questions. For each objective, we can articulate a list of requirements based on experience from past successes and failures, from the remaining uncertainties and areas of scientific debate, and from reasoned assessment of the opportunities to promote significant advancement in climate and climate change research.

Objective 1

Stop the deterioration and improve current observational capability as a first step in building a comprehensive climate observing system. Include climate requirements as a priority in operational systems.

Long-term consistent observations of the key variables that describe the state of the atmosphere, land, and ocean are the foundation for understanding climate. These data are the source of information on the nature and extent of climate variability and are the basis of determining whether climate is changing. Modern observations are the primary means of evaluating climate models.

However, current observational systems are far from adequate in addressing the questions being posed by scientists and policy makers concerning climate change. Virtually all of the data available have been collected for the purpose of weather prediction. Yet these data are being utilized as the key source for examining many critical long-term climate variables. A brief examination of a few of the data sets illustrates the nature of the problem. In situ measurements are the primary sources of land near-surface temperatures and are made available through the World Meteorological Organization's (WMO's) World Weather Watch. The data are not without problems. WMO Resolution 40 seeks to limit the distribution of country-originated research. Extensive areas of tropical land are characterized by poor quality or missing data. Even in the United States we lack a reference temperature network, and no network is dedicated to monitoring decadal homogeneous temperature changes. The United States has had a history of problems related to preserving the homogeneity of maximum and minimum temperature readings. At cooperative observing sites, changes in instrumentation without adequate overlap introduced a serious discontinuity in measurements (Quayle et al., 1991). Automated surface observing network sites located at airports are influenced by urban heat islands and have had (not corrected) daytime overheating problems. There are a number of problems in assessing decadal changes in quantities such as temperature extremes, including station locations without suitable overlaps, urban heat islands and jet exhaust, changes in local conditions, and new instruments not calibrated with previous equipment. Complex adjustments are required to resolve global and regional changes in ocean temperatures measured largely from ships of opportunity. A worldwide network of radiosondes and the microwave sounding unit instrument aboard National Oceanic and Atmospheric Administration (NOAA) polar orbiters are the key to documenting changes in the vertical structure of the atmosphere. WMO and GCOS have developed a network of 140 rawinsonde stations for climate detection. Already, 27 of the 140 stations are not reporting. The former USSR recently reduced its sampling rates. Canada has closed some high-latitude sites. NOAA is considering a reduction of 14-20 sites (NAOS, 1996). Satellite data are also susceptible to problems. Given NOAA policy to launch polar orbiters with minimum overlap, overlap problems tend to occur. The result is a concern (Hurrell and Trenberth, 1998) about the ability of scientists to remove large intersatellite biases from the record of tropospheric temperatures. These examples focus on a single fundamental variable—temperature—but they serve to illustrate the flaws and problems that arise when the answers to climate questions are dependent on an observational system designed for very different purposes.

The climate community has relied on data from observing systems that were not designed to monitor climate. Current priorities of these observing systems do not take into account a variety of climate requirements. Despite these difficulties, the data have been used to document and understand much of what we know today about natural and anthropogenic climate variability and change. Current trends toward reduced data quality, reduced quantity of critical elements, and inadequate information about the manner in which observations are made and processed, jeopardize our ability to document and understand climate variations.

The primary challenge is to develop a permanent climate observing system for monitoring the state of the atmosphere, ocean, land, and hydrologic cycle. The most effective means to accomplish this task will require incremental improvements in the existing observing system. This challenge is multifaceted. First, the major systems developed for operational weather forecasting rarely have the continuity and consistency required for climate research. In many

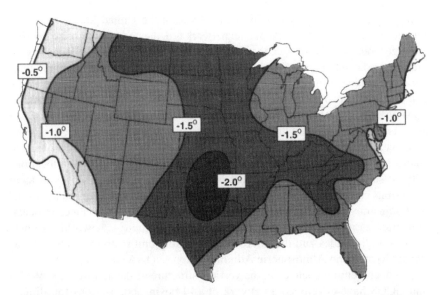

FIGURE II.5.14 Effect of change in average March temperature (°C) resulting from changing the time of observations from 5:00 p.m. to 7:00 a.m. for cooperative climate stations in the United States, the data from which are used detect decade-to-century scale climate change. SOURCE: From Karl et al., 1986.

instances, simple rules of data management (Figure II.5.14) and observation—such as accurate, timely, and regular reporting on the techniques and algorithms used to collect and process data; knowing the differences in observing biases between new and old observing methods prior to eliminating the old observing method; and more effective use of existing data bases—would make a critical difference to climate research.

The collection of U.S. climate-related observations is also inherently a combination of multiagency federal, state, and local efforts. Key elements of the current observational system are vulnerable to relatively small budget cuts in individual agencies or programs, without appropriate recognition of their value to the broader objectives of global change research. These include rural observation sites with long periods of record, portions of the upper-atmosphere sounding network, and portions of the coastal buoy network. The development of a credible climate observing system must be a priority. This requires an integrated observational strategy that is less dependent on individual agency missions or budgets. Such an integrated strategy must include identification of key elements of the current observational system and formulation of an interagency observational plan.

Satellites now collect vast amounts of potentially relevant data about the current atmosphere, hydrosphere, cryosphere, biosphere, and land surface processes. To ensure a long enough data series for decade-to-century time-scale studies, such observations must continue. Satellite measurements must address and minimize the biases associated with drifting orbits that alias the diurnal cycle. New satellites and sensors must overlap the measurements of existing satellites prior to the decommissioning or decay of the latter in order to eliminate inhomogeneities in the climate record (Figure II.5.15). Similarly, the introduction of new instruments on satellites poses tremendous challenges in the development of a homogeneous climate record. Differences in measurements between new and existing instruments must also be resolved prior to discontinuing existing measurements.

Research-based observational systems (e.g., the TOGA observing system, specifically the TAO array) have a difficult and uncertain transition to operational systems, even when the value of observations has clearly been demonstrated. Data collected by the TOGA observing system are valuable to the world's

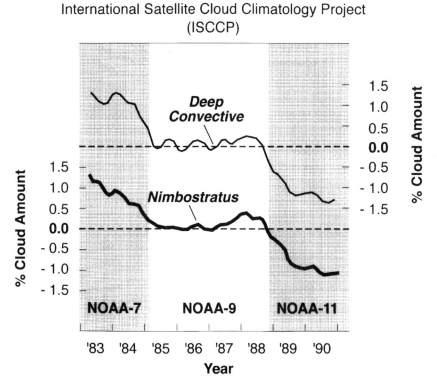

FIGURE II.5.15 Monthly anomalies (50°N to 50°S) in mean cloud amount depicting the biases and drift in measurement associated with different satellite systems even after records were reprocessed for consistent calibration. SOURCE: Klein and Hartmann, 1993. Reprinted with permission of the American Geophysical Union.

operational weather prediction agencies. The array provides surface winds over vast reaches of the Pacific where no other instruments exist. It is essential that the flow of data and information derived from the TOGA observing system be maintained. In the transition to operational systems, research arrays should be evaluated in the context of operational needs to determine whether these systems should be expanded or downsized to maximize operational utilization and efficient use of resources.

Climate issues are inherently global, and a credible, useful observation system must be an international endeavor. In pursuit of this goal, the WMO is working toward defining a set of observations that meet the specific needs of climate system monitoring, climate change detection, and research to improve our understanding of climate variability and change. The WMO effort is designed to create an internationally recognized global climate observing system. Given the importance of internationally agreed upon standards and procedures and the historical role of WMO in organizing weather observations, U.S. participation in the GCOS effort is of considerable importance to climate research.

Current trends toward limited access to international data and information must be overcome (NRC, 1995a). Data and information that quickly lose their market value for operational real-time weather forecasting are still critical for achieving climate research goals and should be made freely available within a short time after their use in operational forecasting. Full and open exchange of data is an important element of the challenge to a climate observing system.

We must take full advantage of existing observing systems by ensuring that operational observations can be utilized for climate research. Long-term, consistent observations of climatologic variables and climate forcing factors are essential for achieving the mission described in this Disciplinary Assessment.

The following **requirements** are essential to achieve this objective:

1. Where feasible, adopt consistent data collection and management rules to ensure the utility of operational and research system measurements for climate research.

2. Develop and adopt interagency plans to ensure the protection of critical long-term observations, to limit gaps in data continuity due to small budget changes in single agencies, and to recognize the value of these observations in a balanced, integrated research program designed to address climate variability and change issues.

3. Maintain major research observation systems, such as the TOGA TAO array, that have demonstrated clear predictive value.

4. Provide strong U.S. support and participation in the development of a GCOS.

5. Ensure full and open international exchange of data and information.

Objective 2

In addition to current observations, enhance the observation and monitoring of key variables, including atmospheric water vapor, ocean temperature, salinity (circulation), surface winds, soil moisture, precipitation (including cloud water and aerosols), snow cover, sea ice thickness, ice sheet topography, and the major forcings of the global climate system (solar output, aerosols, and changes in the land surface).

In developing a comprehensive climate observation strategy, it is important to distinguish between two scientific purposes that observations serve:

1. *Observations Aimed at Elucidating Key Processes Governing the Nature, Timing, Rate, and Geographical Distribution of Climate Variability:* Observation efforts that serve this purpose have to be comprehensive in scope to include all relevant variables that control the process under investigation.

2. *Observations Aimed at Detecting Climate Variability and Change:* Generally, observations that serve this purpose must be carefully selected to maximize the signal of climate change from the noise of the climate system and to ensure that the observed variables are measured with relatively high precision over an extended period of time with very low tolerance for discontinuities in the record.

Observing strategies serving each of these purposes should be individually developed, with an awareness of whatever opportunities for synergism among instruments and platforms become apparent.

Testing climate models requires comparison with long-term observations that are sufficiently comprehensive, and cover enough of the globe, to distinguish among different physical mechanisms. Increased study, through process experiments and monitoring, of water vapor distribution and transport; the transport of heat, salinity, and momentum within the oceans; and the fluxes of energy at the ocean-atmosphere and land-atmosphere surfaces addresses major limitations in current understanding of the coupling between the ocean and the atmosphere. In each case, these elements have been identified as major areas of uncertainty in the analysis of climate model sensitivity and the understanding of processes that govern natural variability.

We must collect a global climate data base of key variables and forcings sufficient to allow a statistical classification of natural variability and the identification of its predictable modes. These observations are also required both to determine how anthropogenic changes to the environment are altering or influencing natural climate variability and its predictable modes and to improve model predictions.

In addition, it is extremely difficult to detect human-induced climate change or to understand and predict natural variability without an adequate assessment of all the primary forcing factors, including solar output, aerosols, greenhouse gases, and changes in the land surface. Determination of the degree to which solar output and/or greenhouse gases are a significant factor in governing the nature of the climate record can be addressed only by improved observations and concerted effort to compare the record with the various forcing factors. At the same time, enhancements to the suite of climate observations must be limited enough that they can be maintained economically.

Practical technologies exist for most of the needed observations. Many are planned as part of major observing systems [e.g., the National Aeronautics and Space Administration (NASA) Earth Observing System (EOS)] or major international research programs [e.g., the Global Energy and Water Cycle Experiment (GEWEX), Climate Variability and Prediction Program (CLIVAR), GOALS–Dec-Cen, and the World Ocean Circulation Experiment (WOCE)]. The challenge will be to form the best total, composite, observational system from this diverse set of efforts and to ensure the continuity and geographic coverage of the data in the face of budgetary constraints and pressure to support activities that promise faster results. Concerted debate and planning for a GCOS are required.

The continuity of climate-related observations is subject to yearly reshaping and descoping resulting from a process of almost continuous budget scrutiny and pressure; thus, continuity is not at all certain. A long-term commitment is particularly uncertain in NASA EOS and other research, rather than operational, agency missions. GEWEX and GOALS are also highly susceptible to budget pressures. Both were developed to address critical elements associated with limitations in understanding moisture and energy fluxes, and to evaluate and enhance our ability to develop seasonal-to-interannual prediction. However, the excitement and dedication of the scientists involved in these major efforts correspond to a period of considerable budgetary uncertainty.

In some areas, technological developments are required to make the needed climate observations practical outside the research mode. The development of a climate observing system, by taking full advantage of current observations and adding the key variables required to describe the climate system (e.g., water vapor) and its forcing factors (e.g., aerosols and solar output), will be a substantial advance (NRC, 1993, 1996a). If observations of the key variables and the major climate forcing factors are not part of a continuous, high-quality observing system, success will be very difficult or impossible. NASA EOS promises to provide new observations of water vapor, precipitation (clouds, aerosols), and solar forcing. Many of these measurements must be "calibrated" against in situ observations, and the in situ instrumentation is inadequate (mechanical rain gauges suffer from aerodynamic biases, and systems to characterize the composition and optical properties of aerosols and clouds are too cumbersome for routine use). In other cases, satellite technologies are immature, and technological improvements are needed before long-term global observations become practical (e.g., ocean salinity, soil moisture, sea ice thickness).

These developments involve both mission agencies such as NASA and NOAA and research support agencies such as the National Science Foundation (NSF). Given the role and involvement of so many agencies with different purposes and responsibilities, the continuity of observations, calibration of new with in situ systems, and maintenance of the breadth of observations can be at risk. Full interagency commitment is needed to maintain a cost-effective and balanced observing system.

The following **requirements** are essential to achieve this objective:

1. Implement and analyze new observations necessary for understanding the processes that couple the components of the Earth system, and improve our understanding of climate variability on decade-to-century time scales.

2. Enhance current operational facilities, with continued implementation of supportive process studies and commitment to long-term Earth observations.

3. Ensure full interagency commitment to both the in situ and the satellite observations necessary to address the major uncertainties in our understanding of the climate system, including a commitment to long-term Earth observations of critical variables such as the major climatic forcing factors.

Objective 3

Utilize historical and paleoclimatic observations to describe the nature of global and regional climate variability.

Modern observations provide a wealth of climate information. However, the period of data collection is insufficient to examine climate variability on longer, decade-to-century time scales. Innumerable studies over the past several decades have demonstrated that data from tree rings, lake sediments, corals, and ice cores provide invaluable records on these scales, and recent work has demonstrated that ocean sediment records from high-deposition-rate areas can also supply quality records of long-term climate variability using existing technology. The continued collection and analysis of these paleoclimatic data are crucial, both for understanding the past history of the Earth's climate and for providing information needed to determine the ability of general circulation models to simulate large-scale climate change. To achieve as broad a representation of natural variability as possible, a substantial global data base has to be developed for each data type and for a variety of physical variables. Furthermore, it must be recognized that in some cases, this data collection must be completed in a timely fashion or the records will no longer be available. In particular, as alpine glaciers and ice caps retreat, a portion of the records they hold is permanently lost unless previously saved by ice core drilling.

Observations from the paleoclimate records require interpretation and synthesis. They are not direct observations of state variables; rather, they provide proxies from which climatic interpretations are derived. Considerable effort must be devoted to improving current techniques of interpreting tree rings, lake sediments, ice cores, and coral records, as well as developing new proxy indicators.

Recorded historical events also provide an important source for reconstructing the climate record of the past few thousand years. These data provide a valuable means to cross-validate physical paleoclimatic data such as tree rings, lake sediments, and coral records.

The following **requirements** are essential to achieve this objective:

1. Widely sample the alpine glaciers and ice caps before this important repository of information on natural variability is lost.
2. Continue efforts to collect and analyze data from around the world from tree rings, lake sediments, corals, and ice cores, and actively pursue high-resolution records from ocean sediments.
3. Focus research efforts on the development and validation of proxy indicators.

Objective 4

Understand the major processes that govern climate variability through analysis of the observed record, correlation with natural and human forcing factors, focused process studies, and construction and analysis of coupled models of the climate system.

The enhanced climate observing program (Objective 2) and the proposed efforts to analyze and synthesize the historical and paleoclimatic record are crucial for identifying the mechanisms that govern climate variability and determining how this variability is related to natural and human forcing factors. When there are strong indications that the observed variability depends crucially on processes that are poorly understood, focused research programs (including field and theoretical studies) should be developed to enhance knowledge of these processes. Present examples include the processes that control the transfer of momentum, heat, and moisture within the atmospheric boundary layer (including low-level clouds and radiation). In the ocean, a portion of the climate variability must be sensitive to the mechanisms that control the exchange of carbon and nutrients between the surface and the permanent thermocline; these mechanisms are also poorly understood. Priorities should be on process studies that reduce the uncertainty in important feedbacks in the climate system (e.g., sea ice and cloud feedbacks) and will lead to improvements in the aggregate parameterization of key small-scale (unresolved) processes (e.g., transfer of water, energy, and carbon between the atmosphere and the land biosphere). The process study research of the last decade has successfully focused on critical regions. For example, TOGA field and modeling campaigns centered on a region of high seasonal-to-interannual climate variability for which a global-scale signal was identified. Regions of strong signal are logical research priorities if we are to fulfill the objective of identifying the mechanisms that govern climate variability.

There will be instances when variability in the climate system is implied solely from the analysis of climate system models and fundamentally involves processes that are poorly resolved. For example, recent work indicates that the simulated climate variability can be extremely sensitive to details of the parameterizations in models, and these sensitivities are accentuated when component models are coupled to create climate system models. In such instances, considerable efforts should be made to improve the parameterizations of unresolved processes.

The development of coupled climate models is integral to understanding the processes that govern climate variability. For example, model studies have demonstrated that ocean-atmosphere interaction is a plausible mechanism for decade-to-century variability. Coupled models also played a major role in enhancing our knowledge of ENSO and in demonstrating predictive skill. In addition, the development of coupled models focuses attention on the physical processes operative at the interfaces of the atmosphere, ocean, biosphere, and cryosphere. Knowledge of the energy and mass fluxes at these interfaces has been repeatedly identified as a major area of uncertainty. For these reasons, the development of coupled climate system models is a priority in climate research. The development of these computationally intensive models requires a strong computational infrastructure and focused effort. The development of models that include explicit representation of atmosphere, ocean, biosphere, and cryosphere systems requires considerable cross-disciplinary communication and collaboration.

The following **requirements** are essential to achieve this objective:

1. Enhance the climate observing system capability, with dedicated monitoring programs, as previously described.
2. Implement focused research initiatives on processes and in regions that are identified as important to understanding variability in the climate system.
3. Enhance the computational infrastructure and the focused efforts to develop climate system models that include explicit representation of atmosphere, ocean, biosphere, and cryosphere.
4. Enhance cross-disciplinary communication and collaboration.

Objective 5

Increase the skill of climate predictions.

Two paradigms for seasonal-to-interannual climate variability prediction are currently in use. The first paradigm employs empirical-statistical methodologies for seasonal and annual forecasts. A second paradigm focuses on the prediction of SST variations in the tropical Pacific and rainfall associated with these variations. This second case involves initializing the upper ocean with in situ measurements taken from the TOGA observing system, assimilating these observations into an ocean model to obtain the initial ocean conditions for the prediction, and then coupling the ocean to an initialized atmosphere and allowing the coupled system to evolve to a later prediction time. Since rainfall is so highly coupled to SST variations, the system effectively gives a prediction of rainfall. These predictions are then used by nations bordering the tropical Pacific for applications in various economic sectors.

The skill of climate prediction can be increased in a number of different ways. Where skill has already been demonstrated, the problem is to increase it by improvements in data quality and quantity; improvements in the coupled models; improvements in data assimilation procedures; and continuous evaluations of the prediction system by making regular and systematic predictions with constant comparisons to verifying data.

Where predictability has not been demonstrated, indications of predictability must first be sought (e.g., by observed correlations of climatic parameters with SST) and then demonstrated in a model simulation mode. The proper initializing data and the correct assimilation procedures can then be examined in a model context. If predictability is indicated in this mode, the system is initialized and evaluated with whatever real data exist. The process is cumbersome and requires a significant amount of computer resources in the early stages and the establishment of in situ observations in the later stages. The demonstration of useful predictability then leads to the next step: the establishment of an observing system and the implementation of a regular and systematic forecast cycle in order to exploit the skill of prediction.

The prediction effort requires a preponderance of research in the early stages and becomes more and more "operational" as prediction systems are built and regularized. Prediction for the tropical Pacific has passed through its earliest research stages and now is nearing a transition to a more stable and permanent status. Crucial to this transition is the maintenance of the TOGA observing system until its density and quality of measurements can be thoroughly assessed by observing system evaluation experiments and tested against the skill of prediction and the expansion of effort to larger regions.

The strategy for research must initially (1) exploit the predictability of the ENSO cycle to the fullest extent and then (2) focus on seeking and developing predictive skill associated with the global coupled system, including other time scales not directly related to ENSO.

Expansion of this skill of prediction on seasonal-to-interannual time scales to larger regions of the globe is the major emphasis of the newly established GOALS research program, a core program of the WCRP as part of CLIVAR. It seeks to increase the skill of prediction in the tropical Pacific in a research mode and then expand predictability to larger regions of the globe in a phased manner. The first phase begins with the tropics, based on the simplifying features of the interactions of heat sources in the tropics that determine most features of the tropical circulation. Established correlations of ENSO with monsoons form the guideposts for this initial expansion. Tantalizing hints of correlation of ENSO with climatic conditions on the West Coast, northwest corner, and southeastern part of the United States are already leading to experimental predictions of rainfall over the North American continent. Land-ocean-atmosphere coupling will play an increasing role as research is expanded beyond the tropics.

Expansions of the skill of prediction to time scales beyond a year are conjectural but are based on the slow time scales of the ocean relative to the atmosphere: if the ocean can be initialized, the coupled dynamics of the atmosphere-ocean system may control the way water reaches the surface enough to retain some imprint of the initial conditions against the inevitable noise superimposed on the system by the atmosphere. Proving predictability under these circumstances requires a greater understanding of slow motions in the ocean, in particular how the interior communicates with the surface of the ocean and then the atmosphere: This forms the research content of the Dec-Cen program, the other component of CLIVAR.

Characterization of the level of predictability of the seasonal-to-interannual time-scale variability in the climate system beyond current tropical Pacific SSTs will be a significant sign of success. This achievement may lead to regular (operational) seasonal-to-interannual prediction with a level of skill that can be utilized internationally. It depends on the development of facilities for integrated assessment, on the objectives of research, and on disseminating forecast information (including uncertainties) and working with nations in terms of how to use it. Fulfillment of these objectives will be a major contribution to addressing societal needs.

A well-defined record of decade-to-century variability derived from historical and paleoclimatic records and the identification, through exploration with coupled climate models and analysis of observations, of the fields and geographical distributions where decade-to-century variations may have predictability, will be a significant advance in our understanding of climate variability.

The following **requirements** are essential to achieve this objective:

1. Maintain major research observation systems that have clear value for improved climate prediction. A key example is the TOGA observing system.

2. Support the development and implementation of a comprehensive research program to study and advance seasonal-to-interannual prediction. Such a program is currently the objective of the WCRP's GOALS.

3. Support the development and implementation of a comprehensive re-

search program to study the mechanisms for decadal-to-centennial variability and the implications for longer time-scale predictability. Currently, planning for this element is incorporated in the Dec-Cen and anthropogenic climate change components of the WCRP.

Objective 6

Continue to improve the analysis and predictive skill of the degree to which humans are affecting climate, including changes in variability and the probability of extreme events.

Analysis of how humans can potentially affect climate and its variability is carried out with a hierarchy of global climate models and observational data sets. Studies with these models indicate that the nature of global and regional climate is in danger of changing due to human activities, most notably in response to increases in greenhouse gases, aerosols, and changes in land use. However, the nature and timing of this change are uncertain. The prediction of future climate change is problematic, in part, because of an inadequate understanding of climate variability, the difficulty of predicting future greenhouse gas and aerosol concentrations, and a limited understanding of the behavior of the coupled climate system. Current climate predictions based on projected increases in greenhouse gases and aerosols indicate the potential for large and rapid climate change relative to the historical record. Improved knowledge of the fully coupled climate system can lead to an enhanced predictive capability that could support societal efforts to adjust to, forestall, or even eliminate some of the negative impacts of projected climate change. This enhanced ability to predict future climate will have a positive impact on economic vitality and national security.

The research of the last decade has clearly identified a number of key factors that require a reduction in uncertainty if progress is to be made in climate prediction:

First, the current observational system does not measure all of the key global factors that force climate change. For example, despite years of debate about the role of solar variations in explaining observed climate fluctuations, we lack a long-term, consistent, calibrated measure of solar input to the Earth system. Similarly, measures of global aerosol concentrations and character are inadequate to assess its role in climate. Without an enhanced climate observing system, such debates are likely to continue without satisfactory resolution.

Second, substantial debate concerning the nature of climate sensitivity to increases in carbon dioxide stems from uncertainties in the measurement of water vapor in the upper troposphere and in the nature of climate-water vapor feedbacks. The nature of this debate demands improved measurement of water vapor.

Third, much of the uncertainty involves ocean-atmosphere coupling, land-vegetation-atmosphere coupling, sea ice modeling, and cloud-climate interactions. Process studies that combine the use of fine-scale regional models, field programs, and diagnostic analysis to bridge the spatial and temporal gaps between observations and typical scales of climate models offer great promise of improving model parameterizations. Diagnostic analysis of paleoclimate and historical data sets can also increase understanding of processes involved in climate change. These studies carried out for a number of large-scale conditions will lead to generalized parameterizations for a range of physical processes (e.g., clouds, sea ice). Reduced uncertainty in modeling surface energy budgets through improved cloud parameterizations will increase the reliability of coupled atmosphere-ocean-land modeling. Furthermore, increased resolution of ocean models will enhance understanding of the coupled system. Systematic analysis of these various climate components should reduce climate drift of the coupled system.

Fourth, experience with weather forecasting models suggests that increased spatial resolution results in improved prediction. In addition, the aspects of climate and climate change prediction of greatest relevance to humans and to ecosystems are those that impact water, water resources, weather hazards, agricultural yields, and human health. Most GCM simulations are at spatial scales that are too coarse for credible climate impact analysis. Increased spatial resolution must be matched with better physical representations.

Fifth, model-data comparison is critical to diagnose and improve climate model predictions. In many cases, the suite of satellite and in situ data sets has been underutilized in efforts to validate climate models. Further, observations from the industrial period represent too short a time span for satisfactory model validation. Greater confidence in model predictions will be gained through efforts to reproduce industrial, preindustrial, and paleoclimatic data sets.

In addition, WCRP efforts to compare climate models based on standard sets of climate simulations through the AMIP (Atmospheric Model Intercomparison Project) process has resulted in increased scrutiny of model parameterizations. The success of this effort has resulted in paleoclimatic intercomparison projects, land surface parameterization comparisons, and intercomparison of limited-area mesoscale models. Continued effort to intercompare models and their parameterization will continue to provide substantial benefit.

Finally, increased coordination of climatic research has the potential to yield significant efficiencies. For decades, we have developed observational strategies, promoted and completed process studies and field campaigns, developed a host of atmospheric and oceanic models, and produced impact analyses of climate change based on model output. As yet, however, the path from a proposed new observational strategy or field campaign through to the development of improved model parameterizations or improved application is often not articulated clearly. The cost, in human and financial resources, of major observational systems and field campaigns is sufficient justification for developing clearly articulated strategies for climate research.

The development of more physically based parameterizations for clouds (including their interaction with radiation), coupled atmosphere-ocean models that do not rely on flux corrections to simulate current and historical climates, and multiple examples of coupled Earth system models that adequately represent the major components of the Earth system will be evidence of significant progress in efforts to project future changes in the climate system, including its response to human activities. The efforts to develop a more comprehensive observing system and to construct more comprehensive climate system models should lead to demonstrated progress in reducing uncertainties in the prediction of human-induced climate change.

The following **requirements** are essential to achieve this objective:

1. Develop an enhanced climate observing system capability, with dedicated monitoring programs, as described previously.

2. Focus on key opportunities for reducing major uncertainties in climate models, including improved observations of water vapor and greater understanding of climate-water vapor feedbacks and improved representation of atmospheric chemistry and indirect chemistry-climate interactions.

3. Develop focused process studies with the objective of addressing key uncertainties associated with boundary layer processes and vertical convection; improved linkages coupling the atmosphere, oceans, and land surface; and more explicit representation of land surface processes, including vegetation and soil characteristics.

4. Improve the opportunities to develop coupled models, and enhance efforts at model-observation and model-model comparisons that give particular attention to simulating the observed changes due to changes in solar irradiance, aerosol loadings, and greenhouse gas concentrations.

5. Focus effort on improving the credibility and usefulness of climate model predictions at spatial scales relevant to analysis of the responses of ecosystems, socioeconomic systems, and human health to climate change predictions.

6. Improve the reconstruction, simulation, diagnostic studies, and analysis of data sets from the industrial, preindustrial, and paleoclimatic periods in order to increase confidence in model predictions.

7. Develop clearly articulated linkages between strategies for observation, analysis, model development, and application of predictions to evaluating consequences of climate change.

Objective 7

Enhance the linkages between climate model predictions and aspects of the Earth system of immediate relevance to humans (e.g., extreme

> **Box II.5.1**
> **Necessities for the Twenty-First Century**
>
> Three clear issues emerge from the discussion of insights gained from research over the past few decades and the scientific and societal drivers of climate and climate change research. These necessities are as follows:
>
> 1. Document and understand the mechanisms of natural variability on time scales of seasons to centuries, and assess the predictability of natural variability.
> 2. Develop prediction, application, and evaluation capability when useful skill is demonstrated.
> 3. Project the response of the climate system to human activities.

weather events, growing seasons, agricultural yields, and spread of diseases), for the purposes of helping society realize maximum benefit from whatever skill is demonstrated.

Climate variations can have substantial economic impacts, as demonstrated on interannual time scales by the effects of ENSO variation on countries bordering the tropical Pacific. Yet current predictions of climate change generally are not accurate enough or at an appropriate spatial resolution to help provide detailed estimates of future impacts on natural ecosystems, agricultural yields, energy use, emergence and transmission of infectious diseases, and other human activities.

Given the magnitude of natural variability on longer time scales and the potentially large impact of human activities on climate, a particularly important objective is to provide reliable regional climate predictions with better characterization of probabilities of extreme events.

The nature of the responses of human societies to change depends on human behavior, demographics, vulnerability, and a host of other factors. It is therefore evident that a comprehensive assessment of the impacts of predicted climate changes will require close cooperative studies by social and physical scientists. Such assessment could be used in the formulation of policies that maximize benefits to society. These issues are summarized in Box II.5.1.

The following **requirements** are essential to achieve these necessities:

1. Develop and construct high-resolution, regional climate models along with empirical methods for producing estimates of climate change characteristics of immediate relevance to humans.

2. Develop mechanisms that promote formal interaction between physical scientists and social scientists, by working on common problems to improve the applications and assessments of climate change impacts.

PRIORITIES FOR CLIMATE RESEARCH

Climate research objectives and their associated requirements can be summarized as four major priorities:

1 Build a permanent climate observing system.
2. Extend the instrumented climate record through the development of integrated historical and proxy data sets.
3. Continue and expand diagnostic efforts and process study research to elucidate key climate variability and change processes.
4. Construct and evaluate models that are increasingly comprehensive, incorporating all major components of the climate system.

These four priorities offer a general framework, whereas the objectives and requirements described previously characterize more specific opportunities to promote significant advancement in climate and climate change research. To some, the list of requirements outlined in the previous section may appear overly ambitious and without priority. However, a comprehensive climate research program that serves societal needs is clearly within our grasp. In many cases, programs required to achieve the objectives outlined in this report are in place. In other cases, changes in requirements can be implemented with minimum budgetary impact. In still other cases, objectives can be fulfilled by increased collaboration and closer interagency planning and linkages. However, even some of the more logical, minimal-impact issues appear to be problematic. For example, in terms of the requirement for continuity and quality as part of the climate observing system, current policies verge on becoming a national and international embarrassment. Addressing these issues must be a priority. Finally, with careful planning to achieve greater efficiencies, the full spectrum of climate objectives should be realizable. There are two primary areas in which greater efficiency has the potential to allow an expanded, and more successful research agenda. The first involves convergence of satellite systems in order to credibly and carefully address both research and mission needs. The second area involves greater coordination of major field and process study campaigns in order to serve multiple scientific objectives. Although each of the listed requirements has substantial merit, we recognize that improvements and augmentations of U.S. climate research programs must still be paced, based on budgetary and other considerations. Consequently, the list of requirements described in the previous section is repeated below but within a prioritized framework. This prioritized framework is based on a relatively simple perspective. Improvements that have minimal budgetary impact but substantial merit should be implemented without hesitation. Requirements with significant programmatic or budgetary implications should have identifiable levels of priority or clear trade-offs with current efforts.

Build a Permanent Climate Observing System

Requirements with Minimal Budgetary Impact

- Where feasible, adopt consistent data collection and management rules to ensure the utility of operational and research system measurements for climate research.
- Develop and adopt interagency plans to ensure the protection of critical long-term observations, to limit gaps in continuity due to small budget changes in single agencies and to recognize the value of these observations in a balanced, integrated research program.
- Provide strong U.S. support and participation in the development of a global climate observing system.
- Ensure full and open international exchange of data and information.

Requirements with Significant Budgetary or Programmatic Impact

- Maintain major research observation systems, such as the TOGA TAO array, that have demonstrated clear predictive value.
- Focus on key opportunities for reducing major uncertainties in climate models, including improved observations of water vapor.
- Ensure full interagency commitment to both the in situ and the satellite observations necessary to address the major uncertainties in our understanding of the climate system, including a commitment to long-term Earth observations of critical variables such as the major climatic forcing factors.

The TOGA TAO array is already in existence and has demonstrated value for both research and operational forecasts; thus, its maintenance is a top priority for building a permanent observing system. At issue is moving the costs from research budgets to operational budgets. Operational studies, through four-dimensional assimilation studies, should provide some guidance as to the importance of the current characteristics of station density and distribution, enabling an assessment of minimum costs to operational agencies.

Current plans for the National Polar-orbiting Operational Environmental Satellite System (NPOESS), with NASA contributions of advanced sounder instruments, offer the potential to satisfy both operational and research needs for improved observations of water vapor. These plans support this requirement under current budgets that save dollars through NOAA-Department of Defense-NASA collaboration.

Critical remaining issues are (1) to ensure sufficient overlap of instruments in space to develop a long-term record and (2) to provide credible measurement of all major climate forcing factors. The addition of commitments for global aerosol measurement and solar energy input to the Earth system should be priorities.

Extend the Instrumented Climate Record Through Development of Integrated Historical and Proxy Data Sets

Requirements with Minimal Budgetary Impact

- Widely sample the alpine glaciers and ice caps before this important repository of information on natural variability is lost.
- Continue efforts to collect and analyze data from around the world from tree rings, lake sediments, corals, and ice cores, and actively pursue high-resolution records from ocean sediments.
- Focus research efforts on development and validation of proxy indicators.

Continue and Expand Diagnostic Efforts and Process Study Research to Elucidate Key Climate Variability and Change Processes

Requirements with Minimal Budgetary Impact

- Enhance cross-disciplinary communication and collaboration.
- Develop clearly articulated linkages between strategies for observation, analysis, model development, and application of predictions to evaluating consequences of climate change.

Requirements with Significant Budgetary or Programmatic Impact

- Implement focused research initiatives on processes and in regions that are identified as important for understanding variability in the climate system.
- Implement and analyze new observations necessary to understand the processes that couple the components of the Earth system, and improve our understanding of climate variability on decade-to-century time scales.
- Develop focused process studies with the objective of addressing key uncertainties associated with boundary layer processes and vertical convection; improved linkages coupling the atmosphere, oceans, and land surface; and more explicit representation of land surface processes, including vegetation and soil characteristics.
- Support the development and implementation of a comprehensive research program to study and advance seasonal-to-interannual prediction. Such a program is currently the objective of WCRP's GOALS.

- Support the development and implementation of a comprehensive research program to study the mechanisms of decadal-to-century variability and its implications for longer time scale predictability. Currently, the planning for this element is incorporated in the Dec-Cen and anthropogenic climate change components of the WCRP.

The climate community has already begun to focus attention on critical regions, through the follow-on to TOGA (GOALS), which expands the focus from the tropical Pacific to the tropics worldwide; the Dec-Cen portion of CLIVAR, which focuses on important regions for longer-term variability such as the North Atlantic; and GEWEX, which focuses on energy and moisture fluxes, particularly at the land-atmosphere interface. Each of these major programs has well-defined justifications and scientific plans and defined major areas of focused study. Completion of the objectives of these three major WCRP programs and focused study of processes at high latitudes are the top priority for continued process study research that can satisfy many aspects of the research requirements listed above. Research funds should be available following the termination of TOGA efforts for continued support of GOALS. Both GEWEX and Dec-Cen are entrained in U.S. budgets and those of other countries. However, at present, these programs lack sufficient resources to complete their objectives in a timely fashion. This type of problem has plagued programs of the WCRP in the past (NRC, 1992). Three solutions are offered. First, we should continue every effort to provide community-based planning and debate that carefully develops priorities and advises on implementation in an efficient manner. Second, every effort should be applied to develop greater coordination of major field and process study campaigns across WCRP and U.S. efforts in order to serve multiple scientific objectives. This may well provide some of the added resources to enable completion of climate research objectives. Third, we must recognize that these efforts are strong candidates for added support.

Construct and Evaluate Models That Are Increasingly Comprehensive, Incorporating All Major Components of the Climate System

Requirements with Minimal Budgetary Impact

- Improve opportunities and enhance efforts at model observation and model-model comparisons that give particular attention to simulating observed changes associated with solar irradiance, aerosol loadings, and greenhouse gas concentrations.
- Develop mechanisms that promote formal interaction between physical scientists and social scientists, by working on common problems to improve the applications and assessments of climate change impacts.

Requirements with Significant Budgetary or Programmatic Impact

- Enhance the computational infrastructure and the focused efforts to develop climate system models that include explicit representation of atmosphere, ocean, biosphere, and cryosphere.
- Focus on key opportunities for reducing major uncertainties in climate models, including greater understanding of climate-water vapor feedbacks and improved representation of atmospheric chemistry and indirect chemistry-climate interactions.
- Focus effort on improving the credibility and usefulness of climate model predictions at spatial scales relevant to analysis of the responses of ecosystems, socioeconomic systems, and human health to climate change predictions.
- Develop and construct high-resolution, regional climate models along with empirical methods for producing estimates of climate change characteristics of immediate relevance to humans.

The observation and process study research described above are key to reducing major uncertainties in climate models and improving the representation of the atmosphere, ocean, biosphere, and cryosphere interfaces. If these efforts move forward, the major requirement will be (1) to have dedicated computational and human resources for the development of coupled system models, and (2) to develop clearly articulated linkages between strategies for observation, analysis, model development, and application of predictions to evaluating consequences of climate change. Increased computational capability with its associated human resources is critical and costly, but is a priority for additional climate research funding. The second priority, which is complementary to improved coupled system models, is the development of higher-resolution models suitable for producing estimates of climate change of relevance to humans. Increased computational capability, associated with high levels of research effort, is the first step toward fulfilling this requirement.

CROSS-CUTTING REQUIREMENTS

Education

Education must be an important facet of the perspective and activities in climate and climate change research entering the twenty-first century. Three elements are of particular importance. First, the general level of public understanding on issues of climate and climate change is disheartening and clearly limits perceptions of the importance of climate research and the development and acceptance of national and international policy. Outreach, contributions to the popular press, and speaking to the general public must be encouraged and rewarded as mechanisms to increase public knowledge of climate and climate change. Second, the long-term health of research programs and their applications depends on a strong background in math and science, beginning with K-12 education, and a strong interest in atmospheric, oceanic, and related sciences. We must inject the importance and excitement of our disciplines into both K-12 and undergraduate educational programs in order to develop and attract the most capable climate researchers. The climate research community should take an active role in enhancing K-12 education through providing up-to-date materials and participating in teacher training. Third, we must maintain and strengthen graduate programs oriented toward the critical scientific questions that define limitations in our understanding of climate. Graduate training in climate at universities is best enhanced by (1) added interdisciplinary efforts directed toward climate as a discipline, and (2) educational efforts directed toward increasing the skills required to develop large-scale models of the Earth system and the skills needed to develop and maintain observational systems. Current training is often inadequate because much of the focused effort occurs at national laboratories and other nonuniversity facilities.

Institutional Arrangements

Diverse institutional arrangements are required to address climate research needs. Community perceptions are that our institutional arrangements are actually becoming less diverse. Over the past two decades the nature of our institutional arrangements both to fund and to conduct research has become homogenized. Research support has tended increasingly to favor projects with short-term payoffs regardless of whether the research is performed at universities or national laboratories. Both funding mechanisms and institutional evaluation of research (including promotion and salaries) have tended to limit long-term comprehensive projects that lack short-term results. For example, national laboratories have adapted a university faculty-type staff evaluation method based on publications and grants, despite very different missions. The trend of homogenization of our institutions must be reversed in order to promote better opportunities to develop long-term sustained efforts. Such efforts are required to develop and manage data sets and observational systems, and to develop comprehensive models of the climate system. Both funding efforts and the evaluation of research efforts must serve to promote critical projects that do not have annual payoffs, when warranted by the nature of the problem. Funding agencies tend to provide opportunities for individual projects and large-scale research efforts (i.e., centers or named programs). Intermediate-sized teams also are important in interdisciplinary efforts or for the issues needed to solve many climate problems. Funding agencies must be able to provide opportunities for a broad range of projects involving single and multiple investigators. Third, some elements of climate research, in particular the development of increased predictive skill, are more efficiently accomplished with dedicated facilities. Dedicated centers—for example, in climate prediction—must be established when warranted by the potential for increased efficiencies or by their potential to catalyze research efforts. Diversity in our institutions is important to promote efficient, sustained efforts that address the major scientific issues in climate research.

CONTRIBUTIONS TO NATIONAL GOALS AND NEEDS

A robust climate research program is likely to contribute substantially to national goals and needs. Nine major contributions can be identified:

1. operational predictions of interannual climate fluctuations up to one year in the future;
2. detection of natural climate variations on decadal time scales and increased understanding of their causes and impacts;
3. plausible climate change scenarios for regional climate and ecosystem change, suitable for impact analysis;
4. improved estimates of the relative global warming potential of various gases and aerosols, including their interactions and indirect effects of other chemical species;
5. improved ability to determine the regional sources and sinks for atmospheric carbon dioxide;
6. reduction in the range of predictions of the rate and magnitude of global warming over the next century;
7. predictions of anthropogenic interdecadal changes in regional climate, in the context of natural variability;
8. documentation of the level of greenhouse gas-induced global warming and documentation of other climatically significant changes in the global environment; and
9. improved understanding of the interactions of human societies with the global environment, enabling quantitative analyses of existing and anticipated patterns of change.

References

Aber, J.D., K.J. Nadelhoffer, P. Steudler, and J.M. Melillo. 1989. Nitrogen saturation in northern forest ecosystems. BioScience 39, 378-386.

Albritton, D., F.C. Fehsenfeld, and A.F. Tuck. 1990. Instrumentation requirements for global atmospheric chemistry. Science 250, 75-81.

Alley, R.B., D.A. Meese, C.A. Shuman, A.J. Gow, K.C. Taylor, P.M. Grootes, J.W.C. White, M. Ram, E.D. Waddington, P.A. Mayewski, and G.A. Zielinski. 1993. Abrupt increase in snow accumulation at the end of the Younger Dryas event. Nature 362, 527-529.

American Thoracic Society. 1996a. Health effects of outdoor air pollution, Part 1. J. Am. Respir. Crit. Care. Med. 153, 3-50.

American Thoracic Society. 1996b. Health effects of outdoor air pollution, Part II. J. Am. Respir. Crit. Care. Med. 153, 477-498.

AMS (American Meteorological Society). 1987. The bachelor's degree in meteorology or atmospheric science. Bull. Am. Meteorol. Soc. 68, 1570.

AMS. 1993. Policy statement: Hurricane detection, tracking and forecasting. Bull. Am. Meteorol. Soc. 74, 1377-1380.

AMS. 1995. Statement: The bachelor's degree in atmospheric science or meteorology. Bull. Am. Meteorol. Soc. 76, 552.

Anfossi, D., S. Sandroni, and S. Viarengo. 1991. Tropospheric ozone in the nineteenth century: The Montalieri series. J. Geophys. Res. 96, 17349-17352.

Atlas, R., A.J. Busalacchi, M. Ghil, S. Bloom, and E. Kalnay. 1987. Global surface wind and flux fields from model assimilation of SEASAT data. J. Geophys. Res.—Oceans 92, 6477-6487.

Auciello, E.P., and R.L. Lavoie. 1993. Collaborative research activities between National Weather Service operational offices and universities. Bull. Am. Meteorol. Soc. 74, 625-629.

Baker, W.E., G.D. Emmitt, F. Robertson, R.M. Atlas, J.E. Molinari, D.E. Bowdle, J. Paegle, R.M. Hardesty, R.T. Menzies, T.N. Krishnamurti, R.A. Brown, M.J. Post, J.R. Anderson, A.C. Lorenc, and J. McElroy. 1995. Lidar-measured winds from space: A key component for weather and climate prediction. Bull. Am. Meteorol. Soc. 76, 869-888.

Baldocchi, D., R. Valenti, S. Running, W. Oechel, and R. Dahlman. 1996. Strategies for measuring and modelling carbon dioxide and water vapor fluxes over terrestrial ecosystems. Global Change Biology 2, 159-168.

Battisti, D.S., and E.S. Sarachik. 1995. Understanding and predicting ENSO. For the IUGG Quadriennal Report, Contributions in Oceanography, Revs. Geophys. (Supp.), 1367-1376.

Bekki, S., and J.A. Pyle. 1993. 2-D assessment of the impact of aircraft sulphur emissions on the stratospheric sulphate aerosol layer. J. Geophys. Res. 97, 15839-15847.

Beljaars, A.C.M., P. Viterbo, M.J. Miller, and A.K. Betts. 1996. The anomalous rainfall over the United States during July 1993—Sensitivity to land-surface parameterization and soil-moisture. Monthly Weather Rev. 124, 362-383.

Bender, M.A., I. Ginis, and Y. Kurihara. 1993. Numerical simulation of tropical-cyclone ocean interaction with a high resolution coupled model. J. Geophys. Res. 98, 23245-23268.

Benjamin, M.M., and B.D. Honeyman. 1992. Trace metals. In Global Biogeochemical Cycles, S.S. Butcher et al. (eds.). Academic Press, New York.

Berger, U., and M. Dameris. 1993. Cooling of the upper atmosphere due to CO_2 increases: A model study. Ann. Geophysicae 11, 809-819.

Berz, G.A. 1998. Annual Review of Natural Catastrophes 1997. Münchener Rück Munich Re Topics, D-80791, Munich, Germany. 20 pp.

Bilitza, D. 1991. History and uses of the ionospheric, thermospheric, and atmospheric data base. J. Geomag. Geoelectr. 43, 901-909.

Brasseur, G.P., and C. Granier. 1992. Mt. Pinatubo aerosol, chlorofluorocarbons, and ozone depletion. Science 257, 1239-1242.

Brickhouse, N.W., C.S. Carter, and K.C. Scantlebury. 1990. Women and chemistry: Shifting the equilibrium toward success. J. Chem. Educ. 67, 116-118.

Bruce, J.P. 1994. Natural disaster reduction and global change. Bull. Am. Meteorol. Soc. 75, 1831-1835.

Burpee, R.W., J.L. Franklin, S.J. Lord, R.E. Tuleya, and S.D. Aberson. 1996. The impact of Omega dropwindsondes on operational hurricane track forecast models. Bull Am. Meteorol. Soc. 77, 925-933.

Businger, J.A., and S.P. Oncley. 1990. Flux measurement with conditional sampling. J. Atmos. Ocean. Technol. 7, 349-352.

Businger, S., S.R. Chiswell, M. Bevis, J. Duan, R.A. Anthes, C. Rocken, R.H. Ware, M. Exner, T. VanHove, and F.S. Solheim. 1996. The promise of GPS in atmospheric profiling. Bull. Am. Meteorol. Soc. 77, 5-18.

Cane, M., and S.E. Zebiak. 1987. Predictability of El Niño events using a physical model. Pp. 153-182 in Atmospheric and Oceanic Variability, H. Cattle (ed.). Royal Meteorological Society Press, London, U.K.

Cardelino, C.A., and W.L. Chameides. 1995. An observation-based model for analyzing ozone precursor relationships in the urban atmosphere. J. Air Waste Manag. Assoc. 45, 161-180.

Chahine, M.T. 1992. The hydrological cycle and its influence on climate. Nature 359, 373-380.

Chameides, W.L., and D.D. Davis. 1982. Global tropospheric chemistry, Special report. Chem. Eng. News 60, 38-52.

Chameides, W.L., R.W. Lindsay, J.L. Richardson, and C.S. Kiang. 1988. The role of biogenic hydrocarbons in urban photochemical smog: Atlanta as a case study. Science 241, 1473-1475.

Changnon, S.A., D. Changnon, E.R. Fosse, D.C. Hoganson, R.J. Roth, Sr., and J.M. Totsch, 1997: Effects of recent weather extremes on the insurance industry: major implications for the atmospheric sciences. Bull. Am. Meteor. Soc., vol. 78, pp 42.

Chapman, R.E. 1992. Benefit-Cost Analysis for the Modernization and Associated Restructuring of the National Weather Service. National Institute of Standards and Technology, NISTIR 4867. Department of Commerce, Washington, D.C.

Charlson, R.E., J.E. Lovelock, M.O. Andreae, and S.G. Warren. 1987. Oceanic phytoplankton, atmospheric sulfur, cloud albedo and climate. Nature 326, 655-661.

Charlson, R.J., J. Langner, and H. Rodhe. 1990. Sulphate aerosol and climate. Nature 348, 22.

Charlson, R.J., J. Langner, H. Rodhe, C.B. Leovy, and S.G. Warren. 1991. Perturbation of the Northern Hemispheric radiative balance by backscattering from anthropogenic sulfate aerosols. Tellus 43AB, 152-163.

Charlson, R.J., S.E. Schwartz, J.M. Hales, R.D. Cess, J.A. Coakley, Jr., J.E. Hansen, and D.J. Hofmann. 1992. Climate forcing by anthropogenic aerosols. Science 255, 423-430.

Chen, M.W., M. Schulz, L.R. Lyons, and D.J. Gorney. 1993. Stormtime transport of ring current and radiation belt ions. J. Geophys. Res.-Space Physics 98, 3835-3849.

China MAP Project. 1997. The Yangtze Delta of China as an Evolving Metro-Agro-Plex, China-MAP. China-MAP Project Office-USA, School of Earth and Atmospheric Sciences, Georgia Institute of Technology, Atlanta. 15 pp.

Cicerone, R.J., S. Stolarski, and S. Walters. 1974. Stratospheric ozone destruction by man-made chlorofluoromethanes. Science 185, 1165-1167.

Clemesha, B.R., C.M. Simonich, and P.P. Batista. 1992. A long-term trend in the height of the atmospheric sodium layer: Possible evidence for global change. Geophys. Res. Lett. 19, 457-460.

Colwell, R.R., and A. Huq. 1994. Environmental reservoir of *Vibrio cholerae*: The causative agent of cholera. Ann. N.Y. Acad. Sci. 740, 44-54.

COMET. 1995. Announcing short-term rental of the COMET forecaster's multimedia library. Bull. Am. Meteorol. Soc. 76, 1964.

Cooper, K.D., L. Oberhelman, T.A. Hamilton, O. Baadsgaard, M. Terhune, G. LeVee, T. Anderson, and H. Koren. 1992. UV exposure reduces immunization rates and promotes tolerance to epicutaneous antigens in humans: Relationship to dose, CD1a-DR+ epidermal macrophage induction, and Langerhans cell depletion. Proc. Natl. Acad. Sci. USA 89, 8497-8501.

Crutzen, P.J. 1974. Estimates of possible future ozone reductions from continued use of fluorochloromethanes (CF_2Cl_2, $CFCl_3$). Geophys. Res. Lett. 1, 205-208.

Delecluse, P., M. Davey, Y. Kitamura, S.G.H. Philander, M. Suarez, and L. Bengtsson. 1998. Coupled general circulation modeling of the tropical Pacific. J. Geophys. Res. (in press).

Draxler, R.R., and J.L. Hefter. 1989. Across North American Tracer Experiment (ANATEX), Vol. I. Description, Ground Level Sampling at Primary Sites, and Meteorology. NOAA Technical Memorandum, ERL ARL-167. NOAA, Washington, D.C.

Duce, R.A. 1986. The impact of atmospheric nitrogen, phosphorous, and iron species on marine biological productivity. Pp. 497-530 in The Role of Air-Sea Exchange in Geochemical Cycling, P. Buat-Ménard (ed.). Published in cooperation with NATO Scientific Affairs Division. D. Reidel Publishing Co., Kluwer Academic Publishers, Norwell, Mass. 537+ pp.

Duce, R.A., P.S. Liss, J.T. Merrill, E.L. Atlas, P. Buat-Ménard, B.B. Hicks, J.M. Miller, J.P. Prospero, R. Arimoto, T.M. Church, W. Ellis, J.N. Galloway, L. Hansen, T.D. Jickells, A.H. Knap, K.H. Reinhardt, B. Schneider, A. Soudine, J.J. Tokos, S. Tsunogai, R. Wollast, and M. Zhou. 1991. The atmospheric input of trace species to the world ocean. Global Biogeochemical Cycles 5, 193-259.

Dutton, J.A. 1992. The atmospheric sciences in the 1990s: Accomplishments, challenges, and imperatives. Bull. Am. Meteorol. Soc. 73, 1549-1562.

Eddy, J. 1976. Maunder Minimum. Science 192, 1189-1202.

Emanuel, K., D. Raymond, A. Betts, L. Bosart, C. Bretherton, K. Droegemeir, B. Farrell, J.M. Fritsch, R. Houze, M. LeMone, D. Lilly, R. Rotunno, M. Shapiro, R. Smith, and A. Thorpe. 1995. Report of the first Prospectus Development Team of the U.S. Weather Research Program to NOAA and the NSF. Bull. Am. Meteorol. Soc. 76, 1194-1208.

EPA (Environmental Protection Agency). 1995. National Air Quality and Emission Trends Report, 1994—Annual Report. Office of Air Quality Planning and Standards, EPA, Research Triangle Park, N.C. 101 pp.

Epstein, P.R. 1995. Emerging diseases and ecosystem instability: New threats to public health. Am. J. Public Health 85, 168-172.

Farber, E. (ed.). 1961. Great Chemists. Interscience Publishers, New York. 1642 pp.

Farrell, W.M., T.L. Aggson, E.B. Rodgers, and W.B. Hanson. 1994. Observations of ionospheric electric fields above atmospheric weather systems. J. Geophys. Res. 99, 19475-19483.

Fels, S.B., J.D. Mahlman, M.D. Schwarzkopf, and R.W. Sinclair. 1980. Stratospheric sensitivity to perturbations in ozone and carbon dioxide: Radiative and dynamical response. J. Atmos. Sci. 37, 2265-2297.

Fishman, J., C.E. Watson, J.C. Larsen, and J.A. Logan. 1990. Distribution of tropospheric ozone determined from satellite data. J. Geophys. Res. 95, 3599-3617.

Fishman, J., K. Fakhruzzaman, B. Cros, and D. Nganga. 1991. Identification of widespread pollution in the Southern Hemisphere deduced from satellite analyses. Science 252, 1693-1696.

Fleming, J.R. 1990. Meteorology in America, 1800-1870. The Johns Hopkins University Press, Baltimore, Md. 264 pp.

Fleming, J.R. 1998. History, Climate and Culture. Oxford University Press, New York.

Fleming, R.J. 1996. The use of commercial aircraft as platforms for environmental measurements. Bull. Am. Meteorol. Soc. 77, 1548-1562.

Foukal, P., and J. Lean. 1990. An empirical model of total solar irradiance variation between 1874 and 1988. Science 247, 556-558.

Fritsch, J.M. 1992. Operational meteorological education and training: Some considerations for the future, (correspondence) Bull. Am. Meteorol. Soc. 73, 1843-1846.

Gadsden, M. 1990. A secular change in noctilucent cloud occurrence. J. Atmos. Terr. Phys. 52, 247-251.

Gates, W.L., A. Henderson-Sellers, G.J. Boer, C.K. Folland, A. Kitoh, B.J. McAvaney, F. Semazzi, N. Smith, A.J. Weaver, and Q.-C. Zeng. 1996. Climate models—Evaluation. Pp. 229-284 in Climate Change 1995—The Science of Climate Change, J.T. Houghton, L.G. Meira Filho, B.A. Callander, N. Harris, A. Kattenberg, and K. Maskell (eds.). Contribution of Working Group I to the Second Assessment Report of the Intergovernmental Panel on Climate Change. Cambridge University Press, Cambridge, U.K. 572 pp.

Giorgi, F., and R. Avissar. 1997. Representation of heterogeneity effects in Earth system modeling: Experience from land surface modeling. Rev. Geophys. 35, 413-438.

Giorgi, F., and L.O. Mearns. 1991. Approaches to the simulation of regional climate change: A review. Rev. Geophys. 29, 191-216.

Gleckler, P.J., D.A. Randall, G. Boer, R. Colman, M. Dix, V. Galin, M. Helfand, J. Kiehl, A. Kitoh, W. Lau, X.Y. Liang, V. Lykossov, B. Mcavaney, K. Miyakoda, S. Planton, and W. Stern. 1995. Cloud-radiative effects on implied oceanic energy transports as simulated by atmospheric general-circulation models. Geophys. Res. Lett. 22, 791-794.

Godbold, D.L., E. Fritz, and A. Hutterman. 1988. Aluminum toxicity and forest decline. Proc. Natl. Acad. Sciences USA 85, 388-3892.

Haagen-Smit, A.J. 1952. Chemistry and physiology of Los Angeles smog. Ind. Eng. Chem. 44, 1362.

Hack, J.J. 1998. Analysis of the improvement in implied meridional ocean energy transport as simulated by the NCAR CCM3. J. Climate 11 (in press).

Halpert, M.S., and C.F. Ropelewski. 1992. Surface temperature patterns associated with the Southern Oscillation. J. Climate 5, 577-593.

Hamilton, K., and R.R. Garcia. 1984. Long-period variations in the solar semidiurnal atmospheric tide. J. Geophys. Res. 89, 11705-11710.

Han, Y., and E.R. Westwater. 1995. Remote sensing of tropospheric water vapor and cloud liquid water by integrated cloud-based sensors. J. Atmos. Ocean. Tech. 12, 1050-1059.

Hansen, J.E., and A.A. Lacis. 1990. Sun and dust versus greenhouse gases: An assessment of their relative roles in global climate change. Nature 346, 713-719.

Hansen, J.E., A. Lacis, R. Ruedy, M. Sato, and H. Wilson. 1993a. How sensitive is the world's climate? National Geographic Research and Exploration 9, 142-158.

Hansen, J., W. Rossow, and I. Fung (eds.). 1993b. Long-Term Monitoring of Global Climate Forcings and Feedbacks. NASA Conference Publication 3234, available from NASA Goddard Space Flight Center, Greenbelt, Md.

Harrison, E.F., P. Minnis, B.R. Barkstrom, V. Ramanathan, R.D. Cess, and G.G. Gibson. 1990. Seasonal variation of cloud radiative forcing derived from the Earth Radiation Budget Experiment. J. Geophys. Res. 95, 18687-18703.

Hebert, P., J.D. Jarrell, and M. Mayfield. 1996. The Deadliest, Costliest, and Most Intense United States Hurricanes of This Century. NOAA Technical Memorandum NWS TPC-1, National Hurricane Center, Miami, Fla.

Herman, J.R., P.K. Bharta, J. Ziemke, Z. Ahmad, and D. Larko. 1996. UV-B increases (1979-1992) from decreases in total ozone. Geophys. Res. Lett. 23, 2117-2120.

Holland, G.J., T. McGeer, and H. Youngren. 1992. Autonomous aerosondes for economical atmospheric soundings anywhere on the globe. Bull. Am. Meteorol. Soc. 73, 1987-1998.

Holton, J.R., P.H. Haynes, M.E. McIntyre, A.R. Douglass, R.B. Rood, and L. Pfister. 1995. Stratosphere-troposphere exchange. Rev. Geophys. 33, 403-439.

Houghton, D.D., T.S. Glickman, J. Dannenberg, and S.L. Marsh. 1996. Bull. Am. Meteorol. Soc. 77, 325-333.

Hoyt, D.V., K.H. Schatten, and E. Nesme-Ribes. 1994. A new reconstruction of solar activity, 1610-1993. Pp. 71-98 in The Solar Engine and Its Influence on Terrestrial Atmosphere and Climate, E. Nesme-Ribes (ed.). Springer-Verlag, New York. Published in cooperation with NATO Scientific Affairs Division. 549+ pp.

Huebert, B.N. 1993. Marine aerosol and gas exchange and global atmospheric effects. First IGAC Scientific Conference, Eilat, Israel.

Hurrel, J.W. and K.E. Trenberth. 1998. J. Climate 11, 945-967.

IARC (International Agency for Research on Cancer) 1992. IARC Monographs on the Evaluation of Carcinogenic Risks to Humans: Solar and Ultraviolet Radiation. Monograph 55. IARC, Lyon, France.

IGAC (International Global Atmospheric Chemistry Project). 1995. Southern Hemisphere Marine Aerosol Characterization Experiment (ACE-1). Radiative Effects of Aerosols in the Remote Marine Atmosphere. Final Science and Implementation Plan. Available from T. Bates, NOAA/ PMEL, 7600 Sandpoint Way NE, Seattle, WA 98115 (bates@pmel.noaa.gov).

IPCC (Intergovernmental Panel on Climate Change). 1990. Climate Change—The IPCC Scientific Assessment, J.T. Houghton, G.J. Jenkins, and J.J. Ephraums (eds.). Cambridge University Press, Cambridge, U.K. 365 pp.

IPCC. 1995. Climate Change 1994—Radiative Forcing of Climate Change and an Evaluation of the IPCC IS92 Emission Scenarios, J.T. Houghton, L.G. Meira Filho, J. Bruce, Hoesung Lee, B.A. Callander, E. Haites, N. Harris, and K. Maskell (eds.). Reports of Working Groups I and II of the Intergovernmental Panel on Climate Change, forming part of the IPCC Special Report to the first session of the Conference of the Parties to the UN Framework Convention on Climate Change. Cambridge University Press, Cambridge, U.K. 339 pp.

IPCC. 1996. Climate Change 1995—The Science of Climate Change: Contribution of Working Group I to the Second Assessment Report of the Intergovernmental Panel on Climate Change, J.T. Houghton, L.G. Meira Filho, B.A. Callander, N. Harris, A. Kattenberg, and K. Maskell (eds.). Cambridge University Press, Cambridge, U.K. 572 pp.

Ji, M., A. Leetmaa, and V.E. Kousky. 1996. Coupled model predictions of ENSO during the 1980s and the 1990s at the National Centers for Environmental Prediction. J. Climate 9, 3105-3120.

Johnson, G.J., and S. Tinning. 1995. Effects of UVB radiation on the human eye. In Proceedings of Conference on Human Health and Global Climate Change. National Academy Press, Washington, D.C.

Johnson, S.R., and M.T. Holt. 1997. The values of weather information, Chapter 3 in *Economic Value of Weather and Climate Forecasts,* R.W. Katz and A.H. Murphy (eds.). Cambridge University Press, Cambridge, U.K.

Joselyn, J.A., and E.C. Whipple. 1990. Effects of the space environment on space science. American Scientist 78, 126-133.

Kalkstein, L.S. 1995. Lessons from a very hot summer. Lancet 346, 857-859.

Karl, T.R., C.N. Williams, P.J. Young, and W.M. Wendland. 1986. A model to estimate the time of observation bias associated with monthly mean maximum, minimum and mean temperatures for the United States. J. Clim. Appl. Meteorol. 25, 145-160.

Karl, T.R., G. Quayle, and P.Y. Groisman. 1993. Detecting climate variations and change: New challenges for observing and data management systems. J. Climate 6, 1481-1494.

Katz, R.W., and A.H. Murphy. 1997. Economic Value of Weather and Climate Forecasts. Cambridge University Press, Cambridge, U.K. 222 pp.

Keckhut, P., A. Hauchecorne, and M.L. Chanin. 1995. Midlatitude long-term variability of the middle atmosphere: Trends and cyclic and episodic changes. J. Geophys. Res. 100, 18887-18897.

Kellogg, W. 1977. Results of the AMS Questionnaire of 1975. Bull. Am. Meteorol. Soc. 58, 39-44.

Kerr, R.B., and X. He. 1994. Global change in the exosphere: Evidence from Arecibo Balmer-alpha and radar observations. EOS, 1994 American Geophysical Union Fall Meeting Supplement, 491, November 1.

Kiehl, J.T., and B.P. Briegleb. 1993. The relative roles of sulfate aerosols and greenhouse gases in climate forcing. Science 260, 311-314.

Kiemle, C., M. Kastner, and G. Ehret. 1995. The convective boundary layer structure from lidar and radiosonde measurements during the EFEDA '91 campaign. J. Atmos. Oceanogr. Tech. 12, 771-782.

Kleeman, R., A.M. Moore, and N.R. Smith. 1995. Assimilation of subsurface thermal data into a simple ocean model for the initialization of an intermediate tropical coupled ocean-atmosphere forecast model. Monthly Weather Rev. 123, 3103-3113.

Klein, S.A., and D.L. Hartmann. 1993. Spurious trends in International Satellite Cloud Climatology Project (ISCCP) C2 data set. Geophys. Res. Lett., pp. 455-458.

Kogan, F.N. 1995. Droughts of the late 1980s in the United States as derived from NOAA polar-orbiting satellite data. Bull. Am. Meteorol. Soc. 76, 655-668.

Kunst, A.E., C.W.N. Looman, and J.P. Mackenbach. 1993. Air pollution, lagged effects of temperature and mortality: The Netherlands 1979-81. J. Epidemiology and Community Health 47, 121-126.

Kuo, Y.-H., R.J. Reed, and S. Low-Nam. 1991. Effects of surface energy fluxes during the early development and rapid intensification stages of seven explosive cyclones in the western Atlantic. Monthly Weather Rev. 119, 457-476.

Kuo, Y.-H., Y.-R. Guo, and E.R. Westwater. 1993. Assimilation of precipitable water measurements into a mesoscale numerical model. Monthly Weather Rev. 121, 1215-1238.

Lacis, A.A., D.J. Wuebbles, and J.A. Logan. 1990. Radiative forcing of climate by changes in the vertical distribution of ozone. J. Geophys. Res. 95, 9971-9982.

Landsberg, H.E. 1969. Weather and Health: An Introduction to Biometeorology. Doubleday and Company, New York. 148 pp.

Latif, M., T.P. Barnett, M.A. Cane, M. Flugel, N.E. Graham, H. von Storch, J.-S. Xu, and S.E. Zebiak. 1994. A review of ENSO prediction studies. Climate Dyn. 9, 167-179.

Latif, M., D. Anderson, T. Barnett, M. Cane, R. Kleeman, A. Leetmaa, J.J. O'Brien, A. Rosati, and E. Schneider. 1998. A review of predictability and prediction of ENSO. J. Geophys. Res. (in press).

Lau, N.-C., and M.J. Nath. 1994. A modeling study of the relative roles of the tropical and extratropical SST anomalies in the variability of the global atmosphere-ocean system. J. Climate 7, 1184-1207.

Lefohn, A.S. (ed.). 1992. Surface Level Ozone Exposures and Their Effects on Vegetation. Lewis Publishers, Chelsea, Mich. 366 pp.

LeMone, M.A., and P.L. Waukau. 1982. Women in meteorology. Bull. Am. Meteorol. Soc. 63, 1266-1276.

Lilly, D., and D.J. Perkey. 1976. Sensitivity of mesoscale predictions to mesoscale initial data. Bull. Am. Meteorol. Soc. 57, 171.

Lindquist, O., K. Johansson, M. Aastrup, A. Andersson, L. Bringmark, G. Hovsenius, L. Hakanson, A. Iverfeldt, M. Meili, and B. Timm. 1991. Mercury in the Swedish environment—Recent research on causes, consequences and corrective methods. Water, Air, Soil Pollut. 55, 1-261.

Logan, J.A. 1994. Trends in the vertical distribution of ozone: An analysis of ozonesonde data. J. Geophys. Res. 99, 25553-25585.

Lorenz, E.N. 1963. The predictability of hydrodynamic flow. Trans. N.Y. Acad. Sci. 25, 409-432.

Lyons, W.A. 1994. Low-light video observations of frequent luminous structures in the stratosphere above thunderstorms. Monthly Weather Rev. 122, 1940-1946.

Marenco, A., H. Gouget, P. Nedelec, J.-P. Pages, and F. Karcher. 1994. Evidence of a long-term increase in tropospheric ozone from Pic du Midi data series—Consequences: Positive radiative forcing. J. Geophys. Res. 99, 16617-16632.

Mass, C.F. 1996. Are we graduating too many atmospheric scientists? Bull. Am. Meteorol. Soc. 77, 1255-1267.

Mayr, E. 1982. The Evolution of Biological Thought: Diversity, Evolution, and Inheritance. Belknap Press/Harvard University Press, Cambridge and London. 974 pp.

McCormick, M.P., R.E. Veiga, and W.P. Chu. 1992. Stratospheric ozone profile and total ozone trends derived from the SAGE I and SAGE II data. Geophys. Res. Lett. 19, 269-272.

McCormick, M.P., E.W. Chiou, L.R. McMaster, W.P. Chu, J.C. Larsen, D. Rind, and S. Oltmans. 1993. Annual variations of water vapor in the stratosphere and upper troposphere observed by the Stratospheric Aerosol and Gas Experiment. J. Geophys. Res. 98, 4867-4875.

McPhaden, M.J., A.J. Busalacchi, R. Cheyney, J.-R. Donguy, K.S. Gage, D. Halpern, M. Ji, P. Julian, G. Meyers, G.T. Mitchum, P.P. Niiler, J. Picaut, R.W. Reynolds, N. Smith, and K. Takeuchi. 1998. The Tropical Pacific Global Atmosphere (TOGA) observing system: A decade of progress. J. Geophys. Res. (in press).

Melfi, S.H., and D.N. Whiteman. 1985. Observation of lower atmospheric moisture structure and evolution using a Raman lidar. Bull. Am. Meteorol. Soc. 66, 1288-1292.

Menzel, W.P., and J.F.W. Purdom. 1994. Introducing GOES-I: The first of a new generation of operational environmental satellites. Bull. Am. Meteorol. Soc. 75, 757-781.

Minnis, P. 1994. Radiative forcing by the 1991 Mt. Pinatubo eruption. Sixth Conference on Climate Variations, American Meteorological Society, Nashville, Tenn.

Minnis, P., E.F. Harrison, L.L. Stowe, G.G. Gibson, F.M. Denn, D.R. Doelling, and W.L. Smith, Jr. 1993. Radiative climate forcing by the Mount Pinatubo eruption. Science 259, 1411-1415.

Molina, M.J., and F.S. Rowland. 1974. Stratospheric sink for chlorofluoromethanes: Chlorine atom catalyzed destruction of stratospheric ozone. Nature 249, 810-812.

Morse, S. 1995. Factors in the emergence of infectious diseases. Emerging Infec. Dis. 1, 7-15.

Moura, A.D. 1994. Prospects for seasonal-to-interannual climate prediction and applications for sustainable development. World Meteorological Society Bulletin 43, 207-215.

Murphy, A.H. 1994. Assessing the economic value of weather forecasts: An overview of methods, results, and issues. Meteorological Applications 1, 69-73.

NCTM (National Council of Teachers of Mathematics). 1989. Curriculum and Evaluation Standards for School Mathematics. Commission on Standards for School Mathematics. The Council, Reston, Va. 258 pp.

Neelin, J.D., D.S. Battisti, A.C. Hirst, F.F. Jin, Y. Wakata, T. Yamagata, and S. Zebiak. 1998. ENSO theory. J. Geophys. Res. (in press).

Nicholls, N., G.V. Gruza, J. Jouzel, T.R. Karl, L.A. Ogallo, and D.E. Parker. 1995. Observed climate variability and change. Pp. 133-192 in Climate Change 1995—The Science of Climate Change, J.T. Houghton, L.G. Meira Filho, B.A. Callander, N. Harris, A. Kattenberg, and K. Maskell (eds.). Contribution of Working Group I to the Second Assessment Report of the Intergovernmental Panel on Climate Change. Cambridge University Press, Cambridge, U.K. 572 pp.

NOAA (National Oceanic and Atmospheric Administration). 1996. North American Atmospheric Observing System Program Plan. NOAA, Department of Commerce, Washington, D.C.

NRC (National Research Council). 1984. Global Tropospheric Chemistry: A Plan for Action. National Academy Press, Washington, D.C. 194 pp.

NRC. 1986. Studies in Geophysics—The Earth's Electrical Environment. Geophysics Study Committee. NTIS Order No. PB86-241874. National Academy Press, Washington, D.C. 264 pp.

NRC. 1990. TOGA: A Review of Progress and Future Opportunities. National Academy Press, Washington, D.C. 66 pp.

NRC. 1991. Rethinking the Ozone Problem in Urban and Regional Air Pollution. National Academy Press, Washington, D.C. 489 pp.

NRC. 1992. A Decade of International Climate Research: The First Ten Years of the World Climate Research Programme. National Academy Press, Washington, D.C. 59 pp.

NRC. 1993. Understanding and Predicting Atmospheric Chemical Change, An Imperative for the U.S. Global Change Research Program. National Academy Press, Washington, D.C. 31 pp.

NRC. 1994a. A Space Physics Paradox—Why Has Increased Funding Been Accompanied by Decreased Effectiveness in the Conduct of Space Physics Research. National Academy Press, Washington, D.C. 96 pp.

NRC. 1994b. Toward a New National Weather Service—Weather for Those Who Fly. Committee on National Weather Service Modernization. National Academy Press, Washington, D.C. 100 pp.

NRC. 1994c. GOALS (Global Ocean-Atmosphere-Land System) for Predicting Seasonal-to-Interannual Climate—A Program of Observation, Modeling, and Analysis. National Academy Press, Washington, D.C. 103 pp.

NRC. 1994d. Ocean-Atmosphere Observations Supporting Short-Term Climate Predictions. National Academy Press, Washington, D.C. 51 pp.

NRC. 1995a. Bits of Power—On the Full and Open Exchange of Scientific Data. Committee on Geophysical and Environmental Data. National Academy Press, Washington, D.C. 21 pp.

NRC. 1995b. A Science Strategy for Space Physics. Space Studies Board. National Academy Press, Washington, D.C. 81 pp.

NRC. 1995c. Natural Climate Variability on Decade-to-Century Time Scales. National Academy Press, Washington, D.C. 630 pp.

NRC. 1996a. A Plan for a Research Program on Aerosol Radiative Forcing and Climate Change. National Academy Press, Washington, D.C. 161 pp.

NRC. 1996b. National Science Education Standards. National Academy Press, Washington, D.C. 262 pp.

NRC. 1996c. Learning to Predict Climate Variations Associated with El Niño and the Southern Oscillation—Accomplishments and Legacies of the TOGA Program. National Academy Press, Washington, D.C. 171 pp.

NSF (National Science Foundation). 1986. Coupling, Energetics, and Dynamics of Atmospheric Regions "CEDAR." CEDAR Science Steering Committee, April 1986 (revised April 1987). NSF, Arlington, Va. 40 pp.

NSF. 1988. GEM (Geospace Environment Modeling)—A Program of Solar-Terrestrial Research in Global Geosciences. GEM Steering Committee, May 1988. NSF, Arlington, Va. 33 pp.

NSF. 1990. RISE (Radiative Inputs of the Sun to Earth—A Research Plan for the 1990s on Solar Irradiance Variation. RISE Science Steering Committee, February 1990. NSF, Arlington, Va. 31 pp.

NWS (National Weather Service). 1992. Natural Disaster Survey Report, Hurricane Andrew: South Florida and Louisiana, August 23-26, 1992. NWS, Silver Spring, Md.

OFCM (Office of the Federal Coordinator for Meteorology). 1995. The National Space Weather Program—The Strategic Plan, August 1995, FCM-P30-1995. Office of the Federal Coordinator for Meteorological Services and Supporting Research, Silver Spring, Md. 25 pp.

OFCM. 1997. The National Space Weather Program—The Implementation Plan, January 1997, FCM-P31-1997. Office of the Federal Coordinator for Meteorological Services and Supporting Research, Silver Spring, Md. 93 pp.

Oltmans, S.J., and D.J. Hofmann. 1995. Increase in lower stratospheric water vapour at a midlatitude Northern Hemisphere site from 1981 to 1994. Nature 374, 146-149.

Oltmans, S.J., and H. Levy II. 1994. Surface ozone measurements from a global network. Atmos. Environ. 28, 9-24.

Oltmans, S.J., A.S. Lefohn, H.E. Scheel, J.A. Harris, H. Levy, II, I.E. Galbally, E.-G. Brunke, C.P. Meyer, J.A. Lathrop, B.J. Johnson, D.S. Shadwick, E. Cuevas, F.J. Schmidlin, D.W. Tarasik, H. Claude, J.B. Kerr, and O. Uchino. 1997. Trends in ozone in the troposphere. Geophys. Res. Lett. (in press).

Patz, J.A., P.R. Epstein, T.A. Burke, and J.M. Balbus. 1996. Global climate change and emerging infectious diseases. J. Am. Med. Assoc. 275, 217-223.

Pielke, R.A., Jr. 1995. Hurricane Andrew in South Florida: Mesoscale Weather and Societal Responses, Environmental and Societal Impacts Group. National Center for Atmospheric Research, Boulder, Colo.

Pielke, R.A., Jr., and J. Kimple. 1997. Societal aspects of weather. Report of the Sixth Prospectus Development Team of the U.S. Weather Research Program to NOAA and NSF. Bull. Am. Meteorol. Soc. 78, 867-876.

Pielke, R.A., T.J. Lee, J.H. Copeland, J.L. Eastman, C.L. Ziegler, and C.A. Finley. 1997. Use of USGS-provided data to improve weather and climate simulations. Ecological Applications (in press).

Prather, M.J. 1985. Continental sources of halocarbons and nitrous oxide. Nature 317, 221-225.

Prather, M.J. 1988. European sources of halocarbons and nitrous oxide: Update 1986. J. Atmos. Chem. 6, 375-406.

Prather, M.J., M.B. McElroy, S.C. Wofsy, G. Russell, and D. Rind. 1987. Chemistry of the global troposphere: Fluorocarbons as tracers of air motion. J. Geophys. Res. 92, 6579-6613.

Price, C. 1993. Global surface temperatures and the atmospheric electrical circuit. Geophys. Res. Lett. 20, 1363-1366.

Quayle, R.G., D.R. Easterling, T.R. Karl, and P.Y. Hughes. 1991. Effects of recent thermometer changes in the cooperative station network. Bull. Am. Meteorol. Soc. 72, 1718-1723.

Ramaswamy, V., R.J. Charlson, J.A. Coakley, J.L. Gras, Harshvardhan, G. Kukla, M.O. McCormick, D. Möller, E. Roeckner, L.L. Stowe, and J. Taylor. 1995. What are the observed and anticipated meteorological and climatic responses to aerosol forcing? Pp. 384-399 in Aerosol Forcing of Climate, R.J. Charlson and J. Heintzenberg (eds.). Wiley and Sons, Chichester, U.K.

Raval, A., and V. Ramanathan. 1989. Observational determination of the greenhouse effect. Nature 342, 758-761.

Reid, G.C. 1991. Solar total irradiance variations and the global sea-surface temperature record. J. Geophys. Res.—Atmospheres 96, 2835-2844.

Ridley, W.P., L.J. Dizikes, and J.M. Wood. 1977. Biomethylation of toxic elements in the environment. Science 197, 329-332.

Rind, D., R. Suozzo, N.K. Balachandran, and M.J. Prather. 1990. Climate change and the middle atmosphere. Part I: The doubled CO_2 climate. J. Atmos. Sci. 47, 475-494.

Rind, D., W.D. Chiou, S. Oltmans, J. Lerner, M.P. McCormick, and L.R. McMaster. 1993. Overview of the Stratospheric Aerosol and Gas Experiment. J. Geophys. Res. 98, 4835-4857.

Roble, R.G., and R.E. Dickinson. 1989. How will changes in carbon dioxide and methane modify the mean structure of the mesosphere and thermosphere? Geophys. Res. Lett. 16, 1441-1444.

Ropelewski, C.F., and M.S. Halpert. 1986. North American precipitation and temperature patterns associated with the El Niño/Southern Oscillation. Monthly Weather Rev. 114, 2352-2362.

Ropelewski, C.F., and M.S. Halpert. 1987. Global and regional scale precipitation patterns associated with the El Niño/Southern Oscillation. Monthly Weather Rev. 115, 1606-1626.

Rosati, A., K. Miyakoda, and R. Gudgel. 1997. The impact of ocean initial conditions on ENSO forecasting with a coupled model. Monthly Weather Rev. 125, 754-772.

Sandroni, D., D. Anfossi, and S. Viarengo. 1992. Surface ozone levels at the end of the nineteenth century in South America. J. Geophys. Res. 97, 2535-2540.

Schimel, D. D. Alves, I. Enting, M. Heimann, F. Joos, D. Raynaud, T. Wigley; M. Prather, R. Derwent, D. Ehhalt, P. Fraser, E. Sanhueza, X. Zhou; P. Jonas, R. Charlson, H. Rodhe, S. Sadasivan; K.P. Shine, Y. Fouquart, V. Ramaswamy, S. Solomon, J. Srinivasan; D. Albritton, R. Derwent, I. Isaksen, M. Lal, and D. Wuebbles. 1996. Radiative forcing of climate change. Pp. 65-131 in Climate Change 1995—The Science of Climate Change, J.T. Houghton, L.G. Meira Filho, B.A. Callander, N. Harris, A. Kattenberg, and K. Maskell (eds.). Contribution of Working Group I to the Second Assessment Report of the Intergovernmental Panel on Climate Change. Cambridge University Press, Cambridge, U.K. 572 pp.

Schulze, E.-D. 1989. Air pollution and forest decline in a spruce (*Picea abies*) forest. Science 244, 776-783.

Sentman, D.D., and E.M. Wescott. 1993. Observations of upper atmospheric optical flashes recorded from an aircraft. Geophys. Res. Lett. 20, 2857-2860.

Serafin, R. 1991. Study on observational systems—A review of meteorological and oceanographic education in observational techniques and the relationship to national facilities and needs. Bull. Am. Meteorol. Soc. 72, 815-826.

Serafin, R., B. Heikes, D. Sargeant, W. Smith, E. Takle, D. Thomson, and R. Wakimoto. 1991. Study of observational systems: A review of meteorological and oceanographic observational techniques and the relationship to national facilities and needs. Bull. Am. Meteorol. Soc. 72, 815-826.

Shannon, J.D., and E.C. Voldner. 1995. Modeling atmospheric concentrations of mercury and deposition to the Great Lakes. Atmos. Environ. 29, 1649-1661.

Shay, L.K., P.G. Black, A.J. Mariano, J.D. Hawkins, and R.L. Elsberry. 1992. Upper ocean response to Hurricane Gilbert. J. Geophys. Res. 97, 20227-20248.

Shepard, L.J. 1993. Lifting the Veil: The Female Face of Science. Shambala Press, Boston and London. 329 pp.

Shine, K.P., Y. Fouquart, V. Ramaswamy, S. Solomon, and J. Srinivasan. 1995. Radiative forcing. Pp. 163-203 in Climate Change 1994—Radiative Forcing of Climate Change and an Evaluation of the IPCC IS92 Emission Scenarios, J.T. Houghton, L.G. Meira Filho, J. Bruce, Hoesung Lee, B.A. Callander, E. Haites, N. Harris, and K. Maskell (eds.). Reports of Working Groups I and II of the Intergovernmental Panel on Climate Change, forming part of the IPCC Special Report to the first session of the Conference of the Parties to the UN Framework Convention on Climate Change. Cambridge University Press, Cambridge, U.K. 339 pp.

Shine, K.P., Y. Fouquart, V. Ramaswamy, S. Solomon, and J. Srinivasan. 1996. Radiative forcing of climate change. Pp. 65-131 in Climate Change 1995—The Science of Climate Change, J.T. Houghton, L.G. Meira Filho, B.A. Callander, N. Harris, A. Kattenberg, and K. Maskell (eds.). Contribution of Working Group I to the Second Assessment Report of the Intergovernmental Panel on Climate Change. Cambridge University Press, Cambridge, U.K. 572 pp.

Shope, R.E. 1991. Global climate change and infectious diseases. Environ. Health Perspect. 96, 171-174.
Silverman, S.M. 1992. Secular variation of the aurora for the past 500 years. Rev. Geophys. 30, 333-351.
Simpson, J., and M.A. LeMone. 1974. Women in meteorology. Bull. Am. Meteorol. Soc. 55, 122-131.
Skoog, B.G., J.I.H. Askne, and G. Elgered. 1982. Experimental determination of water vapor profiles from ground-based radiometers at 12.0 and 31.4 GHz. J. Appl. Meteorol. 21, 394-400.
Smith, W.L., H.E. Revercomb, H.B. Howell, H.M. Woolf, R.O. Knuteson, R.G. Decker, M.J. Lynch, E.R. Westwater, R.G. Strauch, K.P. Moran, B. Stankov, M.J. Falls, J. Jordan, M. Jacobsen, W.F. Dabberdt, R. McBeth, G. Albright, C. Paneitz, G. Wright, P.T. May, and M.T. Decker. 1990. GAPEX: A ground-based atmospheric profiling experiment. Bull. Am. Meteorol. Soc. 71, 310-318.
Solomon, S., R.R. Garcia, F.S. Rowland, and D.J. Wuebbles. 1986. On the depletion of Antarctic ozone. Nature 321, 755-758.
Solomon, S., R.W. Sanders, R.R. Garcia, and J.G. Keys. 1993. Increased chlorine dioxide over Antarctica caused by volcanic aerosols from Mount Pinatubo. Nature 363, 245-248.
Soon, W.H., S.L. Baliunas, and Q. Zhang. 1994. A technique for estimating long-term variations of solar total irradiance: Preliminary estimates based on observations of the Sun and solar-type stars. Pp. 133-144 in The Solar Engine and Its Influence on Terrestrial Atmosphere and Climate, E. Nesme-Ribes (ed.). Springer-Verlag, New York. Published in cooperation with NATO Scientific Affairs Division.
Staehelin, J., and W. Schmid. 1991. Trend analysis of tropospheric ozone concentrations utilizing the 20-year data set of ozone balloon soundings over Payerne. Atmos. Environ. 9, 1739-1749.
Staehelin, J., J. Thudium, R. Buehler, A. Volz-Thomas, and W. Graber. 1994. Trends in surface ozone concentrations at Arosa (Switzerland). Atmos. Environ. 28, 75-87.
Stephens, G.L. 1990. On the relationship between water vapor over the oceans and sea surface temperature. J. Climate 3, 634-645.
Stephens, P.L., and C. Kazarosian. 1992. Results of the AMS membership survey. Bull. Am. Meteorol. Soc. 73, 486-495.
Stowe, L.L., R.M. Carey, and P.P. Pellegrino. 1992. Monitoring the Mt. Pinatubo aerosol layer with NOAA-11 AVHRR data. Geophys. Res. Lett. 19, 159-162.
Tans, P.P., P.S. Bakwin, and D.W. Guenther. 1996. A feasible Global Carbon Cycle Observing System: A plan to decipher today's carbon cycle based on observations. Global Change Biology 2, 309-318.
Tarasick, D.W., D.I. Wardle, J.B. Kerr, J.J. Bellefleur, and J. Davies. 1994. Tropospheric ozone trends over Canada: 1980-1993. Geophys. Res. Lett. 22, 409-412.
Taubenheim, J., G. von Kossart, and G. Entzian. 1990. Evidence of CO_2-induced progressive cooling of the middle atmosphere derived from radio observations. Adv. Space Res. 10, 171-174.
Taylor, F.W., C. Fröhlich, J. Lecolle, M. Strecker. 1987. Analysis of partially emerged corals and reef terraces in the central Vanuatu arc—Comparison of contemporary coseismic and nonseismic with quaternary vertical movements. J. Geophys. Res.—Solid Earth and Planets 92, 4905-4933.
Taylor, H.R., S.K. West, F.S. Rosenthal, B. Munoz, H.S. Newland, H. Abbey, and E.A. Emmett. 1988. Effect of ultraviolet radiation on cataract formation. N. Engl. J. Med. 319, 1429-1433.
Theon, J.S. 1994. The Tropical Rainfall Measuring Mission (TRMM). Advances in Space Research 14, 159-165.
Thomas, G.E. 1991. Mesospheric clouds and the physics of the mesopause region. Rev. Geophys. 29, 553-575.
Trenberth, K.E., G.W. Branstator, D. Karoly, A. Kumar, N.-C. Lau, and C. Ropelewski. 1998. Global atmospheric diagnostics and modeling for TOGA. J. Geophys. Res. (in press).

U.S. Department of Commerce. 1990. Fifty Years of Population Change Along the Nation's Coasts, 1960-2010. National Ocean Survey, National Oceanic and Atmospheric Administration, Washington, D.C.

U.S. Department of Commerce. 1992. Natural Disaster Survey Report, Hurricane Andrew: South Florida and Louisiana, August 23-26, 1992. National Weather Service, National Oceanic and Atmospheric Administration, Silver Spring, Md.

U.S. Department of Commerce. 1994. Natural Disaster Survey Report, Superstorm of March 1993. National Weather Service, National Oceanic and Atmospheric Administration, Silver Spring, Md.

USGCRP (U.S. Global Change Research Program). 1996. Our Changing Planet: The FY 1997 U.S. Global Change Research Program: A Report. A Supplement to the President's Fiscal Year 1997 Budget. U.S. National Science and Technology Council, Subcommittee on Global Change Research. Available from Global Change Research Information Offices, Washington, D.C. 162 pp.

USGCRP. 1997. Our Changing Planet: The FY 1998 U.S. Global Change Research Program: A Report. A Supplement to the President's Fiscal Year 1998 Budget. U.S. National Science and Technology Council, Subcommittee on Global Change Research. Available from Global Change Research Information Offices, Washington, D.C. 118 pp.

Van Dijk, H.F.H., M.H.J. deLouw, J.G.M. Roelofs, and J.J. Verburgh. 1990. Impact of artificial, ammonium-enriched rainwater on soils and young coniferous trees in a greenhouse, Part II: Effects on the trees. Environ. Pollut. 63, 41-59.

Vitousek, P.M., L.R. Walker, L.D. Whiteaker, and P.A. Matson. 1993. Nutrient limitations to plant growth during primary succession in Hawaii Volcanoes National Park. Biogeochemistry 23, 197-215.

Volz-Thomas, A., and D. Kley. 1988. Evaluation of the Montsouris series of ozone measurements in the nineteenth century. Nature 332, 240-242.

Vong, R.J., J.T. Sogmon, and S.F. Mueller. 1991. Cloud water deposition to Appalachian forests. Environ. Sci. Technol. 25, 1014-1021.

Wang, W.C., and N.D. Sze. 1980. Coupled effects of atmospheric N_2O and O_3 on the Earth's climate. Nature 286, 589-590.

Wang, W.C., M.P. Budek, X.Z. Liang, and J.T. Hiehl. 1991. Inadequacy of effective CO_2 as a proxy in simulating the greenhouse effect of other radiatively active gases. Nature 350, 573-577.

Ware, M., M. Exner, D. Feng, M. Gorbunov, K. Hardy, B. Herman, Y. Kuo, T. Meehan, W. Melbourne, C. Rocken, W. Schreiner, S. Sokolovskiy, F. Solheim, X. Zou, R. Anthes, S. Businger, and K. Trenberth. 1996. GPS sounding of the atmosphere from low Earth orbit: Preliminary results. Bull. Am. Meteorol. Soc. 77, 19-40.

Webster, P.J., and R. Lukas. 1992. TOGA COARE: The Coupled-Ocean Atmosphere Response Experiment. Bull. Am. Meteorol. Soc. 73, 1377-1416.

Weeks, M.E. 1968. Discovery of the Elements. Journal of Chemical Education, Easton, Pa. 896 pp.

Weinberg, A.M. 1963. Criteria for scientific choice. Minerva 1, 159-171.

Wennberg, P.O., R.C. Cohen, R.M. Stimpfle, J.P. Koplow, J.G. Anderson, R.J. Salawitch, D.W. Fahey, E.L. Woodbridge, E.R. Keim, R.S. Gao, C.R. Webster, R.D. May, D.W. Toohey, L.M. Avallone, M.W. Proffitt, M. Loewenstein, J.R. Podolske, K.R. Chan, and S.C. Wofsy. 1994. Removal of stratospheric O_3 by radicals: In situ measurements of OH, HO_2, NO, NO_2, ClO, BrO. Science 266, 398-404.

WHO (World Health Organization). 1992. Global Health Situations and Projections, Estimates. WHO, Geneva, Switzerland.

WHO. 1996. Climate Change and Human Health. WHO, Geneva, Switzerland.

Williams, E.R. 1992. The Schumann resonance: A global tropical thermometer. Science 256, 1184-1187.

WMO (World Meteorological Organization). 1989. Fourteenth Status Report on Implementation. Publication 714. WMO, Geneva, Switzerland.
WMO. 1995. Scientific Assessment of Ozone Depletion, 1994. Global Ozone Research and Monitoring Project. Report No. 37. WMO, Geneva, Switzerland.
Zebiak, S.E., and M.A. Cane. 1987. A model El Niño/Southern Oscillation. Monthly Weather Rev. 115, 2262-2278.
Zevin, S.F., and K.L. Seitter. 1994. Results of survey of society membership: Demographics. Bull. Am. Meteorol. Soc. 75, 1855-1866.

Appendixes

APPENDIX A

Acronyms and Abbreviations

ACE	Aerosol Characterization Experiment
ACE-1	(Southern Hemisphere Marine) Aerosol Characterization Experiment
ACRIM	Active Cavity Radiometer Irradiance Monitor
ADEOS	Advanced Earth Observing System
AF	Air Force
AMIP	Atmospheric Model Intercomparison Project
AMS	American Meteorological Society
AOML	Atlantic Oceanographic and Meteorological Laboratory
AOT	aerosol optical thickness
ARM	Atmospheric Radiation Measurement (Program)
ASOS	Automated Surface Observation System
AVHRR	Advanced Very High Resolution Radiometer (satellite instrument)
AWIPS	Automated Weather Interactive Processing System
BASC	Board on Atmospheric Science and Climate
CAAA-90	Clean Air Act Amendments of 1990
CAAI	Committee on Atmospheric Applications and Information
CAPE	convective available potential energy
CAPS	Center for the Analysis and Prediction of Storms
CASH	Commercial Aviation Sensing Humidity (Program)

CASR	Committee on Atmospheric Services and Research
CBL	convective boundary layer
CCM	Community Climate Model
CCN	cloud condensation nuclei
CDNC	cloud droplet number concentration
CEDAR	Coupling, Energetics, and Dynamics of Atmospheric Regions
CEES	Committee on Earth and Environmental Sciences
CENR	Committee on Environment and Natural Resources
CERN	Conseil Européen pour la Recherche Nucléaire
CFC	chlorofluorocarbon
CLIVAR	Climate Variability and Prediction Program
CME	coronal mass ejection
COARE	Coupled Ocean-Atmosphere Response Experiment (TOGA)
COMET	Cooperative Program for Operational Meteorology, Education, and Training
CSSP	Committee on Solar and Space Physics
CSTR	Committee on Solar-Terrestrial Research
Dec-Cen	climate variability on decade-to-century time scales (WCRP)
DIAL	differential absorption lidar
DMS	dimethyl sulfide
DOC	U.S. Department of Commerce
DOD	U.S. Department of Defense
DOE	U.S. Department of Energy
DOI	U.S. Department of the Interior
DU	Dobson unit
EMEP	Cooperative Programme for the Monitoring and Evaluation of Long Range Air Pollutants in Europe
ENSO	El Niño/Southern Oscillation
EOS	Earth Observing System
EPA	U.S. Environmental Protection Agency
ERBE	Earth Radiation Budget Experiment
ERS-1	European Remote Sensing Satellite
EUV	extreme ultraviolet
FAA	Federal Aviation Administration
FCCSET	Federal Coordinating Council for Science, Engineering, and Technology
FY	Fiscal Year
GCM	general circulation model
GCOS	global climate observing system

GDP	Gross Domestic Product
GEM	Geospace Environment Modeling
GEWEX	Global Energy and Water Cycle Experiment
GISS	Goddard Institute for Space Studies
GLOBE	Global Learning and Observing to Benefit the Environment
GOALS	Global Ocean-Atmosphere-Land System
GOES	Geostationary Operational Environmental Satellite
GONG	Global Oscillation Network Group
GPS	global positioning system
GTS	global telecommunication system
HCFC	hydrochlorofluorocarbon
HF	high frequency
HIS	high-resolution interferometer sounder
ICAS	Interdepartmental Committee for Atmospheric Sciences
IGAC	International Global Atmospheric Chemistry (project)
IMF	interplanetary magnetic field
IN	ice nucleus
IOM	Institute of Medicine
IPCC	Intergovernmental Panel on Climate Change
IS	incoherent scatter
ISCCP	International Satellite Cloud Climatology Project
JMA	Japan Meteorological Agency
KPNO	Kitt Peak National Observatory
LAWS	Laser Atmospheric Wind Sounder
LEARN	Laboratory Experience in Atmospheric Research at NCAR
LES	large eddy simulation
LTER	Long-Term Ecological Research
MBA	masters of business administration
MCS	mesoscale convective system
MF	medium frequency
MPP	massively parallel processor
MRF	Medium-Range Forecasting
MST	mesosphere-stratosphere-troposphere
MUF	maximum usable frequency
NAAQS	National Ambient Air Quality Standard
NAE	National Academy of Engineering

NARSTO	North American Research Strategy on Troposphere Ozone
NAS	National Academy of Sciences
NASA	National Aeronautics and Space Administration
NCAR	National Center for Atmospheric Research
NCEP	National Centers for Environmental Prediction
NCLAN	National Crop Loss Assessment Network
NCTM	National Council of Teachers of Mathematics
NDSC	Network for Detection of Stratospheric Change
NESDIS	National Environmental Satellite, Data, and Information Service
NEXRAD	Next Generation Weather Radar
NHC	nonmethane hydrocarbon
NIMBUS	series of experimental environmental research satellites
NMC	National Meteorological Center
NOAA	National Oceanic and Atmospheric Administration
NRC	National Research Council
NSF	National Science Foundation
NSTC	National Science and Technology Council
NSWP	National Space Weather Program
NWP	numerical weather prediction
NWS	National Weather Service
OFCM	Office of the Federal Coordinator for Meteorology
ONR	Office of Naval Research
OSSE	observing system simulation experiment
OSTP	Office of Science and Technology Policy
PCB	polychlorinated biphenyl
PM	particulate matter
PSC	polar stratospheric cloud
QBO	quasi-biennial oscillation
RASS	radioacoustic sounding system
RISE	Radiative Inputs of the Sun to Earth
SAGE	Stratospheric Aerosol and Gas Experiment
SAR	Subcommittee on Atmospheric Research
SBUV	Solar Backscatter Ultraviolet Spectrometer
SEASAT	sea satellite (U.S. oceanographic satellite)
SEP	solar energetic particle
SGS	subgrid scale
SMM	Solar Maximum Mission

SOHO	Solar and Heliospheric Observatory
SSM/I	Special Sensor Microwave/Imager
SST	sea surface temperature
SSTA	sea surface temperature anomoly
STE	stratosphere-troposphere exchange
SUNRISE	Sun's Radiative Inputs from Sun to Earth Program
TAO	Tropical Atmosphere Ocean (array) (TOGA)
TOGA	Tropical Ocean Global Atmosphere (program)
TOMS	Total Ozone Mapping Spectrometer
TRMM	Tropical Rainfall Measurement Mission
UARS	Upper Atmosphere Research Satellite
UAV	unmanned aerospace vehicle
UCAR	University Corporation for Atmospheric Research
UK	United Kingdom
UN	United Nations
USDA	U.S. Department of Agriculture
USGCRP	U.S. Global Change Research Program
USWRP	U.S. Weather Research Program
UV	ultraviolet
VLF	very low frequency
VOC	volatile organic compound
WCRP	World Climate Research Programme
WFO	Weather Forecast Office
WMO	World Meteorological Organization
WOCE	World Ocean Circulation Experiment
WSR-88D	Weather Service Radar (U.S. National Weather Service) 1988 Doppler Weather Radar System
WWW	World Weather Watch (U.N.)
WWW	World Wide Web
XBT	expendable bathythermograph
YMP	CRAY supercomputer model

APPENDIX

B

Listing of Reports of the Committee on Atmospheric Sciences and the Board on Atmospheric Sciences and Climate Since 1958

Research and Education in Meteorology. 1958.
Proceedings of the Scientific Information Meeting on Atmospheric Sciences. 1959.
Meteorology on the Move. 1960.
The Status of Research and Manpower in Meteorology. 1960.
Interaction Between the Atmosphere and the Oceans. 1961.
The Atmospheric Sciences, 1961-1971. 1962.
Atmospheric Ozone Studies: An Outline for an International Observation Program. 1966.
The Feasibility of a Global Observation and Analysis Experiment. 1966.
Weather and Climate Modification: Problems and Prospects. 1966.
Atmospheric Exploration by Remote Probes. 1969.
Education Implications of the Global Atmospheric Research Program. 1969.
The Atmospheric Sciences and Man's Needs: Priorities for the Future. 1971.
Weather and Climate Modification: Problems and Progress. 1973.
Atmospheric Chemistry: Problems and Scope. 1975.
Long-Range Weather Forecasting. 1975.
Report of the Committee on Atmospheric Sciences Ad Hoc Panel to Review the NASA Earth Radiation Budget Program. 1976.
The Atmospheric Sciences: Problems and Applications. 1977.
Planning and Management of Atmospheric Research Programs. 1977.
Severe Storms: Prediction, Detection, and Warning. 1977.
Long-Range Weather Forecast Evaluation. 1979.

APPENDIX B   347

Atmospheric Precipitation: Prediction and Research Problems. 1980.
The Atmospheric Sciences: National Objectives for the 1980s. 1980.
Current Mesoscale Meteorological Research in the United States. 1981.
Changing Climate. Report of the Carbon Dioxide Assessment Committee. 1983.
El Niño and the Southern Oscillation—A Scientific Plan. 1983.
Low-Altitude Wind Shear and Its Hazard to Aviation. Joint study with Aeronautics and Space Engineering Board. 1983.
Report of the Research Briefing Panel on Atmospheric Sciences. Committee on Science, Engineering, and Public Policy. 1983.
Global Tropospheric Chemistry: A Plan for Action. Global Tropospheric Chemistry Panel. 1984
National Solar-Terrestrial Research Program. Committee on Solar-Terrestrial Research. 1984.
An Ocean Climate Research Strategy. Ferris Webster, Senior Fellow, National Research Council. 1984.
Research Recommendations for Increased U.S. Participation in the Middle Atmosphere Program. Panel on the Middle Atmosphere Program, Committee on Solar-Terrestrial Research. 1984.
Solar-Terrestrial Data Access, Distribution, and Archiving. Report prepared jointly by the Committee on Solar and Space Physics, Space Science Board, and the Committee on Solar-Terrestrial Research, Board on Atmospheric Sciences and Climate. 1984.
Research Briefings. Report of the Research Briefing Panel on Weather Prediction Technologies. 1985.
Atmospheric Climate Data—Problems and Promises. Panel on Climate-Related Data. 1986.
The National Climate Program—Early Achievements and Future Directions. Report of the Woods Hole Workshop, July 15-19, 1985. 1986.
Siting and Utilization of a 30-Station Wind Profiler Network. Letter Report by the Panel on Mesoscale Research. 1986.
U.S. Participation in the TOGA Program—A Research Strategy. Tropical Ocean and Global Atmosphere (TOGA) Program. 1986.
Current Issues in Atmospheric Change. Summary and Conclusions of a Workshop, October 30-31, 1986. 1987.
Long-Term Solar-Terrestrial Observations. Panel on Long-Term Observations, Committee on Solar-Terrestrial Research. 1988.
Meteorological Support for Space Operations: Review and Recommendations. Panel on Meteorological Support for Space Operations. 1988.
Field of Solar Physics: Review and Recommendations for Ground-Based Solar Research. Committee on Solar Physics. 1989.
Ozone Depletion, Greenhouse Gases, and Climate Change. Proceedings of a Joint Symposium by the Board on Atmospheric Sciences and Climate and the Committee on Global Change. 1989.

Advancing the Understanding and Forecasting of Mesoscale Weather in the United States. Committee on Meteorological Analysis, Prediction, and Research. 1990.

TOGA: A Review of Progress and Future Opportunities. Advisory Panel for the Tropical Ocean and Global Atmosphere (TOGA) Program. 1990.

The Department of Energy's Atmospheric Chemistry Program, A Critical Review. 1991.

Four-Dimensional Model Assimilation of Data—A Strategy for the Earth System Sciences. 1991.

Prospects for Extending the Range of Prediction of the Global Atmosphere. 1991.

Rethinking the Ozone Problem in Urban and Regional Air Pollution. With the NRC Board on Environmental Studies and Toxicology. 1991.

"Climatological Considerations of the National Weather Service Modernization Program." In: Toward A New National Weather Service, Second Report of the NRC National Weather Service Modernization Committee. 1992.

Coastal Meteorology—A Review of the State of the Science. 1992.

A Decade of International Climate Research—The First Ten Years of the World Climate Research Program. 1992.

Understanding and Predicting Atmospheric Chemical Change—An Imperative for the U.S. Global Change Research Program. 1993.

Atmospheric Effects of Stratospheric Aircraft—An Evaluation of NASA's Interim Assessment. 1994.

Estimating Bounds on Extreme Precipitation Events—A Brief Assessment. 1994.

GEWEX and GCIP: Initial Review of Concepts and Objectives. Letter Report. October 11, 1994.

GOALS (Global Ocean-Atmosphere-Land System)—for Predicting Seasonal-to-Interannual Climate. 1994.

Ocean-Atmosphere Observations Supporting Short-Term Climate Predictions. 1994.

ONR Research Opportunities in Upper Atmospheric Sciences. With the Naval Studies Board and the Space Studies Board. 1994.

A Space Physics Paradox—Why Has Increased Funding Been Accompanied by Decreased Effectiveness in the Conduct of Space Physics Research? 1994.

Natural Climate Variability on Decade-to-Century Time Scales. 1995.

Organizing U.S. Participation in GOALS (Global Ocean-Atmosphere-Land System). 1995.

"Report of the Atmospheric Sciences Data Panel." In: Study of the Long-Term Retention of Selected Scientific and Technical Records of the Federal Government. 1995.

Learning to Predict Climate Variations Associated with El Niño and the Southern Oscillation—Accomplishments and Legacies of the TOGA Program. 1996.

A Plan for a Research Program on Aerosol Radiative Forcing and Climate Change. 1996.

Index

A

Access to electronic data, move to limit, 5, 50-51
ACE-1 Science and Implementation Plan, 164
ACRIM. *See* Active Cavity Radiometer Irradiance Monitor
Active Cavity Radiometer Irradiance Monitor (ACRIM), 257-258
Adaptive observation strategies, 171, 188-190
Advection, 192-193
Aerometric Information Retrieval System, 127
Aerosol climatology, designing and deploying networks to document, 164
Aerosol physics, 63
Aerosol Radiative Forcing and Climate Change, 162
Aerosols. *See also* Atmospheric aerosols
   and atmospheric chemistry, 40
   chemical and physical properties of, 129-131
   direct radiative forcing of climate by, 73-74
   and environmental quality, 23
   and interactions with other atmospheric phenomena, 7, 86-87
   predicting size distributions of, 75-76
Agencies. *See* Federal government and agencies

Agricultural planning, value of predictions to, 26
Aircraft. *See also* Commercial aircraft
   atmospheric effects of, 203, 220-222, 224-225
   remote piloted, 36, 153, 223
Air quality
   forecasting of, 3, 42-43, 138
   improving predictive numerical models for, 2, 7, 134
Air quality monitoring, 2, 134, 150-151
American Meteorological Society (AMS), 177
AMIP. *See* Atmospheric Model Intercomparison Project
AMS. *See* American Meteorological Society
Anthropogenic influences, 246-247, 289-296
   ability to predict, 255, 324
   affecting lower atmosphere, 8, 22-24, 106
   affecting stratospheric processes, 211-212
   driving global chemical change, 117
   separating from solar, 259-263
Appleton, Sir Edward, 210
Appleton anomalies, 236
ARM. *See* Atmospheric Radiation Measurement (ARM) program
Army Signal Corps, 21
Artificial intelligence (AI), 98

ASOS. *See* Automated Surface Observation System
Atmospheric aerosols, 7, 129-131, 162-166
 designing and deploying networks to document aerosol climatology, 164
 designing and implementing intensive field programs for, 164-165
 designing and implementing new suites of measurement technologies for tropospheric aerosols, 163-164
 developing predictive model capability for, 165-166
 and global warming, 292, 293
 maintaining and expanding stratospheric aerosol measurement capability, 162-163
Atmospheric boundary layer, 63, 172-173
 resolving interactions at, 37-41
 and studies, 88
Atmospheric chemistry
 Environmentally Important Atmospheric (chemical) Species, 108, 140-168
 infrastructure, 135-139
 mission, 112-114
 recent insights, 114-121
 recommended research strategies for, 121-132
 summary, 7
Atmospheric components, interactions with other Earth system components, 3, 184-185
Atmospheric dynamics, 169-198
 recommended research, 7-8, 173-175
 small-scale, 6, 63
Atmospheric electricity, 6-7, 63, 67
 and interactions with other atmospheric phenomena, 7, 65
 investigating global electrical circuit and lightning as measures of stability and temperature, 78
 mechanisms of charge separation in clouds, 78
 nature and sources of middle-atmosphere discharges, 78, 249
 production of NO_x by lightning, 79
 recommended research strategies for, 67, 77-79
Atmospheric emissions, rapidly increasing, 4, 45
Atmospheric forecasting. *See* Weather forecasting
Atmospheric information
 developing a strategy for providing, 46-50
 preserving free and open exchange of, 5, 50-51
 prospects for, 48-49
Atmospheric information services
 distributed, 49-50
 funding for, 56-58
 optimizing, 50
Atmospheric Model Intercomparison Project (AMIP), 291, 315
Atmospheric observations. *See* Observations
Atmospheric physics
 atmospheric electricity, 77-79, 93-95
 atmospheric radiation, 71-74
 atmospheric water, clouds, 73, 74-77, 87, 92, 102-103
 boundary layer meteorology, 79-80, 87-89
 cloud physics, 74-77
 instrumentation, 103-106
 mission, 68
 models, improvement and testing, 84-85
 small scale influences on large scale phenomenon, 99-102
Atmospheric potential vorticity, 172
Atmospheric prediction. *See* Weather forecasting
Atmospheric Radiation Measurement (ARM) program, 98
Atmospheric sciences, 1-2, 14-16, 101
 contributions to the national well-being, 17-27, 65, 106, 111, 202, 271, 324
 cost effectiveness of, 169-170
 entering the twenty-first century, 15-16
 history of, 14, 114-116
 imperatives, 2-3, 28-37
 key role of, 14
 oceanography a close partner of, 15
 role in environmental issues, 23-24
Auroral emission, 231
Automated Surface Observation System (ASOS), 21
Automated Weather Interactive Processing System (AWIPS), 21

B

Baroclinicity, effect on boundary layer, 80
BASC. *See* Board on Atmospheric Sciences and Climate

INDEX

Benefits and costs of atmospheric information services, 5, 47, 51-58
Benefits of atmospheric research, 17-27, 47
  enhancing national economic vitality, 24-26
  maintaining environmental quality, 22-24
  protection of life and property, 17-22
  strengthening fundamental understanding, 26-27
Board on Atmospheric Sciences and Climate (BASC), 1, 4, 6, 54, 59
  disciplinary assessments of, 6-9, 28-29
  imperatives of, 2-3, 28-37
  leadership and management planning, 6
  listing of reports of, 346-348
  recommendations of, 3-5, 37-45
Boundary layer meteorology, 6, 65, 100
  effects of inhomogeneity and baroclinicity on boundary layer, 80
  exploiting new remote sensors, 88-89
  interactions of planetary boundary layer, surface characteristics, and clouds, 81
  and interactions with other atmospheric phenomena, 7
  measurements of exchange of water, heat, and trace atmospheric constituents, 80-81
  recommended research strategies for, 67, 79-81
  structure of cloudy boundary layers, 79-80
  turbulence and entrainment, 80

C

CAAA-90. *See* Clean Air Act Amendments of 1990
CAPE. *See* Convective available potential energy
CAPS. *See* Center for the Analysis and Prediction of Storms
Carbon dioxide, 22, 106
CASH. *See* Commercial Aviation Sensing Humidity (CASH) program
CASR. *See* Committee on Atmospheric Services and Research
Catastrophic events, 174, 297
  potential for, 107
Cavendish, Henry, 114
CCN. *See* Cloud condensation nuclei
CDNC. *See* Cloud droplet number concentrations

CEDAR. *See* Coupling, Energetics, and Dynamics of Atmospheric Regions
CEES. *See* Committee on Earth and Environmental Sciences
CENR. *See* Committee on Environment and Natural Resources
Center for the Analysis and Prediction of Storms (CAPS), 180
CFCs. *See* Chlorofluorocarbons
Chaos theory, 41, 98
  outgrowth of meteorology, 27
Charge generation, mechanisms of, 94-96
Charge separation in clouds, mechanisms of, 78
Chemical climatology, documenting, 109
Chemical constituents
  developing new capabilities for observing, 2
  disciplined forecasting for, 3
Chemical instrumentation, continue development and validation of, 159-160
Chemical meteorology system, developing, 138
Chemistry. *See* Atmospheric chemistry
Chlorofluorocarbon (CFC) gases, 26, 123, 206, 222
  and environmental quality, 22, 117-119, 210-211
  longevity of, 255
  substitutes for, 216-217
Circulation systems, quasi-balanced and unbalanced, 172-173
Clean Air Act Amendments of 1990 (CAAA-90), 126, 134-135, 216
Climate, 272-324
  climate monitoring, 2, 281, 307
  climate sensitive enterprises, 25
  climate weather and health, 44
  climatic prediction, increase of skill in, 311-314
  deterioration of current observational capability, 302-306
  enhancing observational capability, 307-309
  historical and paleoclimatic data, use of, 309-310
  improvements in climate prediction, 314-316
  key drivers for research, 297-302
  mission statement, 278
  priorities for climate research, 318-322
  results of research in recent decades, 279-296
    anthropogenic effects, 289-290

decade-to-century variability (DEC-CEN), 283-288, 313-314
  joint effects of greenhouse gas forcing and aerosols, 292-295
  seasonal-to-interannual, 277, 279-283
Climate Variability and Prediction Program (CLIVAR), 308, 313, 321
Clinton administration, 52
CLIVAR. *See* Climate Variability and Prediction Program
Cloud condensation nuclei (CCN), 66
  populations of, 76-77, 87
Cloud droplet number concentrations (CDNCs), 74
Cloud physics, 6, 63, 100
  coverage and radiative properties of clouds, 74
  ice formation in the atmosphere, 75
  improving understanding of precipitation formation, 75, 92-93
  parameterizing subgrid-scale influences of clouds and microphysical processes on cloud models, 76-77
  predicting size distributions of hydrometeors and aerosols affecting radiative transfer, 75-76
  recommended research strategies for, 65-66, 74-77
Clouds
  charge generation in, 94-96
  consequences of, 39-40
  effect on radiation streams, 39
  feedback from, 290
  improved understanding of their roles in climate, 73
  and interactions with other atmospheric phenomena, 7, 85-86
  modeling, 92
  noctilucent, 245, 247
  resolving, 77, 102
  stratocumulus and cirrus, 85-86
Cloudy atmospheres, radiative transfer in, 72
Cloudy boundary layers, structure of, 79-80
CMEs. *See* Coronal mass ejections
Coal burning, 22-23
COARE. *See* Coupled Ocean-Atmosphere Response Experiment
Collaboration
  needed among agencies, 274
  needed among disciplines, 4, 46
COMET. *See* Cooperative Program for Operational Meteorology, Education, and Training
Commercial aircraft, observations from, 30-31, 193
Commercial Aviation Sensing Humidity (CASH) program, 34
Committee on Atmospheric Chemistry, 77
Committee on Atmospheric Services and Research (CASR), listing of reports of, 346-348
Committee on Earth and Environmental Sciences (CEES), 52, 58
Committee on Environment and Natural Resources (CENR), 54-55
  Subcommittee on Air Quality Research, 58
Committee on Solar and Space Physics (CSSP), 199, 204
Committee on Solar-Terrestrial Research (CSTR), 199, 204
Communication systems, space weather effects on, 8, 228, 231-241
Computer models, 14
Computers. *See also* Massively parallel processors (MPPs)
  for atmospheric analysis, 14, 197
  increasingly more powerful, 1, 13, 98-99
Computer-to-computer communication, 5, 47
Computer visualization, 98
Computer workstations, 197
Concentration monitoring networks, maintaining current, 150-151
Condensed-phase chemistry, facilities needed for studying, 7, 139
Confidence
  in climate change predictions, 23
  in forecasts, 1-2, 13
Convection, moist, 82, 88
Convective available potential energy (CAPE), 94
Convective downdrafts, 172
Convective ensemble simulations, 91
Convective heating, 40
Convective momentum transfer, 91
Convective storms, 237
Convective systems, mesoscale, 172
Cooperative Program for Operational Meteorology, Education, and Training (COMET), 180
Cooperative Programme for the Monitoring and Evaluation of Long Range Air Pollutants in Europe (EMEP), 167-168

INDEX

Coordination, needed within atmospheric sciences, 46
Coronal mass ejections (CMEs), 37, 229, 231-233, 238, 242
Cosmic rays, 208
Costs. *See* Benefits and costs of atmospheric information services
Coupled Ocean-Atmosphere Response Experiment (COARE), 91
Coupled systems, seeing components of Earth's environment as, 3
Coupling, Energetics, and Dynamics of Atmospheric Regions (CEDAR), 243
Coupling between chemistry, dynamics, and radiation, 145-147
Courant-Friedrichs-Lewy stability criterion, 193
Coverage and radiative properties of clouds, 74
CRAY supercomputers, YMP model, 197
CSSP. *See* Committee on Solar and Space Physics
CSTR. *See* Committee on Solar-Terrestrial Research
Cyclogenesis, 38
Cyclones, extratropical, 175-177. *See also* Tropical cyclones

D

Data
    acquired for public purposes with public funds, 51
    needed from over oceans, 8
Data assimilation techniques, 171, 188-190
Data denial experiments, 173, 194
Data from satellites and other remote sensors, innovative approaches to analyses of, 72-73
Dec-Cen. *See* Climate, decade-to-century variability (DEC-CEN)
Decision making, incorporating atmospheric information into, 5
Deposition fluxes, developing and evaluating techniques for measuring, 166-167
Differential absorption lidar (DIAL), 194
Digital communication, for aviation weather and flight planning capabilities, 48
Digital computers, for atmospheric analysis, 14
Dimethyl sulfide (DMS), oceanic production of, 76
Disaster statistics, 21

Disciplinary assessments, 6-9
Disciplined forecast process, 3-4, 41-43
Disease vectors, affected by weather and climate, 15, 44
Distributed atmospheric information services, implications of, 49-50
DMS. *See* Dimethyl sulfide
Dobson unit (DU), 124
DOC. *See* U.S. Department of Commerce
DOD. *See* U.S. Department of Defense
DOE. *See* U.S. Department of Energy
DOI. *See* U.S. Department of the Interior
Doppler laser, combining with Global Positioning System (GPS), 89
Doppler weather radar, 36, 68, 178-179
    network, 21
Dropsonde tracking, 193
DU. *See* Dobson unit

E

Earth Observing System (EOS), 32, 98, 202, 217, 224, 308
Earth Radiation Budget Experiment (ERBE), 257-258, 290
The Earth's Electrical Environment, 77
Ecosystem exposure monitoring networks, designing and implementing, 167-168
Ecosystem exposure systems, 7, 110
Ecosystems, 15
Eddy correlation method, 153
Electricity. *See also* Atmospheric electricity
    and Benjamin Franklin, 14
Electronic data, move to limit access to, 5, 50-51
El Niño events, 42, 278
    changes in weather patterns associated with, 18, 183
    value of predictions to agricultural planning, 26
El Niño/Southern Oscillation (ENSO) cycle, 38-39, 85, 180, 273, 277, 279-283, 296-301, 311-313, 317
EMEP. *See* Cooperative Programme for the Monitoring and Evaluation of Long Range Air Pollutants in Europe
Emissions to the atmosphere, rapidly increasing, 4, 45
Energy budget for Earth, 35
Ensemble forecasting, 171, 183, 187-188
ENSO. *See* El Niño/Southern Oscillation cycle

Entrainment, 80, 100
Environmental health, 2
Environmentally Important Atmospheric
    (chemical) Species, 7, 108-109, 112-
    114, 122, 132-135
  atmospheric aerosols, 7, 129-131, 162-166
  developing holistic and integrated
      understanding of, 7
  greenhouse gases, 7, 123-126, 147-157, 300
  nutrients, 7, 132, 166-168
  photochemical oxidants, 7, 126-129, 157-161
  stratospheric ozone, 7, 122-123, 140-147
  toxics, 7, 132, 166-168
Environmental management systems, 7, 111
  assessing efficacy of, 7, 109, 134-135
Environmental quality
  aerosols, 23
  chlorofluorocarbon (CFC) gases, 22, 117-
      119, 210-211
  and global change, 22-23
  greenhouse gases, 22-23
  long-term consequences of chemical
      emissions, 117-119
  maintaining, 22-24
  ozone, 22
Environmental shear, 91
EOS. *See* Earth Observing System
EPA. *See* U.S. Environmental Protection
    Agency
ERBE. *See* Earth Radiation Budget Experiment
ERS-1. *See* European Remote Sensing Satellite
European Centre for Medium Range Weather
    Forecasts, 184
European Remote Sensing Satellite (ERS-1),
    36, 196
EUV. *See* Extreme ultraviolet
Expendable bathythermographs (XBTs), 281
Experimental forecasts, initiating, 3
Expert systems, 32, 35
Exposure assessment networks, deploying, 137
Extreme ultraviolet (EUV) radiation, 206, 263-
    264

F

FAA. *See* Federal Aviation Administration
Fatalities, 18-21, 65
FCCSET. *See* Federal Coordinating Council for
    Science, Engineering, and Technology
Federal government and agencies
  access to atmospheric information, 50-51
  development of new observational
      capabilities, 33-34
  discipline of forecasting, role in, 41-43
  emerging issues, 43
  funding, *see* Federal funding of atmospheric
      research and operations
  historical roles, 21
  interactions at boundaries, 37-38
  planning and management, 58-59
  protection of life and property, 17-18
  role in observations, 29-30
Federal Aviation Administration (FAA), 24, 34
Federal Coordinating Council for Science,
    Engineering, and Technology
    (FCCSET), 52, 58
Federal Coordinator for Meteorological
    Services and Supporting Research, 4,
    46
Federal Council for Science and Technology, 52
Federal funding of atmospheric research and
    operations, 52-58
  by agency, 56
  by categories, 55
  historical, 53, 57
  for information services, 56
  for operations, 57
Field programs, designing and implementing
    intensive, 164-165
Field studies, carrying out process-oriented, for
    algorithm development and
    evaluation, 168
"Fire weather," forecasting, 173, 186-187
Fiscal Year (FY) expenditures, 56
Flash floods, forecasting, 184-185
Flight planning capabilities, by digital
    communication, 48
Flows, surface-induced, 83
Flux measurements
  conducting multiyear, over different
      ecosystems, 151-152
  from oceans, improving methods for, 153
Forecasting. *See* Climate forecasting; Weather
    forecasting
Fossil fuels, consumption of, 23
Franklin, Benjamin, 14
Frontal cyclones, mesoscale, 173
Fundamental condensed phase processes, 111
Fundamental understanding of the atmosphere,
    26-27
Fuzzy logic, 32
FY. *See* Fiscal Year

INDEX
355

G

Gas exchange, conducting large-scale studies of, 152
GCM. *See* General circulation models
GCOSs. *See* Global climate observing systems
GDP. *See* Gross Domestic Product
GEM. *See* Geospace Environment Modeling
General circulation models, atmospheric (GCMs), 66-67, 77, 260, 292, 315
 construction and evaluation of, 321
 Earth-ocean coupling of, 299, 311
 parameterizing, 84-85
 progress in, 300-301
Geomagnetic storms, 231-232
Geophysical fluid flow, fundamental problem of, 40-41
Geospace Environment Modeling (GEM), 243
Geostationary Operational Environmental Satellite (GOES), 181
GEWEX. *See* Global Energy and Water Cycle Experiment
GISS. *See* Goddard Institute for Space Studies
Global changes
 affecting lower atmosphere, 8
 affecting middle and upper atmosphere, 203
 and environmental quality, 22-23
Global climate observing systems (GCOSs), 274, 306, 308
Global electrical circuit, as measure of stability and temperature, 78, 93-94
Global Energy and Water Cycle Experiment (GEWEX), 45, 308, 321
Global observing system, 110
Global Ocean-Atmosphere-Land System (GOALS), 275, 308, 312, 320-321
Global Oscillation Network Group (GONG), 267-268
Global Positioning System (GPS), 194-195
 accuracy of, 235
 observations from, 31
 radio occultation technique with, 37
Global rawinsonde network, halting deterioration in, 8, 173-174
Global stratospheric sulfate layer, 162
Global telecommunication system (GTS), 281
Global transport system, 14
GOALS. *See* Global Ocean-Atmosphere-Land System
Goddard Institute for Space Studies (GISS), 288
GOES. *See* Geostationary Operational Environmental Satellite

GONG. *See* Global Oscillation Network Group
GPS. *See* Global positioning system
Gravity waves, 90
Great Salinity Anomaly, 283
Greenhouse forcing of climate, 294
Greenhouse gases, 7, 22-23, 116, 123-126, 147-157, 289
 conducting large-scale studies of gas exchange, 152
 conducting multiyear flux measurements over different ecosystems, 151-152
 conducting surface-based measurements near source regions, 152
 devising new systems to make accurate concentration measurements, 153-154
 expanding monitoring networks to include vertical profile measurements, 151
 improving and developing models, 154
 improving methods of measuring fluxes from oceans, 153
 maintaining current concentration monitoring networks, 150-151
 primary, 150
 water vapor, 156-157
Gross domestic product (GDP), contributions made by weather and climate information, 24-25
GTS. *See* Global telecommunication system

H

Halogen Occultation Experiment, 157
Heaviside, Oliver, 210
Helioseismology, 267-268
Heterogeneous chemistry, 111, 224
 facilities needed for studying, 7, 139
HF. *See* High frequency
High frequency (HF) events, 184, 234
Holistic research strategy, need to develop, 135
Human health
 affected by weather and climate, 15
 research recommended in, 44
 space weather effects on, 8
Hurricane Andrew, 17-18, 21
Hurricane forecasting, 178-180
 delineating optimal measurement system combinations for, 8
 greatest opportunity to save lives and property, 178
Hurricane statistics, 19

Hydrological cycle
  improving understanding of, 171
  and interactions with other atmospheric phenomena, 7
Hydrometeors, predicting size distributions of, 75-76

I

ICAS. *See* Interdepartmental Committee for Atmospheric Sciences
Ice formation in the atmosphere, 75, 87
Ice nucleus (IN) population, 76
IGAC. *See* International Global Atmospheric Chemistry (IGAC) project
IMF. *See* Interplanetary magnetic field
Immune system, affected by ultraviolet (UV) radiation, 15
Incoherent scatter (IS), 251
Incorporating atmospheric information into decision making, weather-dependent enterprises, 5
Infectious diseases, affected by weather and climate, 15, 44
Inferential observation-based studies, 161
Information. *See* Atmospheric information
Infrastructure
  initiatives needed, 135-139, 200
  modeling, 13
  needed to advance research in atmospheric chemistry, 7, 110-111
  observational, 13
Inhomogeneity, effect on boundary layer, 80
Institutional arrangements for climate research, 323
Instrument development programs, 7, 111, 139
Integrated assessments, support, 161
Integrated field campaigns, continue implementation of, 160-161
Integrating observing systems
  to assimilate new forms of data, 32
  with increased computing power, 31
  with modeling efforts, 31
  through international collaboration, 32
  using information organizing systems, 32
  using multiple data bases, 32
Intelligent systems, 98
Interactions
  among atmospheric phenomena of different scales, 3, 89-92
  among atmospheric phenomena of different sorts, 7
  between atmosphere and other Earth system components, 3
  complexity of, 63-64
  land-atmosphere, 184-185
  long-term, 38-39
  modeling studies of, 3, 252, 254
  nonlinear, 41
  observational studies of, 3
  of planetary boundary layer, surface characteristics, and clouds, 81
  resolving, 37-41
  surface, 38
  theoretical studies of, 3
  water substance, 65
Interdepartmental Committee for Atmospheric Sciences (ICAS), 52-53
Interdisciplinary studies needed, 4, 43-45
  in climate, weather, and health, 4
  in management of water resources in changing climate, 4
  in rapidly increasing emissions to the atmosphere, 4
Intergovernmental Panel on Climate Change (IPCC), 23, 292
International Global Atmospheric Chemistry (IGAC) Project, 87
International Research Institute for Climate Prediction, 42
International Solar-Terrestrial Program, 243
Interplanetary magnetic field (IMF), 37, 204
Interplanetary space, 204-205
Intrastratospheric transport, 146
Ionosphere, 8, 206, 210, 233, 234, 240, 251-254
IPCC. *See* Intergovernmental Panel on Climate Change

K

Kennelly, Arthur E., 210
Kitt Peak National Observatory (KPNO), 269

L

Lagrangian experiments, 165, 192-193
Land-atmosphere interaction, 184-185
Large eddy simulation (LES) models, 80
Large-scale models
  effects of moist convection in, 82

INDEX   357

   incorporation of surface-induced flows into, 83
Laser Atmospheric Wind Sounder (LAWS) instrument, 196
Laser systems, for atmospheric analysis, 14
Lavoisier, Antoine-Laurent, 114
LAWS. *See* Laser Atmospheric Wind Sounder
Leadership and management, 4-5, 46-59
LES. *See* Large eddy simulation
Lidar systems, 36, 88-89. *See also* Differential absorption lidar
Life and property
   need for forecasts and warnings, 18-21
   protection of, 17-22
Life sciences, 15
Lightning
   and Benjamin Franklin, 14
   global monitoring of, 78
   as measure of stability and temperature, 78, 93-94
   propagation of, 94-96

M

Magnetic fields, 37, 226
Magnetic storms, 209
Magnetosphere, 205
Magnetospheric storms, 231
Marconi, Guglielmo, 210
Massively parallel processors (MPPs), 190, 197
Maunder Minimum period, 265
Maximum usable frequency (MUF), 234
MCS. *See* Mesoscale convective system
Measurements. *See also* Concentration measurements; Flux measurements, Observing systems
   central importance of, 198
   conducting surface-based near source regions, 152
   of exchange of water, heat, and trace atmospheric constituents, 80-81
   improving capabilities for making, 6, 103-106
Measurement systems
   satellite-based, 195
   surface exchange, 7, 138
Measurement technologies
   for critical gas- and condensed-phase species, 143-145
   for tropospheric aerosols, designing and implementing new suites of, 163-164

Mechanisms of charge separation in clouds, 78
Medium frequency (MF), 251
Mesoscale convective systems (MCSs), 82, 91, 181
Mesosphere-stratosphere-troposphere (MST), 251
Microphysical processes influencing clouds, 182, 202, 224
   parameterizing, 76-77
Microwave Limb Sounder, 157
Middle-atmosphere, 206, 208
   nature and sources of discharges, 78, 94-96
Middle-upper atmosphere
   global change in, 201, 209, 245-255, 254-255
   monitoring inputs to, 251-252
   monitoring sensitive parameters of, 251
Midlatitude cyclones, small-scale features in, 82-83
Mission to Planet Earth satellite program, 252
Model development, 186
Modeling fluxes, 78
Modeling infrastructure, 13
Models and modelling
   in atmospheric chemistry
      in aerosol research, 165, 166
      in chemistry, dynamics and radiation coupling, 147
      in integrated assessments, 161
      long-term biogenic greenhouse gases, 154
      and operational chemical forecasting, 138
      overarching research challenge, 134
      predicted ozone column change, 124
      in toxic and nutrient investigation, 168
   in atmospheric dynamics and weather forecasting
      adaptive observations, 188
      adjoint models, 176
      in atmospheric convection studies, 180-187
      ensemble forecasting, 187-188
      massively parallel processors, used for, 190, 197
      numerical techniques, 191-193
      in orographic effects, 185-186
      parameterization for, 190-191
      for tropical cyclones, 178
   in atmospheric physics
      radiation transfer, 71-72, 84-85

rapidly increasing computational power, 98-99
representation in cyclones, 82-83
in climate and climate change research
  construction and evaluation of comprehensive models, 275, 321
  in coupled atmosphere-ocean research, 290, 301, 310-311
  in decade-to-century variability, 284, 286-289
  in ENSO prediction, 298
  linkages between climate model prediction and human relevance, 316-317
data for improvement of, 30
fundamental aspects, 26
funding for, 55
in upper atmosphere and near-Earth space research
  atmospheric effects of aircraft, 220-221, 224
  interactive radiative-dynamic-chemistry models, 217
  in middle and upper atmosphere research, 252, 254
  in space weather forecasting, 243
  stratospheric-tropospheric interactions, 222, 224, 225
Model vertical coordinates, 192
Moist convection, 88
  effects in large-scale models, 82
Monitoring. *See* Climate monitoring
Monitoring networks. *See* Observing systems
"Montreal Protocol," 22, 216
Motion of tropical cyclones, physics of, 8
Mt. Pinatubo, 218-219, 286, 288
MPPs. *See* Massively parallel processors
MST. *See* Mesosphere-stratosphere-troposphere
MUF. *See* Maximum usable frequency

N

NAAQS. *See* National Ambient Air Quality Standard
NARSTO. *See* North American Research Strategy on Troposphere Ozone
NASA. *See* National Aeronautics and Space Administration
National Aeronautics and Space Administration (NASA), 54, 98, 223, 241, 243, 308-309, 319

National Ambient Air Quality Standard (NAAQS), 126, 131
National Center for Atmospheric Research (NCAR), 36, 97
National Centers for Environmental Prediction (NCEP), 42, 188
National Crop Loss Assessment Network (NCLAN), 168
National economic vitality
  benefits of weather and climate information, 24-26
  enhancing, 24-26
National Lightning Detection Network, 94
National Oceanic and Atmospheric Administration (NOAA), 21, 196, 241, 243, 287, 303, 309, 319
  Aircraft Operations Center, 36
  International Research Institute for Climate Prediction, 42
  National Centers for Environmental Prediction (NCEP), 42
  Office of Global Programs, 34
National Polar-orbiting Operational Environmental Satellite System (NPOESS), 319
National Research Council (NRC), 199, 204, 210
  Committee on Solar and Space Physics (CSSP), 199, 204
  Committee on Solar-Terrestrial Research (CSTR), 199, 204
National Science and Technology Council (NSTC), 58
National Science Foundation (NSF), 223, 241, 243, 309
National Space Weather Program (NSWP), 241-243
National weather information system, rapid changes in, 4-5
National Weather Service (NWS), 21, 47-49
National well-being
  contributions of the atmospheric sciences to, 17-27, 65, 106, 111, 202, 271, 324
  enhancing national economic vitality, 24-26
  maintaining environmental quality, 22-24
  protection of life and property, 17-22
  strengthening fundamental understanding, 26-27
Nature and sources of middle-atmosphere discharges, 78
NCAR. *See* National Center for Atmospheric Research

INDEX 359

NCEP. *See* National Centers for Environmental Prediction
NCLAN. *See* National Crop Loss Assessment Network
NDSC. *See* Network for Detection of Stratospheric Change
Network for Detection of Stratospheric Change (NDSC), 155-157
Newton, Sir Isaac, 14
Next Generation Weather Radar (NEXRAD), 48, 175, 181
NIMBUS-7 experimental environmental research satellite, 217, 257-258
NOAA. *See* National Oceanic and Atmospheric Administration
Noctilucent clouds, 245, 247
Nonlinearity, fundamental problem of, 40-41, 119-121, 184
Nonspherical particles, radiation transfer through a medium containing, 72
North American Atmospheric Observing System, 32
North American Research Strategy on Troposphere Ozone (NARSTO), 159
North American Strategy for Tropospheric Ozone program, 58
"Nowcasting," 201
NO_x, 120, 143-145
  production by lightning, 79, 96-97
NPOESS. *See* National Polar-orbiting Operational Environmental Satellite System
NRC. *See* National Research Council
NSF. *See* National Science Foundation
NSTC. *See* National Science and Technology Council
NSWP. *See* National Space Weather Program
Numerical computer models of the atmosphere, 3-4
Numerical techniques, 191-193
  for advection, 192-193
  model vertical coordinates, 192
Numerical weather prediction (NWP) models, 181, 190
Nutrients, 7, 132, 149, 166-168
  carrying out process-oriented field studies for algorithm development and evaluation, 168
  designing and implementing ecosystem exposure monitoring networks, 167-168
  developing and evaluating techniques for measuring deposition fluxes, 166-167
NWP. *See* Numerical weather prediction (NWP) models
NWS. *See* National Weather Service

O

Observational technologies, improving understanding of interactions among atmosphere, ocean, land, 13, 101
Observations
  adaptive strategies for making, 31
  from commercial aircraft, 30-31, 193
  deterioration of, 173, 302-306
  from the Global Positioning System (GPS), 31
  in near-Earth space, 37
  new opportunities for, 30-31, 100-101
  preserving free and open exchange of, 1, 47, 50-51, 302-306
  in the stratosphere, 36
  of water in the atmosphere, 34-35
  of wind, 35-36
Observing systems
  for atmospheric chemistry research, 110-111, 136-138, 150-154, 155-156, 160-161, 162-165, 167-168
  for atmospheric physics research, 88, 97, 103-105
  for climate and climate change research, 274, 281, 301-306, 308-309, 318-320
  for dynamics and weather forecasting research, 193-197
  for upper atmosphere and near-Earth research, 216-218, 244, 258
Observing system simulation experiments (OSSEs), 2, 33, 173, 177
Oceanography, close partner of atmospheric sciences, 15
Oceans
  critical boundary for atmosphere, 15
  data needed from over, 8, 36
  fluxes over, 38-39
  long-term interactions with, 38-39
Office of the Federal Coordinator for Meteorology (OFCM), 56, 58. *See also* Federal Coordinator for Meteorological Services and Supporting Research

Operational community, interacting with research community, 2
Operational models, 102
Orographic influences on weather, 172, 185-186
Oscillation effects, quasi-biennial, 220
OSSE. *See* Observing system simulation experiment
Overarching Research Challenges, atmospheric chemistry, 129-131
Ozone destruction, 8, 246
Ozone layer. *See* stratospheric ozone
Ozone, tropospheric. *See* tropospheric photochemical oxidants
Ozonsonde program, 155

P

Paleoclimatic records, 299, 309-310
Parameterization, 69-70, 101
Pattern recognition, 98
PCBs. *See* Polychlorinated biphenyls
Phenomena. *See* Atmospheric phenomena
Photochemical oxidants. *See* tropospheric photochemical oxidants. *See also* Smog
Photoionization, 240
Physical processes. *See also* Atmospheric physical processes
  interactions between radiation and, 74
  occurring on subgrid scales in climate models, 84
  parameterizing, 190-191
Physics. *See* Atmospheric physics,
Phytotoxics, 149
Planetary boundary layer, surface characteristics of, 81
A Plan for a Research Program on Aerosol Radiative Forcing and Climate Change, 87
Polarimetric radar, 181-182
Polar stratospheric clouds (PSCs), 118
Polychlorinated biphenyls (PCBs), 166
Power grid operation, space weather effects on, 8, 227
Pre-chlorofluorocarbon era, 141-142
Precipitation, radars for measuring, 14
Precipitation formation, 70, 100
  improved understanding of, 75
Predictability, 186
Prediction. *See* Weather forecasting

Predictive models
  developing capability, 165-166
  improving numerical, 2
  need to develop, 7, 134
Priestley, Joseph, 114
Primary greenhouse gases, 150
Private meteorological sector
  in fashioning the agenda, 59
  in leadership and management, 46
  in preparing predictions, 24
  in providing weather services, 47-49
Process study observation, 102
Production of NO_x by lightning, 79
PSC. *See* Polar stratospheric cloud
Publicly-funded data acquisition, preserving open access to, 5, 50-51

Q

QBO. *See* Quasi-biennial oscillation
Quantification and characterization of critical gas-phase and heterogeneous mechanisms, 147
Quantitative descriptions, developing, 102-103
Quasi-biennial oscillation (QBO), 220, 250
Quasi-geostrophic theory, 83

R

Radars
  early data networks, 21
  measuring precipitation for atmospheric analysis, 14
  measuring wind for atmospheric analysis, 14
Radiation. *See* Atmospheric physics
Radiation transfer models, 71-72
  using observational data, 84-85
Radiation transfer through a medium containing nonspherical particles, 72
Radiative forcing of climate
  instantaneous, 294
  by trace gases and aerosols, 73-74, 293
Radiative transfer, 70
  in cloudy atmospheres, 72
Radioacoustic sounding system (RASS), 195
Radio occultation technique with GPS, 37
Radiosonde networks, 34, 36
  deterioration of, 303
  early, 21
  worldwide, 32

INDEX

Radiosonde observational networks, early, 21
Rainfall events, variability in location of, 35
RASS. *See* Radioacoustic sounding system
Rawinsonde tracking, 193
Recommendations of the Board on Atmospheric Sciences and Climate (BASC), 3-5, 37-45
Remote sensing capabilities
    exploiting, 88-89
    improving, 1, 163
    satellites for atmospheric analysis, 14
Rossby waves, 26, 220

S

SAGE. *See* Stratospheric Aerosol and Gas Experiment
SAR. *See* Subcommittee on Atmospheric Research
Satellites and atmospheric chemistry research
    global chemistry measurement of greenhouse gases on a range of scales, 153
    recommendation for optimal combinations of remote sensing and in situ observations, 173
    satellite inferences of storm-associated rain rates, 179
    satellite measurement of stratospheric aerosol, 162
    small satellites useful to study chemistry at high altitudes, 147
Satellites and atmospheric physics research
    GCM parameterization compared with data from the International Satellite Cloud Climatology Project, 85
    inferring hydrometeor and cloud characteristics from satellite observations, 101
    innovative approaches to the analysis of data from, 72
    process study parameterizations generalized and extrapolated by satellite data, 102
    satellites for characterizing precipitation over the oceans, 103
Satellites in atmospheric dynamics and weather forecasting research
    GPS receiver and satellite transmissions for water vapor measurement, 194-195
    satellite measurement of wind using Doppler lidar, sea surface scatterometers, 195-197
Satellites in climate and climate change research
    intersatellite measurement bias, 303-305
    stratospheric temperatures from satellite measurement, 287, 288
Satellites in upper-atmosphere and near-Earth space research
    satellite measurement of solar irradiance, 257-259
    satellites showing space environment effects, 235
    UARS measurements of chemistry of the stratosphere, 213
    weather satellite damage from space weather disturbances, 227, 228
Scales of flow, resolving interactions among different, 37-41
Schumann resonances, 78
Scientific strategy
    initiatives supporting, 65
    key components of, 64-65
SEASAT (sea satellite) oceanographic satellite, 196
Seasonal climate forecasting, 8, 183-184
Sea surface temperature anomalies (SSTAs), 282
Sea surface temperatures (SSTs), 298, 312
Semigeostrophic theory, 83
Semi-Lagrangian approach, 192-193
Signal processing, 98
Skin cancer, affected by ultraviolet (UV) radiation, 15
"Skycam" operations, 195
Small-scale dynamics, 6
    effects of moist convection in large-scale models, 82
    incorporation of surface-induced flows into large-scale models, 83
    interactions with larger-scale processes, 89-92, 99-102
    recommended research strategies, 67, 81-83
    representation of small-scale features in midlatitude cyclones, 82-83
Small-scale features, dynamical representation of in midlatitude cyclones, 82-83
SMM. *See* Solar Maximum Mission
Smog, 22, 116, 120-121
Society, greater confidence in forecasts, 1, 13
Solar and Heliospheric Observatory (SOHO), 267-268
Solar effects, 219-220
    separating from anthropogenic, 259-263

Solar energetic particles (SEPs), 229, 231
Solar energy output, over a solar cycle, 257-259, 267-268
Solar influences, 210, 256-271
Solar Maximum Mission (SMM) spacecraft, 257
Solar phenomena
  interactions with near-Earth space, 26
  long-term changes in, 268
  near-Sun wind, 37
  streams from flares, 14
Solar-terrestrial system, need for models of, 9
Solar variability, 8
  effects on global climate system, 8
  and global change, 203-204
Space climate, 228-229
Space physics activities, near-Earth, improving predictive numerical models for, 2
Space weather, 209, 225-245
  disturbances, 228
  forecasting, 3, 6, 9, 43, 200-201
  ionospheric, 240
  magnetospheric, 239
  research needed in, 8, 203
Space weather system, 229-234
Special Sensor Microwave/Imager (SSM/I), 35, 178
SST. *See* Sea surface temperature
SSTAs. *See* Sea surface temperature anomalies
STE. *See* Stratosphere-troposphere exchange
Stratosphere
  observations in the, 36, 224
  recommended research strategies for, 202-204
  roles played in climate system, 200-201, 203, 222-223
Stratosphere-troposphere exchange (STE), 221-222
  better characterization of, 225
Stratospheric Aerosol and Gas Experiment (SAGE), 155-156
Stratospheric aerosols, 224
  maintaining and expanding measurement capability, 162-163
Stratospheric aircraft, 8
Stratospheric modeling, 224
Stratospheric ozone, 7, 122-123, 140-147, 202, 213-218, 223-224
  coupling between chemistry, dynamics, and radiation, 145-147
  and environmental quality, 22
  measuring critical gas- and condensed-phase species, 143-145
  monitoring distribution of, 142-143
  quantification and characterization of critical gas-phase and heterogeneous mechanisms, 147
Stratospheric processes, 8, 208-209, 211-225
Stratospheric-tropospheric exchange, 146
Studies needed. *See* Interdisciplinary studies needed
Subcommittee on Atmospheric Research (SAR), 52-53, 58
Subgrid-scale influences of clouds, parameterizing, 76-77
Subgrid scale (SGS), 84
Sulfate concentrations in atmosphere, 106
Sulfate layer, global stratospheric, 162
Sun, 204-205, 229-234
  evaluating state of, 265
Sun-Earth connections, 226
Sunspot records, 261
Sun's Radiative Inputs from Sun to Earth (SUNRISE) program, 243
Surface effects, quantifying and parameterizing, 87-88
Surface exchange measurement systems, 7, 110, 138
Surface-induced flows, incorporation into large-scale models, 83
Surface UV network, 216
  monitoring, 224

T

TAO. *See* Tropical Atmosphere Ocean array
Teamwork, importance of, 6, 323
Technology transfer programs, 7, 111, 139
Terrain scale, 185-186
Time lagging, 187
TOGA. *See* Tropical Ocean Global Atmosphere (TOGA) program
TOGA-TAO array. *See* Tropical Ocean Global Atmosphere-Tropical Atmospheric Ocean array
Topography, continuous scales of, 185-186
Tornado dynamics, 181
Tornado statistics, 19
Toxics, 7, 132, 166-168
  carrying out process-oriented field studies for algorithm development and evaluation, 168

designing and implementing ecosystem exposure monitoring networks, 167-168
developing and evaluating techniques for measuring deposition fluxes, 166-167
Trace chemical species, 86-87, 115. *See also* Environmentally Important Atmospheric (chemical) species
Trace gases, direct radiative forcing of climate by, 73-74
Transport, 70
  global, 14
  intrastratospheric, 146
  turbulent, 100
  vertical, 33, 39
Triggering, 186
TRMM. *See* Tropical Rainfall Measurement Mission
Tropical Atmosphere Ocean (TAO) array, 38, 274, 281, 305
Tropical cyclones, 177-180
  changes in intensity of, 8
  dynamics of, 172
  and interactions with upper ocean layers, 8
  midlatitude, 82-83
  physics of motion of, 8
Tropical Ocean Global Atmosphere (TOGA) program, 42, 91, 274, 281, 298, 305-306, 311, 321
Tropical Ocean Global Atmosphere-Tropical Atmospheric Ocean (TOGA-TAO) array, 34, 38, 319
Tropical Rainfall Measurement Mission (TRMM), 35
Tropopause
  exchange of material through, 224-225
  role in atmospheric dynamics, 172
Troposphere, exchanges with other layers, 3
Tropospheric aerosols, designing and implementing new suites of measurement technologies for, 163-164
Tropospheric photochemical oxidants instrumentation development, measurements documentation assessment, 7, 126-129, 157-161
Tropospheric stability, 78
Turbulence, 64
  and entrainment, 80

U

UARS. *See* Upper Atmosphere Research Satellite
UAV. *See* Unmanned aerospace vehicle
Ultraviolet (UV) radiation, 14, 213-218
  health effects of, 15, 44
  increasing intensity of, 254-255
  solar, 206
  variability in, 262
U.S. Air Force, 241
U.S. Department of Commerce (DOC), 241, 243
U.S. Department of Defense (DOD), 241, 243, 319
U.S. Department of Energy (DOE), 98, 241
U.S. Department of the Interior (DOI), 241
U.S. Environmental Protection Agency (EPA), 126, 131
U.S. Geological Survey, 241
U.S. Global Change Research Program (USGCRP), 3, 21, 23, 37, 55, 58, 279
U.S. National Climate Program, 279
U.S. Naval Research Laboratory, 210
U.S. Weather Research Program (USWRP), 3, 21, 37, 170
Unmanned aerospace vehicles (UAVs), 98
Upper-atmosphere processes, 199-271
  growing emphasis on prediction of, 14
  recommended strategies for studying, 8-9
  research in, 8-9
  stratospheric processes affecting, 8
Upper Atmosphere Research Satellite (UARS), 157, 217, 224, 257, 262-263
Upper ocean layers, and interactions with tropical cyclones, 8
Upper-troposphere, water vapor in, 64
USGCRP. *See* U.S. Global Change Research Program
USWRP. *See* U.S. Weather Research Program
UV-B radiation, 212-213
UV flux, 217-218

V

Verification, 69-70, 186
Vertical profiles, 195
Vertical transport mechanisms, 33, 39
Volatile organic compounds (VOCs), 122

Volcanic effects, 8, 202, 218-219, 219, 224
Vorticity, potential, 172

W

Water in soil, 184-185
  and hydrology, 174
Water in the atmosphere and on land
  aerosols and clouds, 130-131
  from aircraft, 220
  cloud physics, 74-77
  cloud radiative properties, 85-86
  deposition in precipitation, 167
  distributions in the atmosphere, 102-103
  enhanced observation of water in all forms, 307
  in EOS and GEWEX, 308
  hydrometeors, 75
  ice, 87
  interaction with chemical species, 86-87
  liquid in clouds, 191
  measurements of, 156-157, 182, 194-195, 274
  phase change and atmospheric circulation, 174
  precipitation mechanisms, 92-93
  run off,
  vapor, 34, 174
  water vapor as a greenhouse gas, 149, 290
WCRP. *See* World Climate Research Programme
The Weather Channel, 48
Weather damage, 20-21
Weather-dependent enterprises, incorporating atmospheric information into decision making, 5, 47
Weather fatalities, 18-21, 174
Weather forecasting research, 225. *See also* Climate forecasting
  convection, 180-183
  data acquisition, 193-197
  data manipulation, 188-190
  ensemble forecasting, 187-188
  numerical techniques, 191-193
  recommendations, 173-175
  and storms, 175-180
Weather forecasts, 169-198
  economic benefit of, 24-26
  four-way partnership for providing, 17-18
  initiating experimental, 3, 49
  new systems for providing, 47-48, 171
  spatial scales relevant to, 2-3
  temporal scales relevant to, 2-3
Weather modification, 93-94
Weather satellites, 21
Weather-sensitive enterprises, 25
Weather Service Radar (U.S. National Weather Service) 1988 Doppler Weather Radar System (WSR-88D), 35-36, 181
Weinberg, Alvin M., 15
Wind
  developing new capabilities for observing, 2
  observations of, 35-36
  radars for measuring, 14
WMO. *See* World Meteorological Organization
WOCE. *See* World Ocean Circulation Experiment
World Climate Research Programme (WCRP), 45, 275-276, 279, 313-315, 320-321
World Meteorological Organization (WMO), 302, 306
World Ocean Circulation Experiment (WOCE), 308
World Weather Watch (WWW), 302
World Wide Web (WWW), 48
WSR-88D, 35, 36, 181

X

XBT. *See* Expendable bathythermograph
X-rays, 263-264

Y

YMP. *See* CRAY supercomputers